MONOGRAPHS ON STATISTICS AND APPLIED PROBABILITY

General Editors

D.R. Cox, V. Isham, N. Keiding, T. Louis, N. Reid, R. Tibshirani, and H. Tong

1 Stochastic Population Models in Ecology and Epidemiology *M.S. Barlett* (1960)
2 Queues *D.R. Cox and W.L. Smith* (1961)
3 Monte Carlo Methods *J.M. Hammersley and D.C. Handscomb* (1964)
4 The Statistical Analysis of Series of Events *D.R. Cox and P.A.W. Lewis* (1966)
5 Population Genetics *W.J. Ewens* (1969)
6 Probability, Statistics and Time *M.S. Barlett* (1975)
7 Statistical Inference *S.D. Silvey* (1975)
8 The Analysis of Contingency Tables *B.S. Everitt* (1977)
9 Multivariate Analysis in Behavioural Research *A.E. Maxwell* (1977)
10 Stochastic Abundance Models *S. Engen* (1978)
11 Some Basic Theory for Statistical Inference *E.J.G. Pitman* (1979)
12 Point Processes *D.R. Cox and V. Isham* (1980)
13 Identification of Outliers *D.M. Hawkins* (1980)
14 Optimal Design *S.D. Silvey* (1980)
15 Finite Mixture Distributions *B.S. Everitt and D.J. Hand* (1981)
16 Classification *A.D. Gordon* (1981)
17 Distribution-Free Statistical Methods, 2nd edition *J.S. Maritz* (1995)
18 Residuals and Influence in Regression *R.D. Cook and S. Weisberg* (1982)
19 Applications of Queueing Theory, 2nd edition *G.F. Newell* (1982)
20 Risk Theory, 3rd edition *R.E. Beard, T. Pentikäinen and E. Pesonen* (1984)
21 Analysis of Survival Data *D.R. Cox and D. Oakes* (1984)
22 An Introduction to Latent Variable Models *B.S. Everitt* (1984)
23 Bandit Problems *D.A. Berry and B. Fristedt* (1985)
24 Stochastic Modelling and Control *M.H.A. Davis and R. Vinter* (1985)
25 The Statistical Analysis of Composition Data *J. Aitchison* (1986)
26 Density Estimation for Statistics and Data Analysis *B.W. Silverman* (1986)
27 Regression Analysis with Applications *G.B. Wetherill* (1986)
28 Sequential Methods in Statistics, 3rd edition
G.B. Wetherill and K.D. Glazebrook (1986)
29 Tensor Methods in Statistics *P. McCullagh* (1987)
30 Transformation and Weighting in Regression
R.J. Carroll and D. Ruppert (1988)
31 Asymptotic Techniques for Use in Statistics
O.E. Bandorff-Nielsen and D.R. Cox (1989)
32 Analysis of Binary Data, 2nd edition *D.R. Cox and E.J. Snell* (1989)

33 Analysis of Infectious Disease Data *N.G. Becker* (1989)
34 Design and Analysis of Cross-Over Trials *B. Jones and M.G. Kenward* (1989)
35 Empirical Bayes Methods, 2nd edition *J.S. Maritz and T. Lwin* (1989)
36 Symmetric Multivariate and Related Distributions
K.T. Fang, S. Kotz and K.W. Ng (1990)
37 Generalized Linear Models, 2nd edition *P. McCullagh and J.A. Nelder* (1989)
38 Cyclic and Computer Generated Designs, 2nd edition
J.A. John and E.R. Williams (1995)
39 Analog Estimation Methods in Econometrics *C.F. Manski* (1988)
40 Subset Selection in Regression *A.J. Miller* (1990)
41 Analysis of Repeated Measures *M.J. Crowder and D.J. Hand* (1990)
42 Statistical Reasoning with Imprecise Probabilities *P. Walley* (1991)
43 Generalized Additive Models *T.J. Hastie and R.J. Tibshirani* (1990)
44 Inspection Errors for Attributes in Quality Control
N.L. Johnson, S. Kotz and X, Wu (1991)
45 The Analysis of Contingency Tables, 2nd edition *B.S. Everitt* (1992)
46 The Analysis of Quantal Response Data *B.J.T. Morgan* (1992)
47 Longitudinal Data with Serial Correlation—A state-space approach
R.H. Jones (1993)
48 Differential Geometry and Statistics *M.K. Murray and J.W. Rice* (1993)
49 Markov Models and Optimization *M.H.A. Davis* (1993)
50 Networks and Chaos—Statistical and probabilistic aspects
O.E. Barndorff-Nielsen, J.L. Jensen and W.S. Kendall (1993)
51 Number-Theoretic Methods in Statistics *K.-T. Fang and Y. Wang* (1994)
52 Inference and Asymptotics *O.E. Barndorff-Nielsen and D.R. Cox* (1994)
53 Practical Risk Theory for Actuaries
C.D. Daykin, T. Pentikäinen and M. Pesonen (1994)
54 Biplots *J.C. Gower and D.J. Hand* (1996)
55 Predictive Inference—An introduction *S. Geisser* (1993)
56 Model-Free Curve Estimation *M.E. Tarter and M.D. Lock* (1993)
57 An Introduction to the Bootstrap *B. Efron and R.J. Tibshirani* (1993)
58 Nonparametric Regression and Generalized Linear Models
P.J. Green and B.W. Silverman (1994)
59 Multidimensional Scaling *T.F. Cox and M.A.A. Cox* (1994)
60 Kernel Smoothing *M.P. Wand and M.C. Jones* (1995)
61 Statistics for Long Memory Processes *J. Beran* (1995)
62 Nonlinear Models for Repeated Measurement Data
M. Davidian and D.M. Giltinan (1995)
63 Measurement Error in Nonlinear Models
R.J. Carroll, D. Rupert and L.A. Stefanski (1995)
64 Analyzing and Modeling Rank Data *J.J. Marden* (1995)
65 Time Series Models—In econometrics, finance and other fields
D.R. Cox, D.V. Hinkley and O.E. Barndorff-Nielsen (1996)

66 Local Polynomial Modeling and its Applications *J. Fan and I. Gijbels* (1996)
67 Multivariate Dependencies—Models, analysis and interpretation
   D.R. Cox and N. Wermuth (1996)
68 Statistical Inference—Based on the likelihood *A. Azzalini* (1996)
69 Bayes and Empirical Bayes Methods for Data Analysis
   B.P. Carlin and T.A Louis (1996)
70 Hidden Markov and Other Models for Discrete-Valued Time Series
   I.L. Macdonald and W. Zucchini (1997)
71 Statistical Evidence—A likelihood paradigm *R. Royall* (1997)
72 Analysis of Incomplete Multivariate Data *J.L. Schafer* (1997)
73 Multivariate Models and Dependence Concepts *H. Joe* (1997)
74 Theory of Sample Surveys *M.E. Thompson* (1997)
75 Retrial Queues *G. Falin and J.G.C. Templeton* (1997)
76 Theory of Dispersion Models *B. Jørgensen* (1997)
77 Mixed Poisson Processes *J. Grandell* (1997)
78 Variance Components Estimation—Mixed models, methodologies and applications
   P.S.R.S. Rao (1997)
79 Bayesian Methods for Finite Population Sampling
   G. Meeden and M. Ghosh (1997)
80 Stochastic Geometry—Likelihood and computation
   O.E. Barndorff-Nielsen, W.S. Kendall and M.N.M. van Lieshout (1998)
81 Computer-Assisted Analysis of Mixtures and Applications—
   Meta-analysis, disease mapping and others *D. Böhning* (1999)
82 Classification, 2nd edition *A.D. Gordon* (1999)
83 Semimartingales and their Statistical Inference *B.L.S. Prakasa Rao* (1999)
84 Statistical Aspects of BSE and vCJD—Models for Epidemics
   C.A. Donnelly and N.M. Ferguson (1999)
85 Set-Indexed Martingales *G. Ivanoff and E. Merzbach* (2000)
86 The Theory of the Design of Experiments *D.R. Cox and N. Reid* (2000)
87 Complex Stochastic Systems
   O.E. Barndorff-Nielsen, D.R. Cox and C. Klüppelberg (2001)
88 Multidimensional Scaling, 2nd edition *T.F. Cox and M.A.A. Cox* (2001)
89 Algebraic Statistics—Computational Commutative Algebra in Statistics
   G. Pistone, E. Riccomagno and H.P. Wynn (2001)
90 Analysis of Time Series Structure—SSA and Related Techniques
   N. Golyandina, V. Nekrutkin and A.A. Zhigljavsky (2001)

Analysis of Time Series Structure
SSA and Related Techniques

NINA GOLYANDINA
VLADIMIR NEKRUTKIN
ANATOLY ZHIGLJAVSKY

CHAPMAN & HALL/CRC

Boca Raton   London   New York   Washington, D.C.

Library of Congress Cataloging-in-Publication Data

Golyandina, N. (Nina)
　　Analysis of time series structure : SSA and related techniques /
N. Golyandina, V. Nekrutkin, and A. Zhigljavsky.
　　　　p. cm.—(Monographs on statistics and applied probability)
　　Includes bibliographical references and index.
　　ISBN 1-58488-194-1 (alk. paper)
　　1. Time-series analysis.　　I. Nekrutkin, V.V. (Vladimir
Viktorovich) II. Zhigljavsky, A.A. (Anatoly Aleksandrovich) III.
Title. IV. Series.
　　QA280 .G65 2000
　　519.5′5—dc21　　　　　　　　　　　　　　　　　　　　　　　　00-050442
　　　　　　　　　　　　　　　　　　　　　　　　　　　　　　　　　　　　　CIP

This book contains information obtained from authentic and highly regarded sources. Reprinted material is quoted with permission, and sources are indicated. A wide variety of references are listed. Reasonable efforts have been made to publish reliable data and information, but the author and the publisher cannot assume responsibility for the validity of all materials or for the consequences of their use.

Neither this book nor any part may be reproduced or transmitted in any form or by any means, electronic or mechanical, including photocopying, microfilming, and recording, or by any information storage or retrieval system, without prior permission in writing from the publisher.

The consent of CRC Press LLC does not extend to copying for general distribution, for promotion, for creating new works, or for resale. Specific permission must be obtained in writing from CRC Press LLC for such copying.

Direct all inquiries to CRC Press LLC, 2000 N.W. Corporate Blvd., Boca Raton, Florida 33431.

Trademark Notice: Product or corporate names may be trademarks or registered trademarks, and are used only for identification and explanation, without intent to infringe.

Visit the CRC Press Web site at www.crcpress.com

© 2001 by Chapman & Hall/CRC

No claim to original U.S. Government works
International Standard Book Number 1-58488-194-1
Library of Congress Card Number 00-050442
Printed in the United States of America　1　2　3　4　5　6　7　8　9　0
Printed on acid-free paper

Contents

Preface   ix

Notation   xi

Introduction   1

Part I.   SSA: Methodology   13

1   Basic SSA   15
    1.1   Basic SSA: description   16
    1.2   Steps in Basic SSA: comments   18
    1.3   Basic SSA: basic capabilities   24
    1.4   Time series and SSA tasks   32
    1.5   Separability   44
    1.6   Choice of SSA parameters   53
    1.7   Supplementary SSA techniques   78

2   SSA forecasting   93
    2.1   SSA recurrent forecasting algorithm   95
    2.2   Continuation and approximate continuation   96
    2.3   Modifications to Basic SSA R-forecasting   107
    2.4   Forecast confidence bounds   115
    2.5   Summary and recommendations   127
    2.6   Examples and effects   131

3   SSA detection of structural changes   149
    3.1   Main definitions and concepts   149
    3.2   Homogeneity and heterogeneity   156
    3.3   Heterogeneity and separability   169
    3.4   Choice of detection parameters   189
    3.5   Additional detection characteristics   196
    3.6   Examples   204

Part II. SSA: Theory — **217**

4 Singular value decomposition — **219**
- 4.1 Existence and uniqueness — 219
- 4.2 SVD matrices — 222
- 4.3 Optimality of SVDs — 227
- 4.4 Centring in SVD — 232

5 Time series of finite rank — **237**
- 5.1 General properties — 237
- 5.2 Series of finite rank and recurrent formulae — 243
- 5.3 Time series continuation — 252

6 SVD of trajectory matrices — **257**
- 6.1 Mathematics of separability — 257
- 6.2 Hankelization — 266
- 6.3 Centring in SSA — 268
- 6.4 SSA for stationary series — 276

List of data sets and their sources — **297**

References — **299**

Index — **303**

Preface

This monograph is about a technique of time series analysis which is often called 'singular-spectrum analysis' (SSA). The basic SSA algorithm looks simple, but understanding of what it does and how it fits among the other time series analysis techniques is by no means simple. At least, it was difficult for us: we have spent a few years on this. This book is an account of what we have learned.

Spending so much time on just one technique should be somehow justified. For us, the justification is our belief in the capabilities of SSA: we are absolutely convinced that for a wide range of time series SSA can be extremely useful. More than that, we firmly believe that in the near future no statistical package will be sold without incorporating SSA facilities, and every time series analysis textbook will contain an SSA-related section.

Although not widely known among statisticians and econometrists, SSA has become a standard tool in meteorology and climatology; it is also a well-known technique in nonlinear physics and signal processing. We think that the lack of popularity of SSA among statisticians was mostly due to tradition and the lack of theory of SSA. We should also accept that the main methodological principle of SSA is not really statistical; SSA is more a technique of multivariate geometry than of statistics. In addition to statistics and multivariate geometry, the theory of SSA comprises the elements of signal processing, linear algebra, nonlinear dynamical systems, the theory of ordinary differential and finite-difference equations, and functional analysis. It is thus not surprising that it took a long time for us to achieve some level of understanding of what SSA is.

Despite the fact that the material of the book touches many different fields, a large part of the book is oriented towards a wide circle of readers who need or have an interest in time series analysis.

SSA is essentially a model-free technique; it is more an exploratory, model-building tool than a confirmatory procedure. It aims at a decomposition of the original series into a sum of a small number of interpretable components such as a slowly varying trend, oscillatory components and a 'structureless' noise. The main concept in studying the SSA properties is 'separability,' which characterizes how well different components can be separated from each other.

An important feature of SSA is that it can be used for analyzing relatively short series. On the other hand, asymptotic separation plays a very important role in the theory of SSA. There is no contradiction here because the asymptotic features (which hold as the length of the series N tends to infinity) are found to be met for relatively small N. In practical applications, we typically deal with series of length varying from a few dozen to a few thousand.

Possible application areas of SSA are diverse: from mathematics and physics to economics and financial mathematics, from meteorology and oceanology to social science and market research. Any seemingly complex series with a potential structure could provide another example of a successful application of SSA.

There are a large number of examples in the book. Many of these examples are real-life series from different areas including medicine, physics, astronomy, economics, and finance. These examples are not the most exciting examples of application of SSA; they were not selected to impress the reader. The purpose of the selection was different: the examples serve only for illustrating the methodological and theoretical aspects discussed in the book. Also, each example illustrates a different feature of the method, so that the number of examples can hardly be reduced.

This book could not have been written had we not acquired a particular computer routine realizing SSA (see the Web site *http://vega.math.spbu.ru/caterpillar*). We were very lucky to have had in our team Kirill Braulov from St. Petersburg University who developed the software. We are very grateful to Kirill for his excellent work. We are also very grateful to our other collaborators and colleagues from the Faculty of Mathematics, St. Petersburg University, and especially to Sergei Ermakov, Vladislav Solntsev, Dmitrii Danilov and Alexander Bart, who have participated in a large number of seminars and discussions on the topic. These seminars and discussions were most useful, especially during the initial stage of the work. Also we are grateful to Dmitry Belov (Institute of Physiology, St. Petersburg University) for permission to use his EEG data for one of the examples in the book.

Our Cardiff University colleague, Gerald Gould, has carefully gone through the manuscript and improved the English where necessary; we are much obliged to him for a very important job. Comments from the Chapman & Hall editors have also helped very much in improving the manuscript; we are really thankful to them.

A part of this work has been done in accordance with the grant GR/M21713, "Multivariate methods in change-point detection problems" from the EPSRC. We are very grateful for this support. However, our main gratitude undoubtedly goes to the Procter & Gamble Company, which for many years has been extremely supportive of us. We have worked with a number of very bright and clever people from the company, but first of all we wish to acknowledge Phil Parker and Luigi Ciutti. Their interest in and support for our work have helped us tremendously.

Last but not least, we are very grateful to our families for their patience and understanding during the long period taken to write this book.

Nina Golyandina, Vladimir Nekrutkin, Anatoly Zhigljavsky

St. Petersburg – Cardiff, October 2000

Notation

SVD	singular value decomposition
LRF	linear recurrent formula
SSA	singular-spectrum analysis
c.d.f.	cumulative distribution function
F	time series
N	length of time series
$F_N = (f_0, \ldots, f_{N-1})$	time series of length N
$F_{i,j} = (f_{i-1}, \ldots, f_{j-1})$	subseries of a time series F_N
L	window length
$K = N - L + 1$	number of L-lagged vectors of F_N
X_i	ith L-lagged vector of time series
$\mathbf{X} = [X_1 : \ldots : X_K]$	trajectory matrix with columns X_i
$\mathbf{X}^{\mathrm{T}}$	transposed matrix $\mathbf{X}$
$\mathcal{M}_{L,K}$	linear space of $L \times K$ matrices
$\langle \mathbf{X}, \mathbf{Y} \rangle_{\mathcal{M}}$	inner product of matrices in $\mathcal{M}_{L,K}$
$\|\mathbf{X}\|_{\mathcal{M}}$	Frobenius matrix norm in $\mathcal{M}_{L,K}$
$\mathrm{rank}(\mathbf{X})$	rank of matrix $\mathbf{X}$
$\mathcal{H}$	Hankelization operator
λ_i	ith eigenvalue of the matrix $\mathbf{X}\mathbf{X}^{\mathrm{T}}$
$\mathbf{E}_M$	identical $M \times M$ matrix
$\mathbf{0}_{LK}$	zero $L \times K$ matrix
$\mathbf{0}_M$	zero vector of dimension M
$\mathbf{1}_M$	vector $(1, \ldots, 1)^{\mathrm{T}}$ of dimension M
$\mathbf{R}^M$	Euclidean space of dimension M
$\mathcal{L}$	linear subspace of the Euclidean space
$\dim \mathcal{L}$	dimension of a linear space $\mathcal{L}$
$\mathcal{L}_r$	linear space of dimension r
$\mathrm{span}(P_1, \ldots, P_n)$	linear space spanned by vectors $P_1, \ldots, P_n$
$\mathrm{span}(\mathbf{X})$	linear space spanned by the columns of $\mathbf{X}$
$\mathcal{L}^{(L)} = \mathcal{L}^{(L)}(F_N)$	L-trajectory space of a time series F_N
$\mathrm{dist}(X, \mathcal{L})$	distance from a vector X to a linear space $\mathcal{L}$
$\mathrm{fdim}(F_N)$	difference dimension of a time series F_N
$\mathrm{rank}_L(F_N)$	L-rank of a time series F_N
$\mathrm{rank}(F_N)$	rank of a time series F_N
U_i	ith eigenvector of the SVD of the matrix $\mathbf{X}$

V_i	ith factor vector of the SVD of the matrix $\mathbf{X}$
$\rho^{(L,M)}$	maximal cross-correlation of two series
$\rho_{12}^{(\omega)}$	weighted cross-correlation of two series
$\rho_{12}^{(\Pi)}$	spectral cross-correlation of two series
R_f	covariance function of a stationary series F
m_f	spectral measure of a stationary series F
p_f	spectral density of a stationary series F
Φ_f	spectral function of a stationary series F
Π_f^N	periodogram of a time series F_N
$g(F_1, F_2)$	heterogeneity index of time series F_1, F_2
$\mathbf{G} = \mathbf{G}_{B,T}$	heterogeneity matrix of a time series F
meas	Lebesque measure in $\mathbf{R}$

Introduction

SSA (singular-spectrum analysis) is a novel technique of time series analysis incorporating the elements of classical time series analysis, multivariate statistics, multivariate geometry, dynamical systems, and signal processing. Despite the fact that a lot of probabilistic and statistical elements are employed in the SSA-based methods (they relate to stationarity, ergodicity, principal component and bootstrap techniques), SSA is not a statistical method in terms of classical statistics. In particular, we typically do not make any statistical assumptions concerning either signal or noise while performing the analysis and investigating the properties of the algorithms.

The present book is fully devoted to the methodology and theory of SSA. The main topics are SSA analysis, SSA forecasting, and SSA detection of structural changes. Let us briefly consider these topics.

SSA analysis of time series

The birth of SSA is usually associated with publication of the papers by Broomhead and King (1986a, 1986b) and Broomhead *et al.* (1987). Since then, the technique has attracted a lot of attention. At present, the papers dealing with methodological aspects and applications of SSA number several hundred; see, for example, Vautard *et al.* (1992), Ghil and Taricco (1997), Allen and Smith (1996), Danilov and Zhigljavsky (1997), Yiou *et al.* (2000) and the references therein. An elementary introduction to the subject can be found in the recent book by Elsner and Tsonis (1996).

SSA has proved to be very successful, and has already become a standard tool in the analysis of climatic, meteorological and geophysical time series; see, for example, Vautard and Ghil (1989), Ghil and Vautard (1991), and Yiou *et al.* (1996). It is thus not surprising that among the main journals publishing SSA-related research papers are Journal of Climate, Journal of the Atmospheric Sciences, and Journal of Geophysical Research.

Let us turn to the description of SSA. The basic version of SSA consists of four steps, which are performed as follows. Let $F = (f_0, f_1, \ldots, f_{N-1})$ be a time series of length N, and L be an integer, which will be called the 'window length'. We set $K = N - L + 1$ and define the L-lagged vectors $X_j = (f_{j-1}, \ldots, f_{j+L-2})^{\mathrm{T}}$, $j = 1, 2, \ldots, K$, and the trajectory matrix

$$\mathbf{X} = (f_{i+j-2})_{i,j=1}^{L,K} = [X_1 : \ldots : X_K].$$

Note that the trajectory matrix $\mathbf{X}$ is a Hankel matrix, which means that all the elements along the diagonal $i+j=\mathrm{const}$ are equal. The construction of the trajectory matrix constitutes the first step of the algorithm.

The second step is the singular value decomposition (SVD) of the matrix $\mathbf{X}$, which can be obtained via eigenvalues and eigenvectors of the matrix $\mathbf{S}=\mathbf{X}\mathbf{X}^\mathrm{T}$ of size $L\times L$. This provides us with a collection of L singular values, which are the square roots of the eigenvalues of the matrix $\mathbf{S}$, and the corresponding left and right singular vectors. (The left singular vectors of $\mathbf{X}$ are the orthonormal eigenvectors of $\mathbf{S}$; in SSA literature, they are often called 'empirical orthogonal functions' or simply EOFs. The right singular vectors can be regarded as the eigenvectors of the matrix $\mathbf{X}^\mathrm{T}\mathbf{X}$.) We thus obtain a representation of $\mathbf{X}$ as a sum of rank-one biorthogonal matrices $\mathbf{X}_i$ ($i=1,\ldots,d$), where d ($d\leq L$) is the number of nonzero singular values of $\mathbf{X}$.

At the third step, we split the set of indices $I=\{1,\ldots,d\}$ into several groups $I_1,\ldots,I_m$ and sum the matrices $\mathbf{X}_i$ within each group. The result of the step is the representation

$$\mathbf{X}=\sum_{k=1}^{m}\mathbf{X}_{I_k},\quad\text{where }\mathbf{X}_{I_k}=\sum_{i\in I_k}\mathbf{X}_i.$$

At the fourth step, averaging over the diagonals $i+j=\mathrm{const}$ of the matrices $\mathbf{X}_{I_k}$ is performed. This gives us an SSA decomposition; that is, a decomposition of the original series F into a sum of series

$$f_n=\sum_{k=1}^{m}f_n^{(k)},\quad n=0,\ldots,N-1,\tag{I.1}$$

where for each k the series $f_n^{(k)}$ is the result of diagonal averaging of the matrix $\mathbf{X}_{I_k}$.

The basic scheme of SSA for analysis of time series and some modifications of this scheme are known in the SSA literature cited above. Note that SSA is usually regarded as a method of identifying and extracting oscillatory components from the original series; see, for example, Yiou *et al.* (1996), Ghil and Taricco (1997), Fowler and Kember (1998). The standard SSA literature, however, does not pay enough attention to theoretical aspects which are very important for understanding how to select the SSA parameters and, first of all, the window length L for the different classes of time series. The concept of separability and related methodological aspects and theoretical results provide us with this understanding. It is the study of separability which makes the biggest distinction between our research on SSA analysis and the standard approach to SSA.

The choice of parameters in performing the SSA decomposition (they are the window length L and the way of grouping the matrices $\mathbf{X}_i$) must depend on the properties of the original series and the purpose of the analysis.

The general purpose of the SSA analysis is the decomposition (I.1) with additive components $f_n^{(k)}$ that are 'independent' and 'identifiable' time series; this is

INTRODUCTION

what we mean when we talk about analyzing the structure of time series by SSA. Sometimes, one can also be interested in particular tasks, such as 'extraction of signal from noise,' 'extraction of oscillatory components' and 'smoothing'.

For a properly made SSA decomposition, a component $f_n^{(k)}$ in (I.1) can be identified as a trend of the original series, an oscillatory series (for example, seasonality) or noise. An oscillatory series is a periodic or quasi-periodic series which can be either pure or amplitude-modulated. Noise is any aperiodic series. The trend of the series is, roughly speaking, a slowly varying additive component of the series with all the oscillations removed.

Note that no parametric model for the components in (I.1) is fixed and these components are produced by the series itself. Thus, when analyzing real-life series with the help of SSA one can hardly hope to obtain the components in the decomposition (I.1) as exact harmonics or linear trend, for example, even if these harmonics or linear trend are indeed present in the series (by a harmonic we mean any sine series with some amplitude, frequency and phase). This is an influence of noise and a consequence of the non-parametric nature of the method. In many cases, however, we can get a good approximation to these series.

In the ideal situation the components in (I.1) must be 'independent'. Achieving 'independence' (or 'separability') of the components in the SSA decomposition (I.1) is of prime importance in SSA. From the authors' viewpoint, separability of components in this decomposition is the main theoretical problem in SSA research and the main target in the selection of SSA parameters. Separability of components is the central problem in the book; it is touched upon in virtually every section.

There are different notions of separability (more precisely, L-separability, since the fact of separability depends on the window length L). The most important is weak separability, defined as follows. Provided that the original time series f_n is a sum of m series $f_n^{(k)}$ ($k = 1, \ldots, m$), for a fixed window length L, weak L-separability means that any subseries of length L of the kth series $f_n^{(k)}$ is orthogonal to any subseries of length L of the lth series $f_n^{(l)}$ with $l \neq k$, and the same holds for their subseries of length $K = N - L + 1$. This is equivalent to the fact that there is a way of constructing the SVD of the trajectory matrix $\mathbf{X}$ and grouping the matrices $\mathbf{X}_j$ so that for each k the matrix $\mathbf{X}_{I_k}$ is the trajectory matrix of the series $f_n^{(k)}$.

The demand of exact separability of components is a strict requirement which rarely holds in practice. The notion of approximate separability is more important (and much less restrictive) than the exact one. For a relatively long series, approximate separability of the components is often achieved due to the theoretical concept of asymptotic separability which holds for a rather wide class of components.

To measure the degree of 'separability' of the components in (I.1) we use a number of different characteristics, such as 'spectral correlation coefficient' or 'weighted correlation coefficient'.

Weak separability may not be sufficient to guarantee that a particular SSA decomposition properly reflects the structure of the original time series. Indeed, in the case when two or more of the singular values of the trajectory matrices $\mathbf{X}^{(k)}$ and $\mathbf{X}^{(l)}$ corresponding to two different components $f_n^{(k)}$ and $f_n^{(l)}$ of the original series are equal (in practice, if the singular values are close), then the SVD is not uniquely defined and the two series $f_n^{(k)}$ and $f_n^{(l)}$ are mixed up, so that an additional analysis (such as rotations in the L-dimensional space of the lagged vectors) is required to separate the two series. If there is (approximate) weak separability and all eigenvalues corresponding to different components in (I.1) are sufficiently isolated from each other, then we have (approximate) strong separability, which means that for a proper grouping the SSA decomposition (approximately) coincides with the one assumed.

The absence of approximate strong separability is often observed for series with complex structure. For these series and series of special structure, there are different ways of modifying SSA. Several modifications of the basic SSA technique can be of interest, such as SSA with single and double centring, Toeplitz SSA, and sequential SSA (when the basic scheme is applied several times with different parameters to the residuals from the previous analysis). SSA with centring and Toeplitz SSA are based on particular non-optimal decompositions of the trajectory matrices; they may be useful in analysis of time series of special structure, such as series with linear-like tendencies and stationary-like series.

Toeplitz SSA was suggested in Vautard and Ghill (1989); it is a well known modification of the basic SSA method. By contrast, SSA with double centring of the trajectory matrix is a new version of SSA.

SSA forecasting of time series

The principles of SSA forecasting developed in this book are new with respect to the main-stream SSA approach. Let us now briefly consider the methodological aspects of SSA forecasting.

An important property of the SSA decomposition is the fact that, if the original series f_n satisfies a linear recurrent formula (LRF)

$$f_n = a_1 f_{n-1} + \ldots + a_d f_{n-d} \qquad (I.2)$$

of some dimension d with some coefficients $a_1, \ldots, a_d$, then for any N and L there are at most d nonzero singular values in the SVD of the trajectory matrix $\mathbf{X}$; therefore, even if the window length L and $K = N - L + 1$ are larger than d, we only need at most d matrices $\mathbf{X}_i$ to reconstruct the series.

The fact that the series f_n satisfies an LRF (I.2) is equivalent to its representability as a sum of products of exponentials, polynomials and harmonics, that is as

$$f_n = \sum_{k=1}^{q} \alpha_k(n) e^{\mu_k n} \sin(2\pi \omega_k n + \varphi_k). \qquad (I.3)$$

INTRODUCTION

Here $a_k(n)$ are polynomials, μ_k, ω_k and φ_k are arbitrary parameters. The number of linearly independent terms q in (I.3) is smaller than or equal to d.

SSA forecasting is based on a fact which, roughly speaking, states the following: if the number of terms r in the SVD of the trajectory matrix $\mathbf{X}$ is smaller than the window length L, then the series satisfies some LRF of some dimension $d \leq r$. Certainly, this assertion must not be understood *ad litteram*. However, for infinite series a similar fact can be found in Gantmacher (1998, Chapter XVI, Section 10, Theorem 7). The theorem due to Buchstaber (1994) amplifies these considerations for finite time series; this theorem says that under the above-mentioned conditions the series (with the possible exception of the last few terms) satisfies some LRF. This assertion, however, does not directly lead to a forecasting algorithm, since the last terms of the series are very important for forecasting.

An essential result for SSA forecasting was obtained in Danilov (1997a, 1997b). It can be formulated as follows: if the dimension r of the linear space $\mathfrak{L}_r$ spanned by the columns of the trajectory matrix is less than the window length L and this space is not a vertical space, then the series satisfies a natural LRF of dimension $L - 1$. (If $e_L \notin \mathfrak{L}_r$, where $e_L = (0, 0, \ldots, 0, 1)^\mathrm{T} \in \mathbf{R}^L$, then we say that $\mathfrak{L}_r$ is a 'non-vertical' space.)

If we have a series satisfying an LRF (I.2), then we can obviously continue it for an arbitrary number of steps using the same LRF. It is important that any LRF governing a given series provides the same continuation, and thus we do not necessarily need the LRF with the minimal value of d. Thus, we now know how to continue time series with non-vertical spaces and small ranks of trajectory matrices.

Of course, when we are dealing with real-life time series we can hardly hope to have a time series that is governed by an LRF of small dimension (in terms of SVD, a 'real-life' trajectory matrix with $L \leq K$ has, as a rule, rank L). However, the class of series that can be approximated by the series governed by the LRFs of the form (I.2) or, equivalently, by the (deterministic) time series of the form (I.3) with a small number of terms, is very broad and we can attempt forecasting of these series using an SSA-based forecasting method. We may also be interested in continuing (forecasting) some periodic (perhaps, amplitude-modulated) components of the original series and in forecasting the trend, ignoring noise and all oscillatory components of the series.

The idea of SSA forecasting of a certain time series component is as follows. The selection of a group of $r < \mathrm{rank}\, \mathbf{X}$ rank-one matrices $\mathbf{X}_i$ on the third step of the basic SSA algorithm implies the selection of an r-dimensional space $\mathfrak{L}_r \subset \mathbf{R}^L$ spanned by the corresponding left singular vectors.

If the space $\mathfrak{L}_r$ is non-vertical, then, as was mentioned previously, this space produces the appropriate LRF, which can be used for forecasting (called recurrent forecasting) of the series component, corresponding to the chosen rank-one matrices.

As in the basic SSA, the separability characteristics help in selection of both the window length L and the space $\mathfrak{L}_r$. Moreover, separability is directly related to LRFs: roughly speaking, if two series are separable, then they satisfy certain LRFs.

The SSA recurrent forecasting algorithm can be modified in several ways. For example, we can base our forecast on the Toeplitz SSA or SSA with centring rather than on the basic SSA (the $\mathfrak{L}_r$ is then spanned by the corresponding versions of left singular vectors); in some cases, we can also base the forecast on the LRF of minimal order. Perhaps the most important modification is the so-called SSA vector forecasting algorithm developed in Nekrutkin (1999). The idea of this method is as follows.

For any group of indices I selected at the grouping stage, the application of SSA gives us $K = N - L + 1$ vectors $\widehat{X}_1, \ldots, \widehat{X}_K$ that lie in an r-dimensional subspace $\mathfrak{L}_r$ of $\mathbf{R}^L$. Here r is the number of elements in I, for each j the $\widehat{X}_j$ is the projection of the L-lagged vector X_j onto the subspace $\mathfrak{L}_r$, and the subspace $\mathfrak{L}_r$ is spanned by the r left eigenvectors of the trajectory matrix $\mathbf{X}$ with the indices in the group I. We then continue the vectors $\widehat{X}_1, \ldots, \widehat{X}_K$ for M steps in such a way that (i) the continuation vectors Z_m ($K < m \leq K + M$) belong to the space $\mathfrak{L}_r$ and (ii) the matrix $[\widehat{X}_1 : \ldots : \widehat{X}_K : Z_{K+1} : \ldots : Z_{K+M}]$ is approximately a Hankel matrix. The forecasting series is then obtained by means of diagonal averaging of this matrix.

While the recurrent forecasting algorithm performs the straightforward recurrent continuation of a one-dimensional series (with the help of the LRF so constructed), the vector forecasting method makes the continuation of the vectors in an r-dimensional space and only then returns to the time-series representation. Examples show that vector forecasting appears to be more stable than the recurrent one, especially for long-term forecasting.

Confidence intervals for the forecasts can be very useful in assessing the quality of the forecasts. However, unlike the SSA forecasts themselves (their construction does not formally require any preliminary information about the time series), for constructing confidence bounds we need some assumptions to be imposed on the series and the residual component, which we associate with noise.

We consider two types of confidence bounds; the first one is for the values of the series itself at some future point $N + M$, and the second one is for the values of the signal at this future point (under the assumption that the original series consists of a signal and additive noise). These two types of confidence intervals are constructed in different ways: in the first case, we use the information about forecast errors obtained during the analysis of the series; the second one uses the bootstrap technology.

To build the confidence intervals for the forecast of the entire initial series, we construct the forecasting LRF of dimension $L - 1$ (in the case of the recurrent forecast) and repeatedly apply it to all subseries of the same dimension within the observation period $[0, N-1]$. Then we compare the results with the corresponding values of the series. Under the assumption that the residual series is stationary and

INTRODUCTION

ergodic, we can estimate the quantiles of the related marginal distribution, and therefore build the confidence bounds.

The bootstrap technique is useful for constructing confidence intervals for the signal $F^{(1)}$ at some future time $N + M$ under the assumption that the series $F_N = (f_0, \ldots, f_{N-1})$ is a sum of a signal $F_N^{(1)}$ and noise $F_N^{(2)} = F_N - F_N^{(1)}$. To do that, we first obtain the SSA decomposition $F_N = \widetilde{F}_N^{(1)} + \widetilde{F}_N^{(2)}$, where $\widetilde{F}_N^{(1)}$ (the reconstructed series) approximates $F_N^{(1)}$, and $\widetilde{F}_N^{(2)}$ is the residual series. Assuming that we have a (stochastic) model for the residuals $\widetilde{F}_N^{(2)}$, we then simulate some number S of independent copies $\widetilde{F}_{N,i}^{(2)}$ of the series $F_N^{(2)}$, obtain S series $\widetilde{F}_N^{(1)} + \widetilde{F}_{N,i}^{(2)}$ and get S forecasting results $\widetilde{f}_{N+M-1,i}^{(1)}$. Having obtained the sample $\widetilde{f}_{N+M-1,i}^{(1)}$ ($1 \leq i \leq S$) of the forecasting results, we use it to calculate the empirical lower and upper quantiles of fixed level γ and construct the corresponding confidence interval for the forecast.

Note that the bootstrap confidence bounds can be constructed not only for the SSA forecasts but also for the terms of the SSA decomposition when we are dealing with separation of a signal from noise.

SSA detection of structural changes in time series

We call a time series F_N homogeneous if it is governed by an LRF of order d that is small relative to the length of the series N.

Assume now that the series is homogeneous until some time $Q < N$, but then it stops following the original LRF (this may be caused by a perturbation of the series). However, after a certain time period, it again becomes governed by an LRF. In this case, we have a structural change (heterogeneity) in the series. We may have either a permanent heterogeneity (in this case the new LRF is different from the original one) or a temporary heterogeneity, when both LRFs coincide. Note that even in the latter case, the behaviour of the series after the change is different from the behaviour of the homogeneous (unperturbed) series; for example, the initial conditions for the LRF after the perturbation can be different from the unperturbed initial conditions.

The main idea of employing SSA for detecting different types of heterogeneity is as follows. The results of Section 5.2 imply that for sufficiently large values of the window length L the L-lagged vectors of a homogeneous series span the same linear space $\mathfrak{L}^{(L)}$ independently of N, as soon as N is sufficiently large. Therefore, violations in homogeneity of the series can be described in terms of the corresponding lagged vectors: the perturbations force the lagged vectors to leave the space $\mathfrak{L}^{(L)}$. The corresponding discrepancies are defined in terms of the distances between the lagged vectors and the space $\mathfrak{L}^{(L)}$, which can be determined for different subseries of the original series.

Since, in practice, the series are described by LRFs only approximately, the problem of approximate construction of the spaces $\mathfrak{L}^{(L)}$ arises again. Analogous to the problems of forecasting, the SVD of the trajectory matrices is used for this purpose. As everywhere in the book, the concept of separability plays a very important role when we are interested in detecting changes in components of the series (for example, in the signal, under the presence of additive noise). Unlike the forecasting problems, for studying structural changes in time series, the properties of the SVDs of subseries of the initial series F become of prime importance.

We consider two subseries (say F' and F'') of the series F; we call them 'base subseries' and 'test subseries'. Assume that the lengths of these subseries are fixed and equal to B and T, respectively. Suppose that $B > L$ and $T \geq L$, where L is the window length. Let us make an SVD of the trajectory matrix of the base subseries, select a group of $r < L$ left singular vectors, consider the linear space $\mathfrak{L}'_r$ spanned by these vectors and compute the sum of the squared distances between the space $\mathfrak{L}'_r$ and the L-lagged vectors corresponding to the test subseries. If we normalize this sum by the sum of the squared norms of the L-lagged vectors of the test subseries, then we obtain the so-called heterogeneity index $g = g(F', F'')$ formally defined in Section 3.1. The heterogeneity index $g(F', F'')$ measures the discrepancy between F' and F'' by computing the relative error of the optimal approximation of the L-lagged vectors of the time series F'' by vectors from the space $\mathfrak{L}'_r$.

The main tool used to study structural changes (heterogeneities) in time series is the 'heterogeneity matrix' of size $(N - B + 1) \times (N - T + 1)$. The entries of this matrix are the values of the heterogeneity index $g = g(F', F'')$, where F' and F'' run over all possible subseries of the series F of fixed lengths B and T, respectively.

The columns, rows and some diagonals of the heterogeneity matrix constitute the 'heterogeneity functions'. Change in the indexation system gives us the 'detection functions'; they are more convenient for the purpose of change detection.

We also consider three groups of supplementary detection characteristics. The first group is obtained when we use a different normalization in the expression for the heterogeneity index (rather than using the sum of the squared norms of the L-lagged vectors of the test subseries, we use the sum of the squared terms of the whole series). This renormalization of the heterogeneity index often helps when we monitor changes in monotone series and their components.

The second group of characteristics relates to the series of the roots of the characteristic polynomials of the LRFs that correspond to the SSA decomposition of the base subseries F'. The roots of the characteristic polynomials monitor the dynamics of the linear spaces $\mathfrak{L}'_r$. In particular, this monitoring can be very useful for distinguishing the changes that actually happen in the series from spurious changes that are caused by the fact that abrupt changes in the dynamics of the linear spaces $\mathfrak{L}'_r$ may be related to the changes in the order of the singular values.

The third group of characteristics is basically the moving periodograms of the original series; this group is used to monitor the spectral structure of the original series.

INTRODUCTION

Composition of the book

The book has two parts; they are devoted to the methodology and theory of SSA, respectively. The methodological principles of SSA are thoroughly considered in Part I of the book. This part consists of three chapters, which deal with SSA analysis, SSA forecasting and SSA detection of structural changes, respectively.

SSA analysis of time series is dealt with in Chapter 1. In Section 1.1, the basic algorithm is described. In Section 1.2, the steps of this algorithm are explained and commented on. In Section 1.3, the main capabilities of the basic algorithm are illustrated by a number of real-life examples. In Section 1.4, the major tasks that can be attempted by SSA are formulated and discussed. In Section 1.5, the concept of separability is considered in detail. These considerations play a very important role in the selection of the parameters of SSA, the problem which is dealt with in Section 1.6. In Section 1.7, supplementary SSA techniques, such as SSA with centring and Toeplitz SSA, are considered.

Chapter 2 is devoted to SSA forecasting methodology. In Section 2.1, we formally describe the SSA recurrent forecasting algorithm. In Section 2.2, the principles of SSA forecasting and links with LRFs are discussed. Several modifications of the basic SSA recurrent forecasting algorithm are formulated and discussed in Section 2.3. The construction of confidence intervals for the forecasts is made in Section 2.4. In Section 2.5, we summarize the material of the chapter, and in Section 2.6 we provide several examples illustrating different aspects of SSA forecasting.

The methodology of SSA detection of structural changes in time series is considered in Chapter 3. In Section 3.1, we introduce and discuss the main concepts. In Section 3.2, we consider various violations of homogeneity in time series and the resulting shapes of the heterogeneity matrices and detection functions. In Section 3.3, we generalize the results of Section 3.2 to the case when we are detecting heterogeneities in one of the components of the original series rather than in the series itself (this includes the case when the series of interest is observed with noise). The problem of the choice of detection parameters is dealt with in Section 3.4. In Section 3.5, we consider several additional detection characteristics, and in Section 3.6 we provide a number of examples.

Chapters 4, 5 and 6 constitute the second (theoretical) part of the book, where all the statements of Part I are properly formulated and proved (with the exception of some well-known results where the appropriate references are given).

Chapter 4 considers the singular value decomposition (SVD) of real matrices, which is the main mathematical tool in the SSA method. The existence and uniqueness of SVDs is dealt with in Section 4.1. In Section 4.2, we discuss the structure and properties of the SVD matrices with special attention paid to such features of SVD as orthogonality, biorthogonality, and minimality. In Section 4.3, we consider optimal features of the SVD from the viewpoints of multivariate geometry and approximation of matrices by matrices of lower rank. A number of results on optimality of the standard SVD are generalized in Section 4.4 to the SVD with single and double centring.

Chapter 5 provides a formal mathematical treatment of time series of finite rank; the L-trajectory matrices of these series have rank less than $\min(L, K)$ for all sufficiently large L and K. General properties of such series are considered in Section 5.1. As discussed above, the series of finite rank are related to the series governed by the LRFs; these relations are studied in Section 5.2. The results concerning the continuation procedures are derived in Section 5.3.

In Chapter 6, we make a formal mathematical study of four topics that are highly important for the SSA methodology. Specifically, in Section 6.1 we study weak separability of time series, in Section 6.2 diagonal averaging (Hankelization) of matrices is considered, while centring in SSA is studied in Section 6.3, and specific features of SSA for deterministic stationary sequences are discussed in Section 6.4.

Other SSA and SSA-related topics

On the whole, this book considers many important issues relating to the implementation, analysis and practical application of SSA. There are, however, several other topics which are not covered here. Let us mention some of them.

1. *Multichannel SSA.* Multichannel SSA is an extension of the standard SSA to the case of multivariate time series (see Broomhead and King, 1986b). It can be described as follows. Assume that we have an l-variate time series $f_n = \left(f_n^{(1)}, \ldots, f_n^{(l)}\right)$, where $n = 0, 1, \ldots, N-1$ (for simplicity we assume that the time domain is the same for all the components of the series). Then for a fixed window length L we can define the trajectory matrices $\mathbf{X}^{(i)}$ $(i = 1, \ldots, l)$ of the one-dimensional time series $f_n^{(i)}$. The trajectory matrix $\mathbf{X}$ can then be defined as

$$\mathbf{X} = \begin{pmatrix} \mathbf{X}^{(1)} \\ \cdots \\ \mathbf{X}^{(l)} \end{pmatrix}. \tag{I.4}$$

The other stages of the multichannel SSA procedure are identical to the one-dimensional procedure discussed above with obvious modification that the diagonal averaging should be applied to each of the l components separately. (Multichannel SSA can be generalized even further, for analyzing discrete time random fields and image processing problems; see Danilov and Zhigljavsky, 1997.)

There are numerous examples of successful application of the multichannel SSA (see, for example, Plaut and Vautard, 1994; Danilov and Zhigljavsky, 1997), but the theory of multichannel SSA is yet to be developed. The absence of a theory is the reason why, in the present book, we have confined ourselves to the univariate case only. This case is already difficult enough, and multichannel SSA has additional peculiarities.

Construction of the trajectory matrix in multichannel SSA is not obvious; there are several alternatives to (I.4). The matrix (I.4) seems to be the natural candidate for the trajectory matrix of a multivariate series, but its advantages are not clear. Note also that there is a version of SSA that deals with complex-valued series; it can be considered as a version of multichannel SSA as well. It is, however, not clear how to compare the two-channel SSA with the one-channel complex SSA.

2. *Continuous time SSA.* The basic SSA scheme and most of its variations can be modified for the case of continuous time. There are many significant changes (with respect to the material of the book) that would to be made if one were to try to analyze the corresponding procedure: instead of sums we get integrals, instead of matrices we have linear operators, the SVD becomes the Schmidt decomposition in the corresponding Hilbert space, LRFs become ordinary differential equations, and so on. Note that the theory of generalized continuous time SSA includes the standard discrete time SSA as a particular case. In addition, such a generalization allows us to consider not only embeddings of Hankel type but also many other mappings which transfer functions of one variable to the functions of two variables. Those interested in this approach can find a lot of related material in Nekrutkin (1997).

3. *Use of different window lengths.* The use of different values of the window length is discussed in Section 1.7 in relation to the so-called 'Sequential SSA'. There are some other suggestions in the literature, such as selecting the window length at random (see Varadi *et al.*, 1999) or keeping the ratio L'/N' fixed, where L' is the window length for the subseries of the original series of length $N' = N/k$ which is obtained by sieving the original series (see Yiou *et al.*, 2000). Both methods are suggested for analyzing long series; the latter one is shown to have some similarity with the wavelet analysis of time series.

4. *SSA for sequential detection of structural changes.* The methodology of Chapter 3 aims at a nonsequential (posterior) detection of structural changes in time series. Some of these algorithms can be modified for the more standard change-point problem of sequential detection of change-points. This approach is implemented in Moskvina and Zhigljavsky (2000), where some of the detection algorithms are analyzed as proper statistical procedures. The Web site *http://www.cf.ac.uk/maths/stats/changepoint/* contains more information on the subject and a link to the software that can be downloaded.

Let us mention some other areas related to SSA.

During the last forty years, a variety of techniques of time series analysis and signal processing have been suggested that use SVDs of certain matrices; for surveys see, for example, Marple (1987) or Bouvet and Clergeot (1988). Most of these techniques are based on the assumption that the original series is random and stationary; they include some techniques that are famous in signal processing, such as Karhunen-Loève decomposition and the MUSIC algorithm (for the signal processing references, see, for example, Madisetti and Williams, 1998). Some statistical aspects of the SVD-based methodology for stationary series are considered, for example, in Brillinger (1975, Chapter 9), Subba Rao (1976) and Subba Rao and Gabr (1984).

The analysis of periodograms is an important part of the process of identifying the components in the SSA decomposition (I.1). For example, noise is modeled by aperiodic (chaotic) series whose spectral measures do not have atoms (white noise has constant spectral density). A comparison of the observed spectrum of the residual component in the SSA decomposition with the spectrum of some common time series (these can be found, for example, in Priestley, 1991 and Wei, 1990, Chapter 11) can help in understanding the nature of the residuals and formulation of a proper statistical hypothesis concerning the noise. However, a single realization of a noise series can have a spectrum that significantly differs from the theoretical one. Several simulation-based tests for testing the white noise zero hypothesis against the 'red noise' alternative (i.e., an autoregressive process of the first order) have been devised; the approach is called 'Monte Carlo SSA', see Allen and Smith (1996). This approach has attracted a lot of attention of researchers; for its extension and enhancement see, for example, Paluš and Novotna (1998).

Another area which SSA is related to is nonlinear (deterministic) time series analysis. It is a fashionable area of rapidly growing popularity; see the recent books by Cutler and Kaplan (1997), Kantz and Schreiber (1997), Abarbanel (1996), Tong (1993), and Weigend and Gershenfeld (1993). Note that the specialists in nonlinear time series analysis (as well as statisticians) do not always consider SSA as a technique that could compete with more standard methods; see, for example, Kantz and Schreiber (1997, Section 9.3.2).

It is impossible to discuss all the fields related to SSA. In a certain wide sense, one can consider SSA as a method of approximating the original series (or its component) with the other series governed by an LRF. Then we can consider a long list of publications on the theme, starting with Prony (1795).

On the other hand, the essential feature of SSA is the choice of the optimal basis consisting of the left singular vectors. If we do not restrict ourselves to strong optimality (see the discussion on Toeplitz and centring SSA), then we arrive at a wide class of methods dealing with different bases (including, for example, the wavelet bases) that can be used for the decomposition of the lagged vectors.

As has already been mentioned, in signal processing, nonlinear physics and some other fields, a number of methods are in use that are based on SVDs of the trajectory matrices (as well as other matrices calculated through the terms of time series); these methods are used for different purposes.

Thus, the area of SSA-related methods is very wide. This is one of the reasons why we are confident that the ideas and methodology of SSA described in this book will be useful for a wide circle of scientists in different fields for many years to come.

PART I
SSA: Methodology

CHAPTER 1

Basic SSA

This chapter deals with the basic scheme of SSA and several modifications of it. Only the problem of analysis of the structure of a one-dimensional real-valued time series is considered. Some refined generalizations of the basic scheme adapted to the problems of time series forecasting and homogeneity analysis (including the change-point detection problem) are considered in the subsequent chapters.

Briefly, in this chapter we consider Basic SSA as a model-free tool for time series structure recognition and identification. We do not want to specify the notion 'structure' at the moment but mention that the goal of Basic SSA is a decomposition of the series of interest into several additive components that typically can be interpreted as 'trend' components (that is, smooth and slowly varying parts of the series), various 'oscillatory' components (perhaps with varying amplitudes), and 'noise' components.

In this chapter we do not assign any stochastic meaning to the term 'noise': the concept of a deterministic stationary 'noise' series is generally more convenient for SSA since it deals with a single trajectory of a time series rather than with a sample of such trajectories. Also, it may occur that we are not interested in certain components of the series and can therefore subsume them under the noise components.

Basic SSA performs four steps. At the first step (called the *embedding step*), the one-dimensional series is represented as a multidimensional series whose dimension is called the *window length*. The multidimensional time series (which is a sequence of vectors) forms the *trajectory matrix*. The sole (and very important) parameter of this step is the window length.

The second step, *SVD step*, is the singular value decomposition of the trajectory matrix into a sum of rank-one bi-orthogonal matrices. The first two steps together are considered as the *decomposition stage* of Basic SSA.

The next two steps form the *reconstruction stage*. The *grouping step* corresponds to splitting the matrices, computed at the SVD step, into several groups and summing the matrices within each group. The result of the step is a representation of the trajectory matrix as a sum of several *resultant matrices*.

The last step transfers each resultant matrix into a time series, which is an additive component of the initial series. The corresponding operation is called *diagonal averaging*. It is a linear operation and maps the trajectory matrix of the initial series into the initial series itself. In this way we obtain a decomposition of the initial series into several additive components.

Let us describe these steps formally and discuss their meaning and features.

1.1 Basic SSA: description

Let $N > 2$. Consider a real-valued time series $F = (f_0, \ldots, f_{N-1})$ of length N. Assume that F is a nonzero series; that is, there exists at least one i such that $f_i \neq 0$. Though one can usually assume that $f_i = f(i\Delta)$ for a certain function of time $f(t)$ and a certain time interval Δ, this does not play any specific role in our considerations.

Moreover, the numbers $0, \ldots, N-1$ can be interpreted not only as discrete time moments but also as labels of any other linearly ordered structure. The numbering of the time series values starts at $i = 0$ rather than at the more standard $i = 1$; this is only for convenience of notation.

As was already mentioned, Basic SSA consists of two complementary stages: decomposition and reconstruction.

1.1.1 First stage: decomposition

1st step: Embedding

The *embedding* procedure maps the original time series to a sequence of multidimensional lagged vectors.

Let L be an integer (*window length*), $1 < L < N$. The embedding procedure forms $K = N - L + 1$ *lagged vectors*

$$X_i = (f_{i-1}, \ldots, f_{i+L-2})^{\mathrm{T}}, \quad 1 \leq i \leq K,$$

which have dimension L. If we need to emphasize the dimension of the X_i, then we shall call them *L-lagged vectors*.

The *L-trajectory matrix* (or simply *trajectory matrix*) of the series F:

$$\mathbf{X} = [X_1 : \ldots : X_K]$$

has lagged vectors as its columns. In other words, the trajectory matrix is

$$\mathbf{X} = (x_{ij})_{i,j=1}^{L,K} = \begin{pmatrix} f_0 & f_1 & f_2 & \cdots & f_{K-1} \\ f_1 & f_2 & f_3 & \cdots & f_K \\ f_2 & f_3 & f_4 & \cdots & f_{K+1} \\ \vdots & \vdots & \vdots & \ddots & \vdots \\ f_{L-1} & f_L & f_{L+1} & \cdots & f_{N-1} \end{pmatrix}. \quad (1.1)$$

Obviously $x_{ij} = f_{i+j-2}$ and the matrix $\mathbf{X}$ has equal elements on the 'diagonals' $i + j = $ const. (Thus, the trajectory matrix is a *Hankel matrix*.) Certainly if N and L are fixed, then there is a one-to-one correspondence between the trajectory matrices and the time series.

BASIC SSA: DESCRIPTION

2nd step: Singular value decomposition
The result of this step is the singular value decomposition (SVD) of the trajectory matrix. Let $\mathbf{S} = \mathbf{XX}^\mathrm{T}$. Denote by $\lambda_1, \ldots, \lambda_L$ the *eigenvalues* of $\mathbf{S}$ taken in the decreasing order of magnitude ($\lambda_1 \geq \ldots \geq \lambda_L \geq 0$) and by $U_1, \ldots, U_L$ the orthonormal system of the *eigenvectors* of the matrix $\mathbf{S}$ corresponding to these eigenvalues. Let $d = \max\{i, \text{ such that } \lambda_i > 0\}$.

If we denote $V_i = \mathbf{X}^\mathrm{T} U_i / \sqrt{\lambda_i}$ ($i = 1, \ldots, d$), then the SVD of the trajectory matrix $\mathbf{X}$ can be written as

$$\mathbf{X} = \mathbf{X}_1 + \ldots + \mathbf{X}_d, \tag{1.2}$$

where $\mathbf{X}_i = \sqrt{\lambda_i} U_i V_i^\mathrm{T}$. The matrices $\mathbf{X}_i$ have rank 1; therefore they are *elementary matrices*. The collection $(\sqrt{\lambda_i}, U_i, V_i)$ will be called *i*th *eigentriple* of the SVD (1.2).

1.1.2 Second stage: reconstruction

3rd step. Grouping
Once the expansion (1.2) has been obtained, the grouping procedure partitions the set of indices $\{1, \ldots, d\}$ into m disjoint subsets $I_1, \ldots, I_m$.

Let $I = \{i_1, \ldots, i_p\}$. Then the *resultant matrix* $\mathbf{X}_I$ corresponding to the group I is defined as $\mathbf{X}_I = \mathbf{X}_{i_1} + \ldots + \mathbf{X}_{i_p}$. These matrices are computed for $I = I_1, \ldots, I_m$ and the expansion (1.2) leads to the decomposition

$$\mathbf{X} = \mathbf{X}_{I_1} + \ldots + \mathbf{X}_{I_m}. \tag{1.3}$$

The procedure of choosing the sets $I_1, \ldots, I_m$ is called the *eigentriple grouping*.

4th step: Diagonal averaging
The last step in Basic SSA transforms each matrix of the grouped decomposition (1.3) into a new series of length N.

Let $\mathbf{Y}$ be an $L \times K$ matrix with elements y_{ij}, $1 \leq i \leq L$, $1 \leq j \leq K$. We set $L^* = \min(L, K)$, $K^* = \max(L, K)$ and $N = L + K - 1$. Let $y^*_{ij} = y_{ij}$ if $L < K$ and $y^*_{ij} = y_{ji}$ otherwise.

Diagonal averaging transfers the matrix $\mathbf{Y}$ to the series $g_0, \ldots, g_{N-1}$ by the formula:

$$g_k = \begin{cases} \dfrac{1}{k+1} \displaystyle\sum_{m=1}^{k+1} y^*_{m, k-m+2} & \text{for } 0 \leq k < L^* - 1, \\ \dfrac{1}{L^*} \displaystyle\sum_{m=1}^{L^*} y^*_{m, k-m+2} & \text{for } L^* - 1 \leq k < K^*, \\ \dfrac{1}{N-k} \displaystyle\sum_{m=k-K^*+2}^{N-K^*+1} y^*_{m, k-m+2} & \text{for } K^* \leq k < N. \end{cases} \tag{1.4}$$

The expression (1.4) corresponds to averaging of the matrix elements over the 'diagonals' $i + j = k + 2$: the choice $k = 0$ gives $g_0 = y_{11}$, for $k = 1$ we have

$g_1 = (y_{12} + y_{21})/2$, and so on. Note that if the matrix $\mathbf{Y}$ is the trajectory matrix of some series $(h_0, \ldots, h_{N-1})$ (in other words, if $\mathbf{Y}$ is the Hankel matrix), then $g_i = h_i$ for all i.

Diagonal averaging (1.4) applied to a resultant matrix $\mathbf{X}_{I_k}$ produces the series $\widetilde{F}^{(k)} = (\widetilde{f}_0^{(k)}, \ldots, \widetilde{f}_{N-1}^{(k)})$ and therefore the initial series $f_0, \ldots, f_{N-1}$ is decomposed into the sum of m series:

$$f_n = \sum_{k=1}^{m} \widetilde{f}_n^{(k)}. \tag{1.5}$$

1.2 Steps in Basic SSA: comments

The formal description of the steps in Basic SSA requires some elucidation. In this section we briefly discuss the meaning of the procedures involved.

1.2.1 Embedding

Embedding can be regarded as a mapping that transfers a one-dimensional time series $F = (f_0, \ldots, f_{N-1})$ to the multidimensional series $X_1, \ldots, X_K$ with vectors $X_i = (f_{i-1}, \ldots, f_{i+L-2})^{\mathrm{T}} \in \mathbf{R}^L$, where $K = N - L + 1$. Vectors X_i are called *L-lagged vectors* (or, simply, *lagged vectors*).

The single parameter of the embedding is the *window length L*, an integer such that $2 \leq L \leq N - 1$.

Embedding is a standard procedure in time series analysis. With the embedding being performed, further development depends on the purpose of the investigation.

For specialists in dynamical systems, a common technique is to obtain the empirical distribution of all the pairwise distances between the lagged vectors X_i and X_j and then calculate the so-called correlation dimension of the series. This dimension is related to the fractal dimension of the attractor of the dynamical system that generates the time series. (See Takens, 1981; Sauer, Yorke and Casdagli, 1991, for the theory and Nicolis and Prigogine, 1989, Appendix IV, for the corresponding algorithm.) Note that in this approach, L must be relatively small and K must be very large (formally, $K \to \infty$).

If L is sufficiently large, then one can consider each L-lagged vector X_i as a separate series and investigate the dynamics of certain characteristics for this collection of series. The simplest example of this approach is the well-known 'moving average' method, where the averages of the lagged vectors are computed. There are also much more sophisticated algorithms.

For example, if the initial series can be considered as a locally stationary process, then we can expand each lagged vector X_i with respect to any fixed basis (for instance, the Fourier basis or a certain wavelet basis) and study the dynamics of such an expansion. These ideas correspond to the dynamical Fourier analysis. Evidently, other bases can be applied as well.

The approximation of a stationary series with the help of the autoregression models can also be expressed in terms of embedding: if we deal with the model

$$f_{i+L-1} = a_{L-1}f_{i+L-2} + a_1 f_i + \varepsilon_{i+L-1}, \quad i \geq 0, \tag{1.6}$$

then we search for a vector $A = (a_1, \ldots, a_{L-1}, -1)^{\mathrm{T}}$ such that the inner products (X_i, A) are described in terms of a certain noise series.

Note that these (and many other) techniques that use the embedding can be divided into two large parts, which may be called 'global' and 'dynamical'. The global methods treat the X_i as L-dimensional vectors and do not use their ordering.

For instance, if we calculate the empirical distribution of the pairwise distances between the lagged vectors, then the result does not depend on the order in which these vectors appear. A similar situation occurs for the autoregression model (1.6) if the coefficients a_i are calculated through the whole collection of the lagged vectors (for example, by the least squares method).

This invariance under permutation of the lagged vectors is not surprising since both models deal with stationary-like series and are intended for finding global characteristics of the whole series. The number of lagged vectors K plays the role of the 'sample size' in these considerations, and therefore it has to be rather large. Theoretically, in these approaches L must be fixed and $K \to \infty$.

The situation is different when we deal with the dynamical Fourier analysis and similar methods, and even with the moving averages. Here the order of the lagged vectors is important and describes the dynamics of interest. Therefore, the nonstationary scenario is the main application area for these approaches. As for L and K, their relationship can generally be arbitrary and should depend on the concrete data and the concrete problem.

At any rate, the window length L should be sufficiently large. The value of L has to be large enough so that each L-lagged vector incorporates an essential part of the behaviour of the initial series $F = (f_0, \ldots, f_{N-1})$.

In accordance with the formal description of the embedding step (see Section 1.1.1), the result of this step is a *trajectory matrix*

$$\mathbf{X} = [X_1 : \ldots : X_K]$$

rather than just a collection of the lagged vectors X_i. This means that generally we are interested in the dynamical effects (though some characteristics that are invariant under permutations of the lagged vectors will be important as well).

The trajectory matrix (1.1) possesses an obvious symmetry property: the transposed matrix $\mathbf{X}^{\mathrm{T}}$ is the trajectory matrix of the same series $f_0, \ldots, f_{N-1}$ with window length equal to K rather than L.

1.2.2 Singular value decomposition

Singular value decomposition (SVD) of the trajectory matrix (1.1) is the second step in Basic SSA. SVD can be described in different terms and be used for different purposes. (See Chapter 4 for the mathematical results.) Most SVD features are valid for general $L \times K$ matrices, but the Hankel structure of the trajectory matrix adds a number of specific features. Let us start with general properties of the SVD important for the SSA.

As was already mentioned, the SVD of an arbitrary nonzero $L \times K$ matrix $\mathbf{X} = [X_1 : \ldots : X_K]$ is a decomposition of $\mathbf{X}$ in the form

$$\mathbf{X} = \sum_{i=1}^{d} \sqrt{\lambda_i} U_i V_i^{\mathrm{T}}, \qquad (1.7)$$

where λ_i ($i = 1, \ldots, L$) are eigenvalues of the matrix $\mathbf{S} = \mathbf{X}\mathbf{X}^{\mathrm{T}}$ arranged in decreasing order of magnitudes,

$$d = \max\{i, \text{ such that } \lambda_i > 0\} = \operatorname{rank} \mathbf{X},$$

$\{U_1, \ldots, U_d\}$ is the corresponding orthonormal system of the eigenvectors of the matrix $\mathbf{S}$, and $V_i = \mathbf{X}^{\mathrm{T}} U_i / \sqrt{\lambda_i}$.

Standard SVD terminology calls $\sqrt{\lambda_i}$ the *singular values*; the U_i and V_i are the *left* and *right singular vectors* of the matrix $\mathbf{X}$, respectively. The collection $(\sqrt{\lambda_i}, U_i, V_i)$ is called *ith eigentriple* of the matrix $\mathbf{X}$. If we define $\mathbf{X}_i = \sqrt{\lambda_i} U_i V_i^{\mathrm{T}}$, then the representation (1.7) can be rewritten in the form (1.2), i.e. as the representation of $\mathbf{X}$ as a sum of the elementary matrices $\mathbf{X}_i$.

If all the eigenvalues have multiplicity one, then the expansion (1.2) is uniquely defined. Otherwise, if there is at least one eigenvalue with multiplicity larger than 1, then there is a freedom in the choice of the corresponding eigenvectors. We shall assume that the eigenvectors are somehow chosen and the choice is fixed.

Since SVD deals with the whole matrix $\mathbf{X}$, it is not invariant under permutation of its columns $X_1, \ldots, X_K$. Moreover, the equality (1.7) shows that the SVD possesses the following property of symmetry: $V_1, \ldots, V_d$ form an orthonormal system of eigenvectors for the matrix $\mathbf{X}^{\mathrm{T}} \mathbf{X}$ corresponding to the same eigenvalues λ_i. Note that the rows and columns of the trajectory matrix are subseries of the original time series. Therefore, the left and right singular vectors also have a temporal structure and hence can also be regarded as time series.

SVD (1.2) possesses a number of optimal features. One of these properties is as follows: among all the matrices $\mathbf{X}^{(r)}$ of rank $r < d$, the matrix $\sum_{i=1}^{r} \mathbf{X}_i$ provides the best approximation to the trajectory matrix $\mathbf{X}$, so that $\|\mathbf{X} - \mathbf{X}^{(r)}\|_{\mathcal{M}}$ is minimum.

Here and below the (*Frobenius*) *norm* of a matrix $\mathbf{Y}$ is $\sqrt{\langle \mathbf{Y}, \mathbf{Y} \rangle_{\mathcal{M}}}$, where the *inner product* of two matrices $\mathbf{Y} = (y_{ij})_{i,j=1}^{q,s}$ and $\mathbf{Z} = (z_{ij})_{i,j=1}^{q,s}$ is defined as

$$\langle \mathbf{Y}, \mathbf{Z} \rangle_{\mathcal{M}} = \sum_{i,j=1}^{q,s} y_{ij} z_{ij}.$$

Note that $\|\mathbf{X}\|_{\mathcal{M}}^2 = \sum_{i=1}^{d} \lambda_i$ and $\lambda_i = \|\mathbf{X}_i\|_{\mathcal{M}}^2$ for $i = 1, \ldots, d$. Thus, we shall consider the ratio $\lambda_i / \|\mathbf{X}\|_{\mathcal{M}}^2$ as the characteristic of the contribution of the matrix $\mathbf{X}_i$ in the expansion (1.2) to the whole trajectory matrix $\mathbf{X}$. Consequently, $\sum_{i=1}^r \lambda_i/\|\mathbf{X}\|_\mathcal{M}^2$, the sum of the first r ratios, is the characteristic of the optimal approximation of the trajectory matrix by the matrices of rank r.

Let us now consider the trajectory matrix $\mathbf{X}$ as a sequence of L-lagged vectors. Denote by $\mathfrak{L}^{(L)} \subset \mathbf{R}^L$ the linear space spanned by the vectors $X_1, \ldots, X_K$. We shall call this space the *L-trajectory space* (or, simply, *trajectory space*) of the series F. To emphasize the role of the series F, we use notation $\mathfrak{L}^{(L)}(F)$ rather than $\mathfrak{L}^{(L)}$. The equality (1.7) shows that $\mathcal{U} = (U_1, \ldots, U_d)$ is an orthonormal basis in the d-dimensional trajectory space $\mathfrak{L}^{(L)}$.

Setting $Z_i = \sqrt{\lambda_i} V_i$, $i = 1, \ldots, d$, we can rewrite the expansion (1.7) in the form

$$\mathbf{X} = \sum_{i=1}^d U_i Z_i^\mathrm{T}, \tag{1.8}$$

and for the lagged vectors X_j we have

$$X_j = \sum_{i=1}^d z_{ji} U_i, \tag{1.9}$$

where the z_{ji} are the components of the vector Z_i.

By (1.9), z_{ji} is the ith component of the vector X_j, represented in the basis $\mathcal{U}$. In other words, the vector Z_i is composed of the ith components of lagged vectors represented in the basis $\mathcal{U}$.

Let us now consider the transposed trajectory matrix $\mathbf{X}^\mathrm{T}$. Introducing $Y_i = \sqrt{\lambda_i} U_i$ we obtain the expansion

$$\mathbf{X}^\mathrm{T} = \sum_{i=1}^d V_i Y_i^\mathrm{T},$$

which corresponds to the representation of the sequence of K-lagged vectors in the orthonormal basis $V_1, \ldots, V_d$. Thus, the SVD gives rise to two dual geometrical descriptions of the trajectory matrix $\mathbf{X}$.

The optimal feature of the SVD considered above may be reformulated in the language of multivariate geometry for the L-lagged vectors as follows. Let $r < d$. Then among all r-dimensional subspaces $\mathfrak{L}_r$ of $\mathbf{R}^L$, the subspace $\mathfrak{L}_r^{(0)} \stackrel{\text{def}}{=} \mathfrak{L}(U_1, \ldots, U_r)$, spanned by $U_1, \ldots, U_r$, approximates these vectors in the best way; that is, the minimum of $\sum_{i=1}^K \mathrm{dist}^2(X_i, \mathfrak{L}_r)$ is attained at $\mathfrak{L}_r^{(0)}$. The ratio $\sum_{i=1}^r \lambda_i / \sum_{i=1}^d \lambda_i$ is the characteristic of the best r-dimensional approximation of the lagged vectors.

Another optimal feature relates to the properties of the directions determined by the eigenvectors $U_1, \ldots, U_d$. Specifically, the first eigenvector U_1 determines the direction such that the variation of the projections of the lagged vectors onto this direction is maximum.

Every subsequent eigenvector determines a direction that is orthogonal to all previous directions, and the variation of the projections of the lagged vectors onto this direction is also maximum. Therefore, it is natural to call the direction of ith eigenvector U_i the *ith principal direction*. Note that the elementary matrices $\mathbf{X}_i = U_i Z_i^\mathrm{T}$ are built up from the projections of the lagged vectors onto ith directions.

This view on the SVD of the trajectory matrix composed of L-lagged vectors and an appeal to associations with *principal component analysis* lead to the following terminology. We shall call the vector U_i the ith (principal) *eigenvector*, the vector V_i will be called the *ith factor vector*, and the vector Z_i the *vector of ith principal components*.

1.2.3 Grouping

Let us now comment on the grouping step, which is the procedure of arranging the matrix terms $\mathbf{X}_i$ in (1.2). Assume that $m = 2$, $I_1 = I = \{i_1 \ldots, i_r\}$ and $I_2 = \{1, \ldots, d\} \setminus I$, where $1 \leq i_1 < \ldots < i_r \leq d$.

The purpose of the grouping step is separation of the additive components of time series. Let us discuss the very important concept of separability in detail. Suppose that the time series F is a sum of two time series $F^{(1)}$ and $F^{(2)}$; that is, $f_i = f_i^{(1)} + f_i^{(2)}$ for $i = 0, \ldots, N - 1$. Let us fix the window length L and denote by $\mathbf{X}$, $\mathbf{X}^{(1)}$ and $\mathbf{X}^{(2)}$ the L-trajectory matrices of the series F, $F^{(1)}$ and $F^{(2)}$, respectively.

Consider an SVD (1.2) of the trajectory matrix $\mathbf{X}$. (Recall that if all the eigenvalues have multiplicity one, then this expansion is unique.) We shall say that the series $F^{(1)}$ and $F^{(2)}$ are (weakly) *separable by the decomposition* (1.2), if there exists a collection of indices $I \subset \{1, \ldots, d\}$ such that $\mathbf{X}^{(1)} = \sum_{i \in I} \mathbf{X}_i$ and consequently $\mathbf{X}^{(2)} = \sum_{i \notin I} \mathbf{X}_i$.

In the case of separability, the contribution of $\mathbf{X}^{(1)}$, the first component in the expansion $\mathbf{X} = \mathbf{X}^{(1)} + \mathbf{X}^{(2)}$, is naturally to measure by the share of the corresponding eigenvalues: $\sum_{i \in I} \lambda_i \big/ \sum_{i=1}^{L} \lambda_i$.

The separation of the series by the decomposition (1.2) can be looked at from different perspectives. Let us fix the set of indices $I = I_1$ and consider the corresponding resultant matrix $\mathbf{X}_{I_1}$. If this matrix, and therefore $\mathbf{X}_{I_2} = \mathbf{X} - \mathbf{X}_{I_1}$, are Hankel matrices, then they are necessarily the trajectory matrices of certain time series that are separable by the expansion (1.2).

Moreover, if the matrices $\mathbf{X}_{I_1}$ and $\mathbf{X}_{I_2}$ are close to some Hankel matrices, then there exist series $F^{(1)}$ and $F^{(2)}$ such that $F = F^{(1)} + F^{(2)}$ and the trajectory matrices of these series are close to $\mathbf{X}_{I_1}$ and $\mathbf{X}_{I_2}$, respectively (the problem of finding these series is discussed below). In this case we shall say that the series are *approximately separable*.

Therefore, the purpose of the grouping step (that is the procedure of arranging the indices $1, \ldots, d$ into groups) is to find several groups $I_1, \ldots, I_m$ such that the matrices $\mathbf{X}_{I_1}, \ldots, \mathbf{X}_{I_m}$ satisfy (1.3) and are close to certain Hankel matrices.

Let us now look at the grouping step from the viewpoint of multivariate geometry. Let $\mathbf{X} = [X_1 : \ldots : X_K]$ be the trajectory matrix of a time series F, $F = F^{(1)} + F^{(2)}$, and the series $F^{(1)}$ and $F^{(2)}$ are separable by the decomposition (1.2), which corresponds to splitting the index set $\{1, \ldots, d\}$ into I and $\{1, \ldots, d\} \setminus I$.

The expansion (1.3) with $m = 2$ means that $U_1, \ldots, U_d$, the basis in the trajectory space $\mathfrak{L}^{(L)}$, splits into two groups of basis vectors. This corresponds to the representation of $\mathfrak{L}^{(L)}$ as a product of two orthogonal subspaces (*eigenspaces*) $\mathfrak{L}^{(L,1)} = \mathfrak{L}(U_i, i \in I)$ and $\mathfrak{L}^{(L,2)} = \mathfrak{L}(U_i, i \notin I)$ spanned by $U_i, i \in I$, and $U_i, i \notin I$, respectively.

Separability of two series $F^{(1)}$ and $F^{(2)}$ means that the matrix $\mathbf{X}_I$, whose columns are the projections of the lagged vectors $X_1, \ldots, X_K$ onto the eigenspace $\mathfrak{L}^{(L,1)}$, is exactly the trajectory matrix of the series $F^{(1)}$.

Despite the fact that several formal criteria for separability will be introduced, the whole procedure of splitting the terms into groups (i.e., the grouping step) is difficult to formalize completely. This procedure is based on the analysis of the singular vectors U_i, V_i and the eigenvalues λ_i in the SVD expansions (1.2) and (1.7). The principles and methods of identifying the SVD components for their inclusion into different groups are described in Section 1.6.

Since each matrix component of the SVD is completely determined by the corresponding eigentriple, we shall talk about grouping of the eigentriples rather than grouping of the elementary matrices $\mathbf{X}_i$.

Note also that the case of two series components ($m = 2$) considered above is often more sensibly regarded as the problem of separating out a single component (for example, as a noise reduction) rather than the problem of separation of two terms. In this case, we are interested in only one group of indices, namely I.

1.2.4 Diagonal averaging

If the components of the series are separable and the indices are being split up accordingly, then all the matrices in the expansion (1.3) are Hankel matrices. We thus immediately obtain the decomposition (1.5) of the original series: for every k and n, $\widetilde{f}_n^{(k)}$ is equal to all the entries $x_{ij}^{(k)}$ along the secondary diagonal

$$\{(i, j), \text{ such that } i + j = n + 2\}$$

of the matrix $\mathbf{X}_{I_k}$.

In practice, however, this situation is not realistic. In the general case, no secondary diagonal consists of equal elements. We thus need a formal procedure of transforming an arbitrary matrix into a Hankel matrix and therefore into a series. As such, we shall consider the procedure of *diagonal averaging,* which defines the values of the time series $\widetilde{F}^{(k)}$ as averages of the corresponding diagonals of the matrices $\mathbf{X}_{I_k}$.

It is convenient to represent the diagonal averaging step with the help of the *Hankelization* operator $\mathcal{H}$.

The operator $\mathcal{H}$ acts on an arbitrary $(L \times K)$-matrix $\mathbf{Y} = (y_{ij})$ in the following way (assume for definiteness that $L \leq K$): for $i + j = s$ and $N = L + K - 1$ the element $\widetilde{y}_{ij}$ of the matrix $\mathcal{H}\mathbf{Y}$ is

$$\widetilde{y}_{ij} = \begin{cases} \dfrac{1}{s-1} \sum_{l=1}^{s-1} y_{l,s-l} & \text{for } 2 \leq s \leq L-1, \\ \dfrac{1}{L} \sum_{l=1}^{L} y_{l,s-l} & \text{for } L \leq s \leq K+1, \\ \dfrac{1}{K+L-s+1} \sum_{l=s-K}^{L} y_{l,s-l} & \text{for } K+2 \leq s \leq K+L. \end{cases} \quad (1.10)$$

For $L > K$ the expression for the elements of the matrix $\mathcal{H}\mathbf{Y}$ is analogous, the changes are the substitution $L \leftrightarrow K$ and the use of the transposition of the original matrix $\mathbf{Y}$.

Note that the Hankelization is an optimal procedure in the sense that the matrix $\mathcal{H}\mathbf{Y}$ is closest to $\mathbf{Y}$ (with respect to the matrix norm) among all Hankel matrices of the corresponding size (see Section 6.2). In its turn, the Hankel matrix $\mathcal{H}\mathbf{Y}$ defines the series uniquely by relating the values in the diagonals to the values in the series.

By applying the Hankelization procedure to all matrix components of (1.3), we obtain another expansion:

$$\mathbf{X} = \widetilde{\mathbf{X}}_{I_1} + \ldots + \widetilde{\mathbf{X}}_{I_m}, \quad (1.11)$$

where $\widetilde{\mathbf{X}}_{I_l} = \mathcal{H}\mathbf{X}_{I_l}$.

A sensible grouping leads to the decomposition (1.3) where the resultant matrices $\mathbf{X}_{I_k}$ are almost Hankel ones. This corresponds to approximate separability and implies that the pairwise inner products of different matrices $\widetilde{\mathbf{X}}_{I_k}$ in (1.11) are small.

Since all the matrices on the right-hand side of the expansion (1.11) are Hankel matrices, each matrix uniquely determines the time series $\widetilde{F}^{(k)}$ and we thus obtain (1.5), the decomposition of the original time series.

The procedure of computing the time series $\widetilde{F}^{(k)}$ (that is, building up the group I_k plus diagonal averaging of the matrix $\mathbf{X}_{I_k}$) will be called *reconstruction of a series component* $\widetilde{F}^{(k)}$ *by the eigentriples* with indices in I_k.

1.3 Basic SSA: basic capabilities

In this section we start discussing examples that illustrate basic capabilities of Basic SSA. Note that terms such as 'trend', 'smoothing', 'signal', and 'noise' are used here in their informal, common-sense meaning and will be commented on later.

1.3.1 Trends of different resolution

The example 'Production' (crude oil, lease condensate, and natural gas plant liquids production, monthly data from January 1973 to September 1999) shows the capabilities of SSA in extraction of trends that have different resolutions. Though the series has a seasonal component (and the corresponding component can be extracted together with the trend component), for the moment we do not pay attention to periodicities.

Taking the window length $L = 120$ we see that the eigentriples 1-3 correspond to the trend. Choosing these eigentriples in different combinations we can find different trend components.

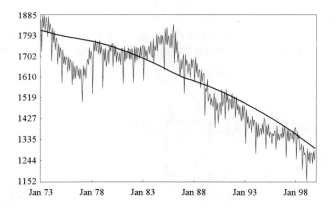

Figure 1.1 *Production: general tendency (rough trend)*.

Figs. 1.1 and 1.2 demonstrate two alternatives in the trend resolution. The leading eigentriple gives a general tendency of the series (Fig. 1.1). The three leading eigentriples describe the behaviour of the data more accurately (Fig. 1.2) and show not only the general decrease of production, but also its growth from the middle 70s to the middle 80s.

1.3.2 Smoothing

The series 'Tree rings' (tree ring indices, Douglas fir, Snake river basin, U.S., annual, from 1282 to 1950), is described in Hipel and McLeod (1994, Chapter 10) with the help of a (3,0)-order ARIMA model. If the ARIMA-type model is accepted, then it is generally meaningless to look for any trend or periodicities. However, we can smooth the series with the help of Basic SSA.

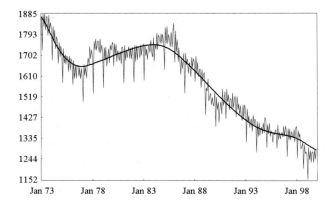

Figure 1.2 *Production: accurate trend.*

Fig. 1.3 shows the initial series and the result of its SSA smoothing, which is obtained by the leading 7 eigentriples with window length 120. Fig. 1.4 depicts the residuals.

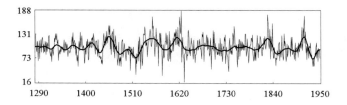

Figure 1.3 *Tree rings: smoothing result.*

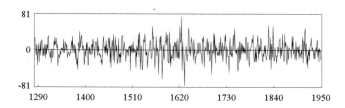

Figure 1.4 *Tree rings: residuals.*

Another example demonstrating SSA as a smoothing technique uses the 'White dwarf' data, which contains 618 point measurements of the time variation of the intensity of the white dwarf star PG1159-035 during March 1989. The data is discussed in Clemens (1994). The whole series can be described as a smooth quasi-periodic curve with a noise component.

Using Basic SSA with window length $L = 100$ and choosing the leading 11 eigentriples for the reconstruction, we obtain the smooth curve of Fig. 1.5 (thick line). The residuals (Fig. 1.6) seem to have no evident structure (to simplify the visualization of the results; these figures present only a part of the series).

Further analysis shows that the residual series can be regarded as a Gaussian white noise, though it does not contain very low frequencies (see the discussion in Section 1.6.1).

Thus, we can assume that in this case the smoothing procedure leads to noise reduction and the smooth curve in Fig. 1.5 describes the signal.

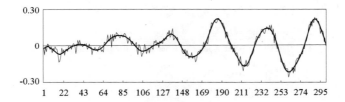

Figure 1.5 *White dwarf: smoothed series.*

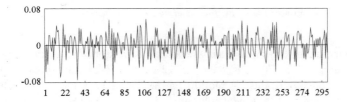

Figure 1.6 *White dwarf: residuals.*

1.3.3 Extraction of seasonality components

The 'Eggs' data (eggs for a laying hen, monthly, U.S., from January 1938 to December 1940, Kendall and Stuart, 1976, Chapter 45) has a rather simple structure: it is the sum of an explicit annual oscillation (though not a harmonic one) and the trend, which is almost constant.

The choice $L = 12$ allows us to extract simultaneously all seasonal components (12, 6, 4, 3, 2.4, and 2-months harmonics) as well as the trend.

The graph in Fig. 1.7 depicts the initial series and its trend (thick line), which is reconstructed from the first eigentriple.

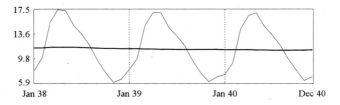

Figure 1.7 *Eggs: initial series and its trend.*

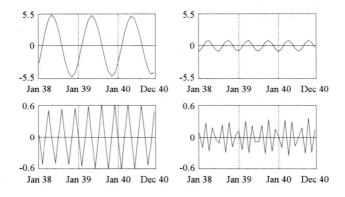

Figure 1.8 *Eggs: four leading seasonal harmonics.*

The four leading seasonal *harmonic components* (briefly, *harmonics*) of the series are depicted in Fig. 1.8; they are: 12-months, 6-months (presented in the same scale), 4-months and 2.4-months harmonics (also in the same scale). The corresponding pairs of the eigentriples are 2-3; 4-5; 6-7, and 8-9. The two weakest harmonics, 3-months and 2-months (10-11 and 12 eigentriples, respectively), are not shown.

1.3.4 Extraction of cycles with small and large periods

The series 'Births' (number of daily births, Quebec, Canada, from January 1, 1977 to December 31, 1990) is discussed in Hipel and McLeod (1994). It shows, in addition to a smooth trend, two cycles of different ranges: the one-year periodicity and the one-week periodicity.

Both periodicities (as well as the trend) can be simultaneously extracted by Basic SSA with window length $L = 365$. Fig. 1.9 shows the one-year cycle of the series added to the trend (white line) on the background of the 'Births' series from 1981 to 1990. Note that the form of this cycle varies in time, though the main two peaks (spring and autumn) remain stable. The trend corresponds to the

BASIC SSA: BASIC CAPABILITIES

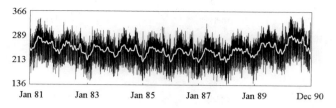

Figure 1.9 *Births: initial time series and its annual periodicity.*

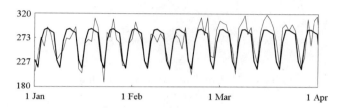

Figure 1.10 *Births: one-week periodicity.*

leading eigentriple, while the one-year periodic component is reconstructed from the eigentriples 6-9 and 12-19.

Fig. 1.10 demonstrates the one-week cycle on the background of the initial series for approximately the first three months of 1977. This cycle corresponds to the eigentriples 2-5 and 10-11.

1.3.5 Extraction of periodicities with varying amplitudes

The capability of SSA in extracting an oscillating signal with a varying amplitude can be illustrated by the example of the 'Drunkenness' series (monthly public drunkenness intakes, Minneapolis, U.S., from January 1966 to July 1978, McCleary and Hay, 1980). The initial series is depicted in Fig. 1.11 (thin line).

Taking $L = 60$ in Basic SSA and reconstructing the series from the fourth and fifth eigentriples, we see (bottom line in Fig. 1.11) an almost pure 12-months periodic component. The amplitude of this annual periodic component approximately equals 120 at the beginning of the observation time. The amplitude then gradually decreases and almost disappears at the end. The amplitude is reduced by a factor of about 10, but the trend in the data is diminished only by a factor of three to four.

1.3.6 Complex trends and periodicities

The 'Unemployment' series (West Germany, monthly, from April 1950 to December 1980, Rao and Gabr, 1984) serves as an example of SSA capability of

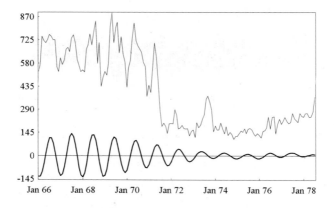

Figure 1.11 *Drunkenness: varying amplitudes.*

extracting complex trends simultaneously with the amplitude-modulated periodicities.

The result of extraction is presented in Fig. 1.12 (the initial series and the reconstructed trend) and in Fig. 1.13 (seasonality).

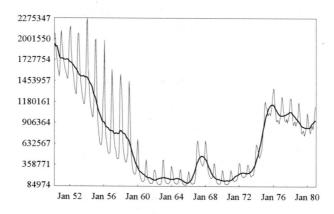

Figure 1.12 *Unemployment: trend.*

The window length was taken as $L = 180$. Since both the trend and the seasonality are complex, many eigentriples are required to reconstruct them. The trend is reconstructed from the eigentriples 1, 2, 5-7, 10, 11, 14, 15, 20, 21, 24, 27, 30, and 33, while the eigentriples with numbers 3, 4, 8, 9, 12, 13, 16-19, 22, 23, 25, 26, 34, 35, 43, 44, 71, and 72 describe the seasonality.

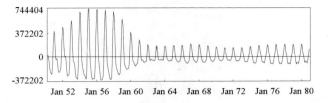

Figure 1.13 *Unemployment: seasonality.*

If we were to take a smaller number of eigentriples for the trend, then we would obtain a less refined description of a smooth, slowly varying component of the series corresponding to a more general tendency in the series.

1.3.7 Finding structure in short time series

The series 'War' (U.S. combat deaths in the Indochina war, monthly, from 1966 to 1971, Janowitz and Schweizer, 1989, Table 10) is chosen to demonstrate the capabilities of SSA in finding a structure in short time series.

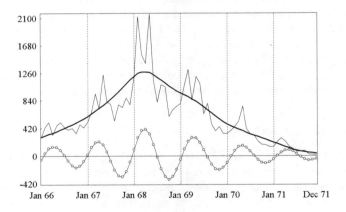

Figure 1.14 *War: trend and annual periodicity.*

Selecting a window length $L = 18$, we can see (Fig. 1.14) that the two leading eigentriples perfectly describe the trend of the series (thick line on the background of the initial data). This trend relates to the overall involvement of U.S. troops in the war.

The third (bottom) plot of Fig. 1.14 shows the component of the initial series reconstructed from the eigentriples 3 and 4. There is little doubt that this is an annual oscillation modulated by the war intensity. This oscillation has its origin

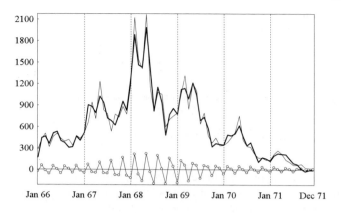

Figure 1.15 *War: quarter periodicity and series approximation.*

in the climatic conditions of South-East Asia: the summer season is much more difficult for any activity than the winter one.

Two other series components, namely that of the quarterly cycle corresponding to the eigentriples 5 and 6 (depicted at the bottom of Fig. 1.15) and the omitted 4-months cycle, which can be reconstructed from the eigentriples 7 and 8, are both modulated by the war intensity and both are less clear for interpretation. Nevertheless, if we add all these effects together (that is, reconstruct the series component corresponding to the eight leading eigentriples), a perfect agreement between the result and the initial series becomes apparent: see Fig. 1.15, top two plots, with the thick line corresponding to the reconstruction.

1.4 Time series and SSA tasks

In the previous section the terms 'trend', 'smoothing', 'amplitude modulation' and 'noise' were used without any explanation of their meaning. In this section we shall provide the related definitions and corresponding discussions. We shall also describe the major tasks that can be attempted by Basic SSA. Examples of application of Basic SSA for solving these tasks have been considered in Section 1.3.

1.4.1 Models of time series and the periodograms

Formally, SSA can be applied to an arbitrary time series. However, a theoretical study of its properties requires specific considerations for different classes of series. Moreover, different classes assume different choices of parameters and expected results. We thus start this section with a description of several classes of time series, which are natural for the SSA treatment, and use these classes to discuss the important concept of (approximate) separability defined earlier in Section 1.2.3. (For the theoretical aspects of separability see Section 6.1.)

Since the main purpose of SSA is a decomposition of the series into additive components, we always implicitly assume that this series is a sum of several simpler series. These 'simple' series are the objects of the discussion below. Note also that here we only consider deterministic time series, including those that can be regarded as 'noise'. Stochastic models of the noise series, in their relation to the separability problem, are discussed in Sections 6.1.3 and 6.3.

(a) Stationary series

The concept of a deterministic stationary time series is asymptotic (rigorous definitions and results on the subject are given in Section 6.4, here we stick to a looser style). Specifically, an infinite series $F = (f_0, f_1, \ldots, f_n, \ldots)$ is called *stationary* if for all nonnegative integers k, m the following convergence takes place:

$$\frac{1}{N} \sum_{j=0}^{N-1} f_{j+k} f_{j+m} \xrightarrow[N \to \infty]{} R_f(k-m), \qquad (1.12)$$

where the (even) function $R_f(n)$ is called the *covariance function* of the series F. The covariance function can be represented as

$$R_f(n) = \int_{(-1/2, 1/2]} e^{i 2 \pi n \omega} m_f(d\omega),$$

where m_f is a measure called the *spectral measure* of the series F.

The form of the spectral measure determines properties of the corresponding stationary series in many respects. For example, the convergence (1.12) implies, loosely speaking, the convergence of the averages

$$\frac{1}{N} \sum_{j=0}^{N-1} f_{j+k} \xrightarrow[N \to \infty]{} 0 \qquad (1.13)$$

for any k if and only if m_f does not have an atom at zero.

Thus, the definition of stationarity is related to the ergodicity not only of the second order, but also of the first order as well. Below, when discussing stationarity, we shall always assume that (1.13) holds, which is the zero-mean assumption for the original series.

If the measure m_f is discrete, then, roughly speaking, we can assume that the stationary series F has the form

$$f_n \sim \sum_k a_k \cos(2\pi \omega_k n) + \sum_k b_k \sin(2\pi \omega_k n), \quad \omega_k \in (0, 1/2], \qquad (1.14)$$

where $a_k = a(\omega_k), b_k = b(\omega_k), b(1/2) = 0$ and the sum $\sum_k (a_k^2 + b_k^2)$ converges. (Note that $a(1/2) \neq 0$ if one of the ω_k is exactly $1/2$.)

Then the measure m_f is concentrated at the points $\pm \omega_k$, $\omega_k \in (0, 1/2)$, with the weights $(a_k^2 + b_k^2)/4$. The weight of the point $1/2$ equals $a^2(1/2)$.

A series of the form (1.14) will be called *almost periodic* (see Section 6.4.2 for the precise definition). *Periodic* series correspond to a spectral measure m_f concentrated at the points $\pm j/T$ ($j = 1, \ldots, [T/2]$) for some integer T. In terms of the representation (1.14), this means that the number of terms in this representation is finite and all the frequencies ω_k are rational.

Almost periodic series that are not periodic are called *quasi-periodic*. For these series the spectral measure is discrete, but it is not concentrated on the nodes of any grid of the form $\pm j/T$. The *harmonic* $f_n = \cos 2\pi \omega n$ with an irrational ω provides an example of a quasi-periodic series.

Aperiodic (in other terminology — *chaotic*) series are characterized by a spectral measure that does not have atoms. In this case one usually assumes the existence of the *spectral density*: $m_f(d\omega) = p_f(\omega)d\omega$. Aperiodic series serve as models for *noise*; they are also considered in the theory of chaotic dynamical systems. If the spectral density exists and is constant, then the aperiodic series is called *white noise*.

Almost periodic and chaotic series have different asymptotic behaviour of their covariance functions: in the aperiodic case this function tends to zero, but the almost periodic series are (generally) characterized by almost periodic covariance functions.

As a rule, real-life stationary series have both components, periodic (or quasi-periodic) and noise (aperiodic) components. (The series 'White dwarf' – Section 1.3.2 – is a typical example of such series.)

Note that it is difficult, or even impossible when dealing with a finite series, to distinguish between a periodic series with a large period and a quasi-periodic series. Moreover, on finite time intervals aperiodic series are hardly distinguished from a sum of harmonics with wide spectrum and small amplitudes.

For a description of finite, but reasonably long, stationary series, it is convenient to use the language of the *Fourier expansion* of the initial series. This is the expansion

$$f_n = c_0 + \sum_{k=1}^{[N/2]} \Big(c_k \cos(2\pi n\, k/N) + s_k \sin(2\pi n\, k/N) \Big), \tag{1.15}$$

where N is the length of the series, $0 \leq n < N$, and $s_{N/2} = 0$ for even N. The zero term c_0 is equal to the average of the series, so that if the series is centred, then $c_0 = 0$.

For a series of a finite length, the *periodogram* of the series is an analogue of the spectral measure. By definition (see Section 6.4.5) the periodogram $\Pi_f^N(\omega)$ of the series $F = (f_0, \ldots, f_{N-1})$ is

$$\Pi_f^N(\omega) = \frac{1}{N} \left| \sum_{n=0}^{N-1} e^{-i2\pi \omega n} f_n \right|^2, \quad \omega \in (-1/2, 1/2]. \tag{1.16}$$

Since the elements of the series F are real numbers, $\Pi_f^N(-\omega) = \Pi_f^N(\omega)$ for $|\omega| < 1/2$, and therefore we can consider only the interval $[0, 1/2]$ for ω. If the series F is represented in the form (1.15), then it is not difficult to show that

$$\Pi_f^N(k/N) = \frac{N}{2} \begin{cases} 2c_0^2 & \text{for } k = 0, \\ c_k^2 + s_k^2 & \text{for } 0 < k < N/2, \\ 2c_{N/2}^2 & \text{for } k = N/2. \end{cases} \quad (1.17)$$

The last case is, of course, possible only when N is odd.

Let us consider the Fourier expansions (1.15) of two series $F^{(1)}$ and $F^{(2)}$ of length N and denote the corresponding coefficients by $c_k^{(j)}$ and $s_k^{(j)}$, $j = 1, 2$. Using the notation

$$d_k = \begin{cases} c_k^{(1)} c_k^{(2)} + s_k^{(1)} s_k^{(2)} & \text{for } k \neq 0 \text{ and } N/2, \\ 2c_k^{(1)} c_k^{(2)} & \text{for } k = 0 \text{ or } N/2, \end{cases} \quad (1.18)$$

we can easily see that the inner product of two series is

$$\left(F^{(1)}, F^{(2)}\right) \stackrel{\text{def}}{=} \sum_{k=0}^{N-1} f_n^{(1)} f_n^{(2)}$$
$$= \frac{N}{2} \left(2d_0 + \sum_{0 < k < N/2} d_k + 2d_{N/2} \right), \quad (1.19)$$

where $d_{N/2} = 0$ for odd N.

This immediately yields that the norm $||F|| = \sqrt{(F, F)}$ of the series (1.15) is expressed through its periodogram as follows:

$$||F||^2 = \sum_{k=0}^{[N/2]} \Pi_f^N(k/N). \quad (1.20)$$

The equality (1.20) implies that the value (1.17) of the periodogram at the point k/N describes the influence of the harmonic components with frequency $\omega = k/N$ into the sum (1.15). Moreover, (1.20) explains the normalizing coefficient N^{-1} in the definition (1.16) of the periodogram.

Some other normalizations of the periodograms are known in literature and could be useful as well. In particular, using below the periodogram analysis of the time series for the purposes of SSA, we shall plot the values of $\Pi_f^N(k/N)/N$ (for fixed k this is called *power* of the frequency k/N), but we shall keep the name 'periodogram' for the corresponding line-plots.

The collection of frequencies $\omega_k = k/N$ with positive powers is called the *support of the periodogram*. If the support of a certain periodogram belongs to some interval $[a, b]$, then this interval is called the *frequency range of the series*.

Asymptotically, for the stationary series, the periodograms approximate the spectral measures (see Theorem 6.4 of Section 6.4.5).

Thus, a standard model of a stationary series is a sum of a periodic (or quasi-periodic) series and an aperiodic series possessing a spectral density.

If the length N of a stationary series is large enough, then the frequencies j/N, which are close to the most powerful frequencies of the almost periodic component of the series, have large values in the periodogram of the series.

For short series the grid $\{j/N,\ j = 0, \ldots, [N/2]\}$ is a poor approximation to the whole range of frequencies $[0, 1/2]$, and the periodogram may badly reflect the periodic structure of the series components.

(b) Amplitude-modulated periodicities

The nature of the definition of stationarity is asymptotic. This asymptotic nature has both advantages (for example, the rigorous mathematical definition allows illustration of all the concepts by model examples) as well as disadvantages (the main one is that it is not possible to check the stationarity of the series using only a finite-time interval of it).

At the same time, there are numerous deviations from stationarity. We consider only two classes of the nonstationary time series which we describe at a qualitative level. Specifically, we consider amplitude-modulated periodic series and series with trends. The choice of these two classes is related to their practical significance and importance for the SSA.

The trends are dealt with in the next subsection. Here we discuss the *amplitude-modulated* periodic signals, that is, series of the form $f_n = A(n)g_n$, where g_n is a periodic sequence and $A(n) \geq 0$. Usually it is assumed that on the given time interval ($0 \leq n \leq N - 1$) the function $A(n)$ varies much more slowly than the low-frequency harmonic component of the series g_n.

Series of this kind are typical in economics where the period of the harmonics g_n is related to seasonality, but the amplitude modulation is determined by long-term tendencies.

An explanation of the same sort is suitable for the example 'War' of Section 1.3.7, where the seasonal component of the combat deaths (Fig. 1.14, bottom line) seems to be modulated by the intensity of the military activities.

Let us discuss the periodogram analysis of the amplitude-modulated periodic signals, for the moment restricting ourselves to the amplitude-modulated harmonic

$$f_n = A(n)\cos(2\pi\omega + \theta), \quad n = 0, \ldots, N - 1. \tag{1.21}$$

As a rule, the periodogram of the series (1.21) is supported on a short frequency interval containing ω. This is not surprising since, for example, for large $\omega_1 \approx \omega_2$ the sum

$$\cos(2\pi\omega_1 n) + \cos(2\pi\omega_2 n) = 2\cos\left(\pi(\omega_1 - \omega_2)n\right)\cos\left(\pi(\omega_1 + \omega_2)n\right)$$

is a product of a slowly varying sequence

$$A(n) = 2\cos\left(\pi(\omega_1 - \omega_2)n\right)$$

and the harmonic with the high frequency $(\omega_1 + \omega_2)/2$.

Note that for $n \leq 1/2(\omega_1 - \omega_2)$ the sequence $A(n)$ is positive and its oscillatory nature cannot be seen for small n.

Fig. 1.16 depicts the periodogram of the main seasonal (annual plus quarterly) component of the series 'War' (Section 1.3.7). We see that the periodogram is supported around two main seasonal frequencies, but is not precisely concentrated at these two points. For the 'War' series, this is caused by the amplitude modulation.

However, the above discussion implies that in the general case the appearance of exactly the same modulation can be caused by two different reasons: either it can be the 'true' modulation, which can be explained by taking into account the nature of the signal, or the modulation is spurious, with its origin in the closeness of the frequencies of the harmonic components of the original series.

The other possible reason of spreading around the main frequencies is the discreteness of the periodogram grid $\{k/N\}$: if a frequency ω of a harmonic does not belong to the grid, then it spreads over it.

Note that since the length of the 'War' series is proportional to 12, the frequencies $\omega = 1/12$ and $\omega = 1/3$, which correspond to annual and quarterly periodicities, fall exactly on the periodogram grid $\{k/36,\ k = 1, \ldots, 18\}$.

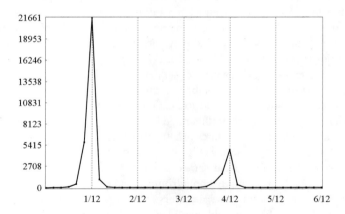

Figure 1.16 *War: periodogram of the main seasonality component.*

Evidently, not only periodic series can be modulated by the amplitude; the same can hold for quasi-periodic and chaotic sequences. However, identification of these cases by means of the periodogram analysis is more difficult.

(c) Trends

It seems that there is no commonly accepted definition of the concept 'trend'. Certainly, the main tendency of the series can be postulated with the help of a parametric model, and subsequent estimation of the parameters would allow us to talk about, say, linear, exponential, or logistic trends; see, for instance, Anderson (1994, Chapter 3.8). Very often the problem of trend approximation is stated directly, as a pure approximation problem, without any worry concerning the tendencies. The most popular approximation is polynomial; see, for example, Otnes and Enochson (1978, Chapter 3.8) or Anderson (1994, Chapter 3.2.1).

For us, this meaning of the notion 'trend' is not suitable just because Basic SSA is a model-free, and therefore nonparametric method.

Under the assumption that the series F is a realization of a certain random discrete-time process, $\xi(n)$, trend is often defined as $\mathbf{E}\xi(n)$, the expectation of the random process (see, for instance, Diggle, 1990, Section 1.4). We cannot use this definition since we are working with only one trajectory and do not have an ensemble of trajectories for averaging.

In principle, the trend of a time series can be described as a function that reflects slow, stable, and systematic variation over a long period of time; see Kendall and Stuart (1976, Section 45.12). The notion of trend in this case is related to the length of the series — from the practical point of view this length is exactly the 'long period of time'.

Moreover, we have already collected oscillatory components of the series into a separate class of (centred) stationary series and therefore the term 'cyclical trend' (see Anderson, 1994, Chapter 4) does not suit us. In general, an appropriate definition of trend for SSA defines the trend as an additive component of the series which is (i) not stationary, and (ii) 'slowly varies' during the whole period of time that the series is being observed (compare Brillinger, 1975, Chapter 2.12).

At this point, let us mention some consequences of this understanding of the notion 'trend'. The most important is the nonuniqueness of the solution to the problem 'trend identification' or 'trend extraction' in its nonparametric setup. This nonuniqueness has already been illustrated by the example 'Production'; see Section 1.3, where Figs. 1.1-1.2 depict two forms of trend: the trend that describes the general tendency of the series (Fig. 1.1) and the detailed trend (Fig. 1.2).

Furthermore, for a finite time series, a harmonic component with a low frequency is practically indistinguishable from a trend (it can even be monotone on a finite time interval). In this case, auxiliary subject-related information about the series can be decisive for the problem of distinguishing trend from the periodicity.

For instance, even though the reconstructed trend in the example 'War' (see Section 1.3.7 and Fig. 1.14) looks like a periodicity observed over a time interval that is less than half of the period, it is clear that there is no question of periodicity in this case.

In the language of frequencies, trend generates large powers in the low-frequency range of the periodogram.

Finally, we have assumed that any stationary series is centred. Therefore, the average of all the terms f_n of any series F is always added to its trend. On the periodogram, a nonzero constant component of the series corresponds to an atom at zero.

Therefore, a general descriptive model of the series that we consider in the present monograph is the additive model where the components of the series are trends, oscillations, and noise components. In addition, the oscillatory components are subdivided into periodic and quasi-periodic, while the noise components are, as a rule, aperiodic series. Both stationarity and amplitude modulation of the oscillatory and noise components are allowed. The sum of all the additive components, except for the noise, will be called the *signal*.

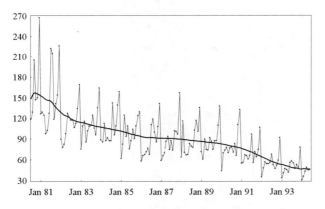

Figure 1.17 *Rosé wine: initial time series and the trend.*

Example 1.1 *Additive components of time series*

Let us consider the 'Rosé wine' series (monthly rosé wine sales, Australia, from July 1980 to June 1994, thousands of litres). Fig. 1.17 depicts the series itself (the thin line) and Fig. 1.18 presents its periodogram.

Fig. 1.17 shows that the series 'Rosé wine' has a decreasing trend and an annual seasonality of a complex form. Fig. 1.18 shows the periodogram of the series; it seems reasonable that the trend is related to large values at the low-frequency range, and the annual periodicity is related to the peaks at frequencies $1/12$, $1/6$, $1/4$, $1/3$, $1/2.4$, and $1/2$. The nonregularity of the powers for these frequencies indicates a complex form of the annual periodicity.

Fig. 1.19 depicts two additive components of the 'Rosé wine' series: the seasonal component (top graph), which is described by the eigentriples 2-11, 13 and the residual series ($L = 84$). The trend component (thick line in Fig. 1.17) is reconstructed from the eigentriples 1, 12, and 14.

Periodogram analysis demonstrates that the expansion of the series into three parts is indeed related to the separation of the spectral range into three regions: low

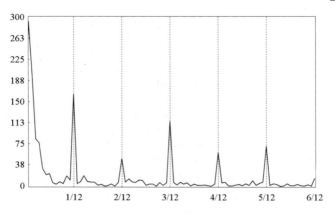

Figure 1.18 *Rosé wine: periodogram for the series.*

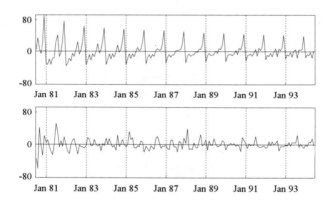

Figure 1.19 *Rosé wine: two components of the series.*

frequencies correspond to the trend (the thick line in Fig. 1.20), the frequencies describing the seasonalities correspond to the periodic component (Fig. 1.20, the thin line), and the residual series (which can be regarded as noise) has all the other frequencies (Fig. 1.21).

The periodograms of the whole series (see Fig. 1.18), its trend and the seasonal component (see Fig. 1.20) are presented in the same scale.

1.4.2 Basic SSA: Classification of the main tasks

Classification of the main tasks, which Basic SSA can be used for, is naturally related to the above classification of the time series and their components. It is, of

TIME SERIES AND SSA TASKS

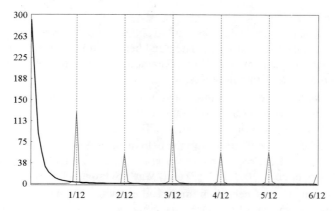

Figure 1.20 *Rosé wine: periodograms of the trend and the seasonal component.*

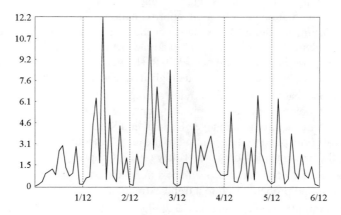

Figure 1.21 *Rosé wine: periodogram of the residuals.*

course, neither rigid nor exact, but it helps to understand features of the method and the main principles for the choice of its parameters in analyzing specific data sets; see Section 1.6.

1. *Trend extraction and smoothing*

   These two problems are in many ways similar and often cannot be distinguished in practice. None of these problems has an exact meaning, unless a parametric model is assumed. Therefore, a large number of model-free methods can be applied to solve each of them. Nevertheless, it is convenient to distinguish trend extraction and smoothing, at least on a qualitative level.

- *Trend extraction*

  This problem occurs when we want to obtain more or less refined non-oscillatory tendency of the series. The trend being extracted, we can investigate its behaviour, approximate it in a parametric form, and consider its continuation for forecasting purposes.

  Note that here we are not interested in whether the residual from the trend extraction has 'structure' (for example, it can contain a certain seasonality) or is a pure noise series.

  Results of trend extraction with the help of Basic SSA can be demonstrated by the examples 'Production' (Section 1.3.1, Figs. 1.1 and 1.2), 'Unemployment' (Section 1.3.6, Fig. 1.12) and 'War' (Section 1.3.7, Fig. 1.14).

  In the language of periodograms, trend extraction means extraction of the low-frequency part of the series that could not be regarded as an oscillatory one.

- *Smoothing*

  Smoothing a series means representing the series as a sum of two series where the first one is a 'smooth approximation' of it. Note that here we do not assume anything like existence of the trend and do not pay attention to the structure of the residuals: for example, the residual series may contain a strong periodicity of small period. In the language of frequencies, to smooth a series we have to remove all its high-frequency components.

  Methods that use weighted moving averages (see Anderson, 1994, Chapter 3.3) or weighted averages depending on time intervals, including the local polynomial approximation (see Kendall and Stuart, 1976, Chapter 36), perfectly correspond to the meaning of the term 'smoothing'. The same is true for the median smoothing; see Tukey (1977, Chapter 7).

  If a series is considered as a sum of a trend and a noise, a smoothing procedure would probably lead to a trend extraction.

  The example 'Snake' (Section 1.3.2, Fig. 1.3) shows the smoothing capabilities of Basic SSA. There is no distinct border between the trend extraction and smoothing, and the example 'Unemployment' (Section 1.3.6, Fig. 1.12) can be considered both for the refined trend extraction and for the result of a certain smoothing.

2. *Extraction of oscillatory components*

   The general problem here is identification and separation of the oscillatory components of the series that do not constitute parts of the trend. In the parametric form (under the assumptions of zero trend, finite number of harmonics, and additive stochastic white noise), this problem is extensively studied in the classical spectral analysis theory (see, for example, Anderson, 1994, Chapter 4).

   The statement of the problem in Basic SSA is specified mostly by the model-free nature of the method. One of the specifics is that the result of Basic SSA extraction of a single harmonic component of a series is not, as a rule, a purely harmonic sequence. This is a general feature of the method, it was thoroughly discussed in the Introduction. From the formal point of view, this means that in practice we deal with an approximate separability rather than with the exact one (see Section 1.5).

Also, application of Basic SSA does not require rigid assumptions about the number of harmonics and their frequencies. For instance, the example 'Births' (Section 1.3.4) illustrates simultaneous extraction of two (approximately) periodic components in daily data (the annual and weekly periodicities).

Certainly, auxiliary information about the initial series always makes the situation clearer and helps in choosing the parameters of the method. For example, the assumption that there might be an annual periodicity in monthly data suggests that the the analyst must pay attention to the frequencies $j/12$ ($j = 1, \ldots, 6$). The presence of sharp peaks on the periodogram of the initial series leads to the assumption that the series contains periodic components with these frequencies.

Finally, we allow the possibility of amplitude modulation for the oscillatory components of the series. In examples 'War' (Section 1.3.7), 'Drunkenness' (Section 1.3.5) and 'Unemployment' (Section 1.3.6) the capabilities of Basic SSA for their extraction have been demonstrated.

The most general problem is that of finding the whole structure of the series, that is splitting it into several 'simple' and 'interpretable' components, and the noise component.

3. *Obtaining the refined structure of a series*

   According to our basic assumption, any series that we consider can be represented as the sum of a signal, which itself consists of a trend and oscillations, and noise.

   If the components of the signal are expressed in a parametric form (see, for instance, Ledemann and Lloyd, 1984, Chapter 18.2) for a parametric model of seasonal effects), then the main problem of the decomposition of the series into its components can be formalized. Of course, for the model-free techniques such as Basic SSA, this is not so.

   The previously discussed problems are similar in the sense that we want to find a particular component of a series without paying much attention to the residuals. Our task is now to obtain the whole structure of the signal, that is to extract its trend (if any), to find its seasonal components and other periodicities, and so on. The residuals should be identified as the noise component.

   Therefore, we must take care of both the signal and noise. The decomposition of the signal into components depends, among the other things, on the interpretability of these components (see the 'War' example of Section 1.3.7).

   As for the residual series, we have to be convinced that it does not contain parts of the signal. If the noise can be assumed stochastic, then various statistical procedures may be applied to test the randomness of the residuals. Due to its simplicity, the most commonly used model of the noise is the model of stochastic white noise. From the practical point of view, it is usually enough to be sure that the residual series has no evident structure.

These considerations work for the formally simpler problem of *noise reduction,* which differs from that discussed above in that here the series is to be split into two components only, the signal and the noise, and a detailed study of the signal is not required.

In practice, this setup is very close to the setup of the problem of 'smoothing', especially when the concept 'noise' is understood in a broad sense.

For a series with a large signal-to-noise ratio, the signal generates sharp peaks at the periodogram of the series. It is sometimes important to remember that if the frequency range of the noise is wide (as for white noise), then the powers of the frequencies, relating to these peaks, include the powers of the same frequencies of the noise. The exception is only when the frequency ranges of the signal and noise are different.

1.5 Separability

As mentioned above, the main purpose of SSA is a decomposition of the original series into a sum of series, so that each component in this sum can be identified as either a trend, periodic or quasi-periodic component (perhaps, amplitude-modulated), or noise.

The notion of separability of series plays a fundamental role in the formalization of this problem (see Sections 1.2.3 and 1.2.4). Roughly speaking, SSA decomposition of the series F can be successful only if the resulting additive components of the series are (approximately) separable from each other.

This raises two problems that require discussion:

1. Assume that we have obtained an SSA decomposition of the series F. How do we check the quality of this decomposition? (Note that the notion of separability and therefore the present question are meaningful when the window length L is fixed.)

2. How, using only the original series, can we predict (at least partially and approximately) the results of the SSA decomposition of this series into components and the quality of this decomposition?

Of course, the second question is related to the problem of the choice of the SSA parameters (window length and the grouping manner). We shall therefore delay the corresponding discussion until the next section. Here we consider the concept of separability itself, both from the theoretical and the practical viewpoints.

1.5.1 Weak and strong separability

Let us fix the window length L, consider a certain SVD of the L-trajectory matrix $\mathbf{X}$ of the initial series F of length N, and assume that the series F is a sum of two series $F^{(1)}$ and $F^{(2)}$, that is, $F = F^{(1)} + F^{(2)}$.

In this case, separability of the series $F^{(1)}$ and $F^{(2)}$ means (see Section 1.2.3) that we can split the matrix terms of the SVD of the trajectory matrix $\mathbf{X}$ into two different groups, so that the sums of terms within the groups give the trajectory matrices $\mathbf{X}^{(1)}$ and $\mathbf{X}^{(2)}$ of the series $F^{(1)}$ and $F^{(2)}$, respectively.

The separability immediately implies (see Section 6.1) that each row of the trajectory matrix $\mathbf{X}^{(1)}$ of the first series is orthogonal to each row of the trajectory matrix $\mathbf{X}^{(2)}$ of the second series, and the same holds for the columns.

Since rows and columns of trajectory matrices are subseries of the corresponding series, the orthogonality condition for the rows (and columns) of the trajectory matrices $\mathbf{X}^{(1)}$ and $\mathbf{X}^{(2)}$ is just the condition of orthogonality of any subseries of length L (and $K = N - L + 1$) of the series $F^{(1)}$ to any subseries of the same length of the series $F^{(2)}$ (the subseries of the time series must be considered here as vectors).

If this orthogonality holds, then we shall say that the series $F^{(1)}$ and $F^{(2)}$ are *weakly separable*. A finer (and more desirable in practice) notion of separability is the notion of strong separability which, in addition to the orthogonality of the subseries of the two series, puts constraints on the singular values of the matrices $\mathbf{X}^{(1)}$ and $\mathbf{X}^{(2)}$. This notion is discussed in Section 1.5.4. Note that if all the singular values of the trajectory matrix $\mathbf{X}$ are different, then the conditions for weak separability and strong separability coincide. Below, for brevity, we shall use the term 'separability' for 'weak separability'.

The condition of (weak) separability can be stated in terms of orthogonality of subspaces as follows: the series $F^{(1)}$ and $F^{(2)}$ are separable if and only if the subspace $\mathfrak{L}^{(L,1)}$ spanned by the columns of the trajectory matrix $\mathbf{X}^{(1)}$, is orthogonal to the subspace $\mathfrak{L}^{(L,2)}$ spanned by the columns of the trajectory matrix $\mathbf{X}^{(2)}$, and similar orthogonality must hold for the subspaces $\mathfrak{L}^{(K,1)}$ and $\mathfrak{L}^{(K,2)}$ spanned by the rows of the trajectory matrices.

Example 1.2 *Weak separability*

Let us illustrate the notion of separability in the language of geometry. Consider the series $F = F^{(1)} + F^{(2)}$ with elements $f_n = f_n^{(1)} + f_n^{(2)}$ where $f_n^{(1)} = a^n$, $f_n^{(2)} = (-1/a)^n$, $a = 1.05$ and $0 \leq n \leq 27$. Let $L = 2$, then the series $F^{(1)}$ and $F^{(2)}$ are (weakly) separable.

The two-dimensional phase diagram of the series F, that is the plot of the vectors $X_{n+1} = (f_n, f_{n+1})^{\mathrm{T}}$, is shown in Fig. 1.22 in addition to both principal directions of the SVD of the trajectory matrix of F. Since the principal directions are determined by the eigenvectors, they are proportional to the vectors $(1, 1.05)^{\mathrm{T}}$ and $(1.05, -1)^{\mathrm{T}}$. The projection of the vectors X_n on the first of these directions fully determines the series $F^{(1)}$ and 'annihilates' the series $F^{(2)}$, while the projection of these vectors on the second principal direction has the opposite effect: the series $F^{(2)}$ is left untouched and $F^{(1)}$ disappears.

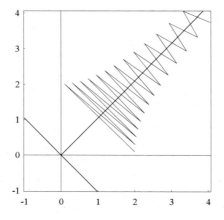

Figure 1.22 *Separability: two-dimensional phase diagram of the series.*

Since a similar phenomenon holds also for the K-dimensional phase diagram of the series ($K = 27$), Fig. 1.22 provides us with a simple geometrical interpretation of separability.

Let us give one more separability condition, which is only a necessary condition (it is not sufficient). This condition is very clear and easy to check.

Let $L^* = \min(L, K)$ and $K^* = \max(L, K)$. Introduce the weights

$$w_i = \begin{cases} i+1 & \text{for } 0 \le i \le L^* - 1, \\ L^* & \text{for } L^* \le i < K^*, \\ N-i & \text{for } K^* \le i \le N - 1. \end{cases} \quad (1.22)$$

Define the inner product of series $F^{(1)}$ and $F^{(2)}$ of length N as

$$\left(F^{(1)}, F^{(2)}\right)_w \stackrel{\text{def}}{=} \sum_{i=0}^{N-1} w_i f_i^{(1)} f_i^{(2)} \quad (1.23)$$

and call the series $F^{(1)}$ and $F^{(2)}$ **w**-*orthogonal* if

$$\left(F^{(1)}, F^{(2)}\right)_w = 0.$$

It can be shown (see Section 6.2) that separability implies **w**-orthogonality. Therefore, if the series F is split into a sum of separable series $F^{(1)}, \ldots, F^{(m)}$, then this sum can be interpreted as an expansion of the series F with respect to a certain **w**-orthonormal basis, generated by the original series itself. Expansions of this kind are typical in linear algebra and analysis.

The window length L enters the definition of **w**-orthogonality; see (1.22). The weights in the inner product (1.23) have the form of a trapezium. If L is small relative to N, then almost all the weights are equal, but for $L \approx N/2$ the influence of the central terms in the series is much higher than of those close to the endpoints in the time interval.

1.5.2 Approximate and asymptotic separability

Exact separability does not happen for real-life series and in practice we can talk only about approximate separability. Let us discuss the characteristics that reflect the degree of separability, leaving for the moment questions relating to the singular values of the trajectory matrices.

In the case of exact separability, the orthogonality of rows and columns of the trajectory matrices $\mathbf{X}^{(1)}$ and $\mathbf{X}^{(2)}$ means that all pairwise inner products of their rows and columns are zero. In statistical language, this means that the noncentral covariances (and therefore, noncentral correlations — the cosines of the angles between the corresponding vectors) are all zero. (Below, for brevity, when talking about covariances and correlations, we shall drop the word 'noncentral'.)

This implies that we can consider as a characteristic of separability of two series $F^{(1)}$ and $F^{(2)}$ the *maximum correlation coefficient* $\rho^{(L,K)}$, that is the maximum of the absolute value of the correlations between the rows and between the columns of the trajectory matrices of these two series (as usual, $K = N - L + 1$).

We shall say that two series $F^{(1)}$ and $F^{(2)}$ are *approximately separable* if all the correlations between the rows and the columns of the trajectory matrices $\mathbf{X}^{(1)}$ and $\mathbf{X}^{(2)}$ are close to zero.

Let us consider other characteristics of the quality of separability. The following quantity (called the *weighted correlation* or *w-correlation*) is a natural measure of deviation of two series $F^{(1)}$ and $F^{(2)}$ from w-orthogonality:

$$\rho_{12}^{(w)} = \frac{\left(F^{(1)}, F^{(2)}\right)_w}{\|F^{(1)}\|_w \|F^{(2)}\|_w}, \qquad (1.24)$$

where $\|F^{(i)}\|_w = \sqrt{\left(F^{(i)}, F^{(i)}\right)_w}$, $i = 1, 2$.

If the absolute value of the w-correlation is small, then the two series are almost w-orthogonal, but, if it is large, then the series are far from being w-orthogonal and are therefore badly separable.

For long (formally, for infinitely long) series it is convenient to introduce the notion of asymptotic separability. Consider two infinite series $F^{(1)}$ and $F^{(2)}$ and denote by $F_N^{(1)}$ and $F_N^{(2)}$ the finite series consisting of the first N elements of the series $F^{(1)}$ and $F^{(2)}$. Assume also that the window length L is a function of the series length N.

We shall say that the series $F^{(1)}$ and $F^{(2)}$ are *asymptotically separable* if the maximum $\rho^{(L,K)}$ of the absolute values of the correlation coefficients between the rows/columns of the trajectory matrices of the series $F_N^{(1)}$ and $F_N^{(2)}$ tends to zero, as $N \to \infty$. The standard behaviour of the window length $L = L(N)$ in the definition of the asymptotic separability is such that $L, K \to \infty$.

From the practical viewpoint, the effect of the asymptotic separability becomes apparent in the analysis of long series and means that two asymptotically separable series are approximately separable for large N.

Section 6.1 contains several analytical examples of both exact and asymptotic separability. These examples show that the class of asymptotically separable series is much wider than the class of series that are exactly separable, and the conditions on the choice of the window length L are much weaker in the case of asymptotic separability.

For instance, exact separability of two harmonics with different periods can be achieved when both periods are divisors of both the window length L and $K = N - L + 1$. This requires, in particular, that the quotient of the periods is rational.

Another example of exact separability is provided by the series $f_n^{(1)} = \exp(\alpha n)$ and $f_n^{(2)} = \exp(-\alpha)\cos(2\pi n/T)$ with an integer T. (We thus deal with the exponential trend and an amplitude-modulated harmonic.) Here separability holds if the window length L and K are proportional to T.

Conditions for asymptotic separability are much weaker. In particular, two harmonics with arbitrary different frequencies are asymptotically separable as soon as L and K tend to infinity. Moreover, under the same conditions the periodic components are asymptotically separable from the trends of a general form (for example, from exponentials and polynomials).

1.5.3 Separability and Fourier expansions

Since separability of series is described in terms of orthogonality of their subseries, and the inner product of series can be expressed in terms of the coefficients in the Fourier expansion (1.19), new separability characteristics related to these expansions can be introduced.

In the expansion (1.19), the terms d_k defined in (1.18) have the meaning of the weights of the frequencies k/N in the inner product (1.19). If all the d_k are zero, then the series are orthogonal and this can be interpreted in terms of the Fourier expansion: each frequency k/N makes a zero input into the inner product.

On the other hand, if $\Pi_{f_1}^N(k/N)\,\Pi_{f_2}^N(k/N) = 0$ for all k, then, since

$$\frac{N}{2}|d_k| \leq \sqrt{\Pi_{f_1}^N(k/N)\,\Pi_{f_2}^N(k/N)}, \qquad (1.25)$$

all the d_k are zero and the orthogonality of the series has the following explanation in terms of the periodograms: in this case the supports of the periodograms of the series $F^{(1)}$ and $F^{(2)}$ do not intersect.

Thus, we can formulate the following sufficient separability condition in terms of periodograms: if for each subseries of length L (and K as well) of the series $F^{(1)}$ the frequency range of its periodogram is disjoint from the frequency range of the periodogram of each subseries of the same length of the series $F^{(2)}$, then the two series are exactly separable.

SEPARABILITY

In this language, the above (sufficient) conditions for separability of two finite harmonic series with different integer periods T_1 and T_2 become obvious: if

$$L = k_1 T_1 = k_2 T_2, \qquad K = m_1 T_1 = m_2 T_2$$

with integer k_1, k_2, m_1, m_2, then for any subseries of length L of the first series its periodogram must contain just one frequency $\omega_1^{(L)} = 1/k_1$, and at the same time the corresponding frequency for the second series must be equal $\omega_2^{(L)} = 1/k_2$. For the subseries of length K, the analogous condition must hold for the frequencies $\omega_1^{(K)} = 1/m_1$ and $\omega_2^{(K)} = 1/m_2$.

For stationary series, the analogous conditions for asymptotic separability are: if the supports of the spectral measures of stationary series are disjoint, then these series are asymptotically separable as $L \to \infty$ and $K \to \infty$ (see Section 6.4.4).

The simplest example of this situation is provided by the sum of two harmonics with different frequencies.

When we deal with an approximate orthogonality of the series rather than with the exact one, we can use the characteristics describing the degree of disjointness of the supports of the periodograms, such that their smallness guarantees the smallness of the correlation coefficient between the series. Indeed, the formulae (1.18), (1.20) and (1.25) yield

$$\left|(F^{(1)}, F^{(2)})\right| \leq \Phi^{(N)}(F^{(1)}, F^{(2)})$$

$$\stackrel{\text{def}}{=} \sum_{k=0}^{[N/2]} \sqrt{\Pi_{f_1}^N(k/N)\, \Pi_{f_2}^N(k/N)} \leq \left\|F^{(1)}\right\| \left\|F^{(2)}\right\|.$$

Therefore,

$$\rho_{12}^{(\Pi)} \stackrel{\text{def}}{=} \frac{\Phi^{(N)}(F^{(1)}, F^{(2)})}{\sqrt{\sum_{k=0}^{[N/2]} \Pi_{f_1}^N(k/N)} \sqrt{\sum_{k=0}^{[N/2]} \Pi_{f_2}^N(k/N)}} \qquad (1.26)$$

can be taken as a natural measure of the spectral orthogonality of the series F_1 and F_2. We shall call this characteristic the *spectral correlation coefficient*. Obviously, the value of the spectral correlation coefficient is between 0 and 1, and the absolute value of the standard correlation coefficient between the series F_1 and F_2 does not exceed $\rho_{12}^{(\Pi)}$.

Therefore, smallness of the spectral correlation coefficients between all the subseries of length L (and K as well) of the series $F^{(1)}$ and $F^{(2)}$ is a sufficient condition for approximate separability of these series.

For this condition to hold, it is not necessary that the series $F^{(1)}$ and $F^{(2)}$ be subseries of stationary series. For example, if one of the series is a sum of a slowly varying monotone trend and a noise, while the other series is a high-frequency oscillatory series, then for a sufficiently large N and $L \lesssim N/2$ we have every reason to expect approximate separability of these two series.

Indeed, when L and K are sufficiently large, the periodogram of any subseries of the first series is mostly supported at the low-frequency range, while the main support of the periodogram of the subseries of the second series is in the range of high frequencies. This implies smallness of the spectral correlation coefficients.

We consider another similar example. Assume that the average values of all the subseries of the series $F^{(1)}$ are large, the average values of the corresponding subseries of $F^{(2)}$ are close to zero, and the amplitudes of all the harmonic components of the series $F^{(1)}$ and $F^{(2)}$ are small. Then it is easy to see that all the spectral correlations (1.26) are small.

This example explains the approximate separation of 'large' signals $F^{(1)}$ from series $F^{(2)}$ that oscillate rapidly around zero (a phenomenon that is regularly observed in practice). Note that the conditions imposed on $F^{(2)}$ in this example are typical for finite subseries of aperiodic series.

Thus, in many cases a qualitative analysis of the periodograms of the series provides an insight into their separability features. At the same time, smallness of the spectral correlation coefficients provides only a sufficient condition for approximate separability and it is not difficult to construct examples where exact separability takes place, but the spectral correlation coefficients are large.

Example 1.3 *Separability and spectral correlation*
Let $N = 399$, $a = 1.005$, $T = 200$ and the terms of the series $F^{(1)}$ and $F^{(2)}$ are

$$f_n^{(1)} = a^{-n}, \quad f_n^{(2)} = a^n \cos(2\pi n/T).$$

It can be shown that the choice of the window length $L = 200$ (hence, $K = 200$) leads to the exact separability of the series.

At the same time, the periodogram analysis does not suggest exact separability: the frequency ranges of the two series significantly intersect and the maximum spectral correlation coefficients between their subseries equals 0.43.

Thus, for a fixed window length L we have several characteristics of the quality of (weak) separability, related to the rows and columns of the trajectory matrices:

1. *Cross-correlation matrices between the rows (and columns) of the trajectory matrices.* If all the correlations are zero, then we have exact separability, while the smallness of their absolute values corresponds to approximate separability.

2. *Weighted correlation coefficient* $\rho_{12}^{(w)}$ *between time series* $F^{(1)}$ *and* $F^{(2)}$ *defined by* (1.24). The equality $\rho_{12}^{(w)} = 0$ is a necessary (but not sufficient) condition for separability. Irrespective of separability, the expansion of a time series onto **w**-uncorrelated or approximately **w**-uncorrelated components is a highly desirable property.

3. *Matrices of spectral correlations between the rows (and columns) of the trajectory matrices.* The vanishing of all the spectral correlations is a sufficient (but not necessary) condition for separability. Smallness of these correlations implies approximate separability. In the latter case, this separability has a useful interpretation in the language of periodograms: the frequency ranges (with

1.5.4 Strong separability

The criteria for (weak) separability of two series for a fixed window length give a solution to the problem that can be stated as follows: 'Does the sum of the SVDs of the trajectory matrices of the series $F^{(1)}$ and $F^{(2)}$ coincide with *one of* the SVDs of the trajectory matrix of the series $F = F^{(1)} + F^{(2)}$?'

Another question, closer to practical needs, can be stated as follows: 'Is it possible to group the matrix terms of *any* SVD of the trajectory matrix $\mathbf{X}$ of the series $F = F^{(1)} + F^{(2)}$, to obtain the trajectory matrices of the series $F^{(1)}$ and $F^{(2)}$?'

If the answer to this question is the affirmative, then we shall say that the series $F^{(1)}$ and $F^{(2)}$ are *strongly separable*. It is clear that if the series are weakly separable and all the singular values of the trajectory matrix $\mathbf{X}$ are different, then strong separability holds.

Moreover, strong separability of two series $F^{(1)}$ and $F^{(2)}$ is equivalent to the fulfillment of the following two conditions: (a) the series $F^{(1)}$ and $F^{(2)}$ are weakly separable, and (b) the collections of the singular values of the trajectory matrices $\mathbf{X}^{(1)}$ and $\mathbf{X}^{(2)}$ are disjoint.

Let us comment on this. Assume that

$$\mathbf{X}^{(1)} = \sum_k \mathbf{X}_k^{(1)}, \quad \mathbf{X}^{(2)} = \sum_m \mathbf{X}_m^{(2)}$$

are the SVDs of the trajectory matrices $\mathbf{X}^{(1)}$ and $\mathbf{X}^{(2)}$ of the series $F^{(1)}$ and $F^{(2)}$, respectively. If the series are weakly separable, then

$$\mathbf{X} = \sum_k \mathbf{X}_k^{(1)} + \sum_m \mathbf{X}_m^{(2)}$$

is the SVD of the trajectory matrix $\mathbf{X}$ of the series $F = F^{(1)} + F^{(2)}$.

Assume now that the singular values corresponding to the elementary matrices $\mathbf{X}_1^{(1)}$ and $\mathbf{X}_1^{(2)}$ coincide. This means that using the SVD of the matrix $\mathbf{X}$ we cannot uniquely identify the terms $\mathbf{X}_1^{(1)}$ and $\mathbf{X}_1^{(2)}$ in the sum $\mathbf{X}_1^{(1)} + \mathbf{X}_1^{(2)}$, since these two matrices correspond to the same eigenvalue of the matrix $\mathbf{X}\mathbf{X}^\mathrm{T}$.

To illustrate this discussion, let us consider a simple example.

Example 1.4 *Weak and strong separability*
Let $N = 3$, $L = K = 2$ and consider the series

$$F^{(1)} = (1, -a, a^2), \quad F^{(2)} = (1, a^{-1}, a^{-2}), \quad a \neq 0.$$

In this case, the matrices $\mathbf{X}^{(1)}$ and $\mathbf{X}^{(2)}$ are

$$\mathbf{X}^{(1)} = \begin{pmatrix} 1 & -a \\ -a & a^2 \end{pmatrix}, \quad \mathbf{X}^{(2)} = \begin{pmatrix} 1 & a^{-1} \\ a^{-1} & a^{-2} \end{pmatrix}.$$

Checking the weak separability of the series is easy. At the same time, the matrices $\mathbf{X}^{(1)}(\mathbf{X}^{(1)})^\mathrm{T}$ and $\mathbf{X}^{(2)}(\mathbf{X}^{(2)})^\mathrm{T}$ have only one positive eigenvalue each, and the singular values of the matrices $\mathbf{X}^{(1)}$ and $\mathbf{X}^{(2)}$ are

$$\sqrt{\lambda^{(1)}} = 1 + a^2, \quad \sqrt{\lambda^{(2)}} = 1 + a^{-2},$$

respectively. Thus, for any $a \neq 1$ the series $F^{(1)}$ and $F^{(2)}$ are strongly separable, but for $a = 1$ these series are only weakly separable; they loose the strong separability.

Indeed, if $a \neq 1$, then the SVD of the trajectory matrix $\mathbf{X}$ of the series $F^{(1)} + F^{(2)}$ is uniquely defined and has the form

$$\mathbf{X} = \begin{pmatrix} 2 & -a + a^{-1} \\ -a + a^{-1} & a^2 + a^{-2} \end{pmatrix}$$
$$= (1 + a^2) U_1 V_1^\mathrm{T} + (1 + a^{-2}) U_2 V_2^\mathrm{T} = \mathbf{X}^{(1)} + \mathbf{X}^{(2)},$$

where $U_1 = V_1 = (1, -a)^\mathrm{T}/\sqrt{1 + a^2}$ and $U_2 = V_2 = (1, a^{-1})^\mathrm{T}/\sqrt{1 + a^{-2}}$.

At the same time, when $a = 1$ there are infinitely many SVDs of the matrix $\mathbf{X}$:

$$\mathbf{X} = \begin{pmatrix} 2 & 0 \\ 0 & 2 \end{pmatrix} = 2 U_1 V_1^\mathrm{T} + 2 U_2 V_2^\mathrm{T},$$

where $\{U_1, U_2\}$ is an arbitrary orthonormal basis in $\mathbf{R}^2$, $V_1 = \mathbf{X}^\mathrm{T} U_1/2$ and $V_2 = \mathbf{X}^\mathrm{T} U_2/2$. From all these SVDs only one is acceptable, the one corresponding to $U_1 = (1, 1)^\mathrm{T}/\sqrt{2}$ and $U_2 = (1, -1)^\mathrm{T}/\sqrt{2}$. This SVD gives us the required decomposition $\mathbf{X} = \mathbf{X}^{(1)} + \mathbf{X}^{(2)}$.

In practice, the lack of strong separability (under the presence of the weak separability, perhaps, approximate) becomes essential when the matrix $\mathbf{X}\mathbf{X}^\mathrm{T}$ has two close eigenvalues. This leads to an instability of the SVD computations. Let us return to the example of the asymptotic separability of the harmonic series $F^{(1)}$ and $F^{(2)}$ with

$$f_n^{(1)} = \cos(2\pi\omega_1 n), \quad f_n^{(2)} = \cos(2\pi\omega_2 n),$$

where $\omega_1 \neq \omega_2$ and $L, K \to \infty$.

As demonstrated in Section 5.1, for all L and K the SVD of the trajectory matrix of each of the series $F^{(1)}, F^{(2)}$ consists of two terms so that the eigen and factor vectors are the harmonic series with the same frequency as the original series (however, the phase of the harmonics may differ from the phase in the original series). Also, the singular values asymptotically (when $L, K \to \infty$) coincide and do not depend on the frequencies of the harmonics.

Moreover, the series $F^{(1)}$ and $F^{(2)}$ are asymptotically separable, and, since these series have the same amplitude, asymptotically all four singular values are equal. Therefore, even when N, L, and K are large, we cannot as a rule separate the periodicities $F^{(1)}$ and $F^{(2)}$ out of the sum $F = F^{(1)} + F^{(2)}$ (if, of course, we do not use special rotations in the four-dimensional eigenspace of the trajectory matrix $\mathbf{X}$ of the series F).

Another way to deal with the case of equal singular values is described in Section 1.7.3. It is based on the simple fact that if the series $F^{(1)}$ are $F^{(2)}$ weakly separable, then we can always find a constant $c \neq 0$ such that the series $F^{(1)}$ and $cF^{(2)}$ are strongly separable.

The presence of close singular values is the reason why SSA often fails to decompose the component consisting of many harmonics with similar weights. If these weights are small, then it may be natural to consider such components as the noise components.

1.6 Choice of SSA parameters

In this section we discuss the role of the parameters in Basic SSA and the principles for their selection. As was mentioned in Section 1.4.1, we assume that the time series under consideration can be regarded as a sum of a slowly varying trend, different oscillatory components, and a noise. The time series analysis issues related to this assumption were discussed in Section 1.4.2.

Certainly, the choice of parameters depends on the data we have and the analysis we have to perform. We discuss the selection issues separately for all the main problems of time series analysis.

There are two parameters in Basic SSA: the first is an integer L, the window length, and the second parameter is structural; loosely speaking, it is the way of grouping.

1.6.1 Grouping effects

Assume that the window length L is fixed and we have already made the SVD of the trajectory matrix of the original time series. The next step is to group the SVD terms in order to solve one of the problems discussed in Section 1.4.2. We suppose that this problem has a solution; that is, the corresponding terms can be found in the SVD, and the result of the proper grouping would lead to the (approximate) separation of the time series components (see Section 1.5).

Therefore, we have to decide what the proper grouping is and how to find the proper groups of the eigentriples. In other words, we need to identify an eigentriple corresponding to the related time series component. Since each eigentriple consists of an eigenvector (left singular vector), a factor vector (right singular vector) and a singular value, this is to be achieved using only the information contained in these vectors (considered as time series) and the singular values.

(a) General issues

We start by mentioning several purely theoretical results about the eigentriples of several 'simple' time series (see Section 5.1).

Exponential-cosine sequences
Consider the series
$$f(n) = Ae^{\alpha n}\cos(2\pi\omega n + \phi), \tag{1.27}$$
$\omega \in [0, 1/2]$, $\phi \in [0, 2\pi)$ and denote $T = 1/\omega$.

Depending on the parameters, the exponential-cosine sequence produces the following eigentriples:

1. *Exponentially modulated harmonic time series with a frequency $\omega \in (0, 1/2)$*
   If $\omega \in (0, 1/2)$, then for any L and N the SVD of the trajectory matrix has two terms. Both eigenvectors (and factor vectors) have the same form (1.27) with the same frequency ω and the exponential rate α. If $\alpha \leq 0$ then for large N, L and $K = N - L + 1$, both singular values are close (formally they asymptotically coincide for $L, K \to \infty$). Practically, they are close enough when L and K are several times greater than $T = 1/\omega$.

2. *Exponentially modulated saw-tooth curve ($\omega = 1/2$)*
   If $\omega = 1/2$ and $\sin(\phi) \neq 0$, then f_n is proportional to $(-e^{\alpha})^n$. In this case for any L the corresponding SVD has just one term. Both singular vectors have the same form as the initial series.

3. *Exponential sequence ($\omega = 0$)*
   If $\omega = 0$ and $\cos(\phi) \neq 0$, then f_n is proportional to $e^{\alpha n}$ and we have an exponential series. For any N and window length L, the trajectory matrix of the exponential series has only one eigentriple. Both singular vectors of this eigentriple are exponential with the same parameter α.

4. *Harmonic series ($\alpha = 0$)*
   If $\alpha = 0$ and $\omega \neq 0$, then the series is a pure harmonic one. The eigenvectors and factor vectors are harmonic series with the same ω. If $\omega \neq 1/2$ and $T = 1/\omega$ is a divisor of K and L, then both singular values coincide.

Polynomial series
Consider a polynomial series of the form
$$f_n = \sum_{k=0}^{m} a_k n^k, \quad a_m \neq 0.$$

1. *General case*
   If f_n is a polynomial of degree m, then the order of the corresponding SVD does not exceed $m + 1$ and all the singular vectors are polynomials too; also their degrees do not exceed m.

2. *Linear series*
   For a linear series
   $$f_n = an + b, \quad a \neq 0,$$
   with arbitrary N and L the SVD of the L-trajectory matrix consists of two terms. All singular vectors are also linear series with the same $|a|$.

CHOICE OF SSA PARAMETERS

Note that the exponential-cosine and linear series (in addition to the sum of two exponential series with different rates) are the only series that have at most two terms in the SVD of their trajectory matrices for any series length N and window length $L \geq 2$ (see the proofs in Section 5.1). This fact helps in their SSA identification as components of more complex series.

Let us now turn to the various different grouping problems and the corresponding grouping principles. We start with mentioning several general rules.

1. *If we reconstruct a component of a time series with the help of just one eigentriple and both singular vectors of this eigentriple have a similar form, then the reconstructed component will have approximately the same form.*

   This means that when dealing with a single eigentriple we can often predict the behaviour of the corresponding component of the series. For example, if both singular vectors of an eigentriple resemble linear series with similar slopes, then the corresponding component is also almost linear. If the singular vectors have the form of the same exponential series, then the trend has a similar form. Harmonic-like singular vectors produce harmonic-like components (compare this with the results for exponential-cosine series presented at the beginning of this section).

   The conservation law under discussion can be extended to incorporate monotonicity (monotone singular vectors generate monotone components of the series) as well as some other properties of time series.

2. *If $L \ll K$ then the factor vector in an eigentriple has a greater similarity with the component, reconstructed from this eigentriple, than the eigenvector.* Consequently we can approximately predict the result of reconstruction from a single eigentriple with the help of its factor vector.

3. *If we reconstruct a series with the help of several eigentriples, and the periodograms of their singular vectors are (approximately) supported on the same frequency interval $[a, b]$, then the frequency power of the reconstructed series will be mainly supported on $[a, b]$.* This feature is analogous to that in item 1 but concerns several eigentriples and is formulated in terms of the Fourier expansions.

4. *The larger the singular value of the eigentriple is, the bigger the weight of the corresponding component of the series.* Roughly speaking, this weight may be considered as being proportional to the singular value.

Now let us turn to the SSA problems.

(b) Grouping for extraction of trends and smoothing

1. *Trends*
According to our definition, trend is a slowly varying component of a time series which does not contain oscillatory components. Assume that the time series itself

is such a component alone. Practice shows that in this case, one or more leading singular vectors will be slowly varying as well. Exponential and polynomial sequences are good examples of this situation.

For a general series F we typically assume that its trend component $F^{(1)}$ is (approximately) strongly separable from all the other components. This means that among the eigentriples of the series F, there are eigentriples that approximately correspond to the SVD components of the series $F^{(1)}$.

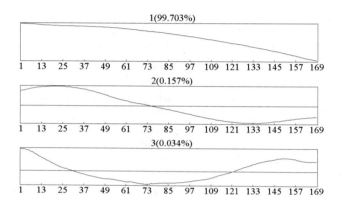

Figure 1.23 *Production: three factor vectors of the 'accurate trend'.*

Thus, *to extract a trend of a series, we have to collect all the elementary matrices related to slowly varying singular vectors.*

The ordinal numbers of these eigentriples depend not only on the trend $F^{(1)}$ itself, but on the 'residual series' $F^{(2)} = F - F^{(1)}$ also. Consider two different extremes. First, let the series F have a strong trend tendency $F^{(1)}$ with a relatively small oscillatory-and-noise component $F^{(2)}$. Then most of the trend eigentriples will have the leading positions in the SVD of the whole series F. Certainly, some of these eigentriples can have small singular values, especially if we are looking for a more or less refined trend.

For instance, in the 'Production' example (Section 1.3.1, Fig. 1.2) a reasonably accurate trend is described by the three leading eigentriples, and the singular value in the third eigentriple is five times smaller than the second one. The corresponding factor vectors are shown at Fig. 1.23.

The other extreme is the situation where we deal with high oscillations on the background of a small and slow general tendency. Here, the leading elementary matrices describe oscillations, while the trend eigentriples can have small singular values (and therefore can be far from the top in the ordered list of the eigentriples).

2. Smoothing

The problem of smoothing may seem similar to that of trend extraction but it has

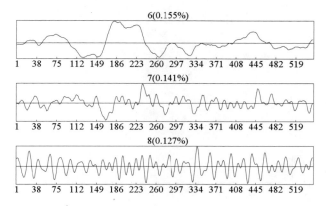

Figure 1.24 *Tree rings: factor vectors with ordinal numbers 6, 7, and 8.*

its own specifics. In particular, we can smooth any time series, even if it does not have an obvious trend component (the example 'Tree rings' in Section 1.3.2 is just one of this sort). That means that for the extraction of a trend, we collect all the eigentriples corresponding to a slowly varying (but not oscillatory) part of the series; at the same time a smoothed component of a series is composed of a collection of the eigentriples whose singular vectors do not oscillate rapidly.

In the 'Tree rings' example (Section 1.3.2, Fig. 1.3) seven leading eigentriples corresponding to low frequencies were chosen for smoothing. Fig. 1.24 demonstrates factor vectors with numbers 6, 7, and 8. The sixth factor vector is slowly varying, and therefore it is selected for smoothing, while the eighth one corresponds to rather high frequencies and is omitted. As for the 7th factor vector, it demonstrates mixing of rather low (0-0.05) and high (approximately 0.08) frequencies. To include all low frequencies, the seventh eigentriple was also selected for smoothing.

For relatively long series, periodogram analysis serves as a good description of the matter. Periodograms (see Figs. 1.25 and 1.26) confirm that smoothing in the 'Tree rings' example splits the frequencies into two parts rather well. The small intersection of the frequency ranges for the result of the smoothing and the residual is due to mixing of some of the frequencies in some of the eigentriples (see the 7th factor vector in Fig. 1.24).

(c) Grouping for oscillations

1. *Harmonic series*

Let us start with a pure harmonic with a frequency ω and a certain phase and amplitude. Since we assume that such a component $F^{(1)}$ in the original series is approximately strongly separable from $F^{(2)} = F - F^{(1)}$, we may hope that two (or one if $\omega = 1/2$) SVD eigentriples of the trajectory matrix generated by

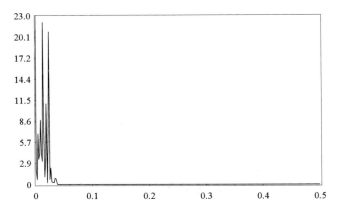

Figure 1.25 *Tree rings: periodogram of smoothed series.*

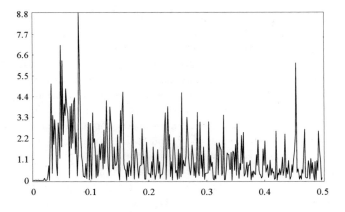

Figure 1.26 *Tree rings: periodogram of smoothing residuals.*

F correspond to $F^{(1)}$. The problem is, therefore, to identify these eigentriples among all other eigentriples generated by F.

Let $\omega \neq 1/2$. As was stated in the example of the exponential-cosine function, the pure harmonic corresponding to (1.27) with $\alpha = 0$ generates an SVD of order two with the singular vectors having the same harmonic form.

Consider the ideal situation where $T = 1/\omega$ is a divisor of the window length L and $K = N - L + 1$. Since T is an integer, it is a period of the harmonic.

Let us take, for definiteness, the left singular vectors (that is, the eigenvectors). In the ideal situation described above, the eigenvectors have the form of sine and cosine sequences with the same T and the same phases. The factor vectors are also of the same form.

Thus, the identification of the components that are generated by a harmonic is reduced to the determination of these pairs. Viewing the pairwise scatterplots of the eigenvectors (and factor vectors) simplifies the search for these pairs; indeed, the pure sine and cosine with equal frequencies, amplitudes, and phases create the scatterplot with the points lying on a circle.

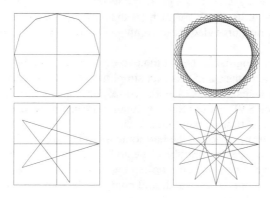

Figure 1.27 *Scatterplots of sines/cosines.*

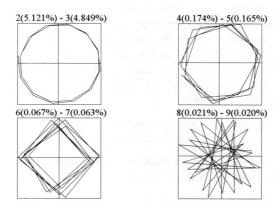

Figure 1.28 *Eggs: scatterplots of paired harmonic factor vectors.*

If $T = 1/\omega$ is an integer, then these points are the vertices of the regular T-vertex polygon. For the rational frequency $\omega = q/p < 1/2$ with relatively prime integers p and q, the points are the vertices of the regular p-vertex polygon. Fig. 1.27 depicts scatterplots of four pairs of sin/cosine sequences with zero phases, the same amplitudes and frequencies $1/12$, $10/53$, $2/5$, and $5/12$.

Small deviations from this ideal situation would imply that the points in the scatterplots are no longer exactly the vertices of the regular p-vertex polygon, although staying reasonably close to them. As an example, Fig. 1.28 provides scatterplots of paired factor vectors in the 'Eggs' example (Section 1.3.3), corresponding to the harmonics with the frequencies $1/12, 2/12, 3/12$ and $5/12 = 1/2.4$.

Therefore, an analysis of the pairwise scatterplots of the singular vectors allows one to visually identify those eigentriples that correspond to the harmonic components of the series, provided these components are separable from the residual component.

In practice, the singular values of the two eigentriples of a harmonic series are often close to each other, and this fact simplifies the visual identification of the harmonic components. (In this case, the corresponding eigentriples are, as a rule, consecutive in the SVD order.) Such a situation typically occurs when, say, both L and K are several times greater than $1/\omega$.

Alternatively, if the period of the harmonic is comparable to N, then the corresponding singular values may not be close and therefore the two eigentriples may not be consecutive. The same effect often happens when the two singular values of a harmonic component are small and comparable with the singular values of the components of the noise.

The series F may contain several purely harmonic components, and the frequency of each one should be identified using the corresponding pair of eigentriples. In easy cases it is a straightforward operation (see, for example, the scatterplot of period 12 at Fig. 1.28). In more complex cases, the periodogram analysis applied to the singular vectors often helps.

One more method of approximate identification of the frequency, which can be useful even for short series, is as follows. Consider two eigentriples, which approximately describe a harmonic component with frequency ω_0. Then the scatterplot of their eigenvectors can be expressed as a two-dimensional curve with Euclidean components of the form

$$x(n) = r(n)\cos(2\pi\omega(n)n + \phi(n))$$
$$y(n) = r(n)\sin(2\pi\omega(n)n + \phi(n)),$$

where the functions r, ω and ϕ are close to constants and $\omega(n) \approx \omega_0$. The polar coordinates of the curve vertices are $(r(n), \delta(n))$ with $\delta(n) = 2\pi\omega(n)n + \phi(n)$.

Since $\Delta_n \stackrel{\text{def}}{=} \delta(n+1) - \delta(n) \approx 2\pi\omega_0$, one can estimate ω_0 by averaging polar angle increments Δ_n ($n = 0, \ldots, L-1$). The same procedure can be applied to a pair of factor vectors.

If the period of the harmonic is equal to 2, that is $\omega = 1/2$, then the situation is simpler since in this case the SVD of this harmonic consists of only one eigentriple, and the corresponding eigenvector and factor vector have a saw-tooth form. Usually the identification of such vectors is easy.

2. Grouping for identification of a general periodic component

Consider now the more comprehensive case of extraction of a general periodic component $F^{(1)}$ out of the series F. If the integer T is the period of this component, then according to Section 1.4.1,

$$f_n^{(1)} = \sum_{k=1}^{[T/2]} a_k \cos(2\pi kn/T) + \sum_{k=1}^{[T/2]} b_k \sin(2\pi kn/T). \quad (1.28)$$

Hence (see Section 5.1), there are at most $T-1$ matrix components in the SVD of the trajectory matrix of the series $F^{(1)}$, for any window length $L \geq T-1$. Moreover, for large L the harmonic components in the sum (1.28) are approximately strongly separable, assuming their powers $a_k^2 + b_k^2$ are all different.

In this case, each of these components produces either two (for $k \neq T/2$) or one (for $k = T/2$) eigentriples with singular vectors of the same harmonic kind.

Therefore, under the assumptions:

(a) the series $F^{(1)}$ is (approximately) strongly separable from $F^{(2)}$ in the sum $F = F^{(1)} + F^{(2)}$ for the window length L,

(b) all the nonzero powers $a_k^2 + b_k^2$ in the expansion (1.28) are different, and

(c) L is large enough,

we should be able to approximately separate all the eigentriples, corresponding to the periodic series $F^{(1)}$ in the SVD of the trajectory matrix of the whole series F.

To perform this separation, it is enough to identify the eigentriples that correspond to all the harmonics with frequencies k/T (this operation has been described above in the section 'Grouping for oscillations: Harmonic series') and group them.

For instance, if it is known that there is a seasonal component in the series F and the data is monthly, then one must look at periodicities with frequencies $1/12$ (annual), $1/6$ (half-year), $1/4$, $1/3$ (quarterly), $1/2.4$, and $1/2$. Of course, some of these periodicities may be missing.

In the example 'Eggs' (see Section 1.3.3), all the above-mentioned periodicities are present; they can all be approximately separated for the window length 12.

If some of the nonzero powers in (1.28) coincide, then the problem of identification of the eigentriples in the SVD of the series F has certain specifics. For instance, suppose that $a_1^2 + b_1^2 = a_2^2 + b_2^2$. If L is large, then four of the singular values in the SVD of the trajectory matrix of the series F are (approximately) the same, and the corresponding singular vectors are linear combinations of the harmonics with two frequencies: $\omega_1 = 1/T$ and $\omega_2 = 2/T$.

Therefore, the components with these frequencies are not strongly separable. In this case, the periodogram analysis of the singular vectors may help a lot; if their periodograms have sharp peaks at around the frequencies ω_1 and ω_2, then the corresponding eigentriples must be regarded as those related to the series $F^{(1)}$.

Figs. 1.29 and 1.30 depict the periodograms of the sixth and the eighth eigenvectors from the example 'Rosé wine' (Section 1.4.1, Fig. 1.17). Due to the closeness of the singular values in the eigentriples 6-9 (the eigenvalue shares for these eigentriples lie between 0.349% and 0.385%) the harmonics (mostly, with the fre-

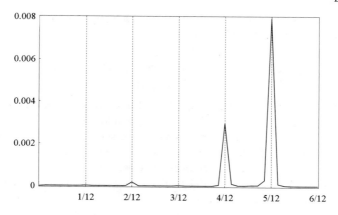

Figure 1.29 *Rosé wine: periodogram of the sixth eigenvector.*

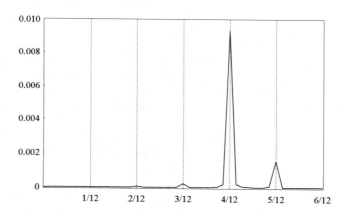

Figure 1.30 *Rosé wine: periodogram of the eighth eigenvector.*

quencies 4/12 and 5/12) are being mixed up, and this is perfectly reflected in the periodograms provided.

3. Modulated periodicities

The case of amplitude-modulated periodicity is much more complex since we do not assume the exact form of modulation. However, the example of the exponentially modulated harmonic (1.27) with $\alpha \neq 0$ shows that sometimes identification of the components of such signals can be performed. Let us start with this series. Being one of the simplest, the exponentially modulated harmonic can be considered as an additive component of some econometric series describing an

CHOICE OF SSA PARAMETERS

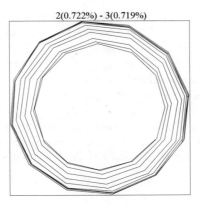

Figure 1.31 *Hotels: spiral in the scatterplot of factor vectors.*

exponential growth associated with an exponentially modulated seasonal oscillation.

On the whole, the situation here is analogous to the case of a pure harmonic series. If the frequency is not $1/2$ and the window length L is large, then we have two eigentriples with approximately equal singular values, both being characterized by the singular vectors of the same exponential-cosine shape. Therefore, the scatterplots of these pairs of eigen/factor vectors have the form of a spiral and visually are easily distinguishable.

For instance, Fig. 1.31 depicts the scatterplot of the second and third factor vectors corresponding to the annual periodicity with increasing amplitude in the series 'Hotels' for $L = 48$ (description of this series can be found in Section 1.7.1).

If the period of the harmonic is 2, then the series is $f_n = (-a)^n$, where $a = e^\alpha$. The singular vectors of the single eigentriple, created by this series, have exactly the same shape as the modulated saw-tooth curve. Of course, visual identification of this series is also not difficult.

If the series modulating the pure harmonic is not exponential, then the extraction of the corresponding components is much more difficult (see the theoretical results of Section 5.1 concerning the general form of infinite series that generate finite SVDs of their trajectory matrices). Let us, however, describe the situation when the identification of the components is possible.

As we already mentioned in Section 1.4.1, if the amplitude of the modulated harmonic $F^{(1)}$ varies slowly, then the range of frequencies is concentrated around the main frequency, which can clearly be seen in the periodogram of this modulated harmonic. If the window length L and $K = N - L + 1$ are large, then all the singular vectors of this series will have the same property.

Therefore, if the series $F^{(1)}$ is (approximately) strongly separable from a series $F^{(2)}$ in the sum $F = F^{(1)} + F^{(2)}$, then one can expect that the frequency interval of $F^{(2)}$ has a small (in terms of powers) intersection with the frequency interval of the modulated harmonic $F^{(1)}$. Thus, by analyzing periodograms of the singular vectors in all the eigentriples of the series F, one can hope to identify the majority of those that (approximately) describe $F^{(1)}$.

The situation is similar when an arbitrary periodic (not necessarily harmonic) signal is being modulated. In this case, each term in the sum (1.28) is multiplied by the function modulating the amplitude, and every frequency k/T gives rise to a group of neighbouring frequencies. Under the condition that $F^{(1)}$ and $F^{(2)}$ are strongly separable, one should look for singular vectors in the SVD of the series F such that their periodograms are concentrated around the frequencies k/T.

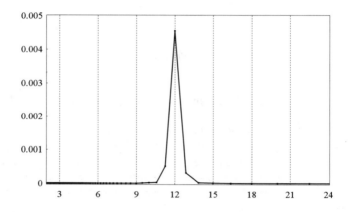

Figure 1.32 *Unemployment: periodogram of the 4th eigenvector (in periods).*

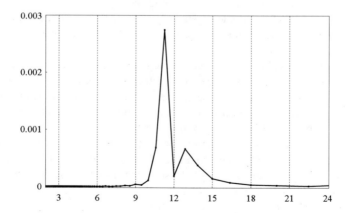

Figure 1.33 *Unemployment: periodogram of the 12th eigenvector (in periods).*

In the example 'Unemployment' (Section 1.3.6) the modulated annual periodicity generates, in particular, the pairs of eigentriples 3-4 and 12-13. Figs. 1.32 and 1.33 depict the periodograms of the fourth and the twelfth eigenvectors, describing the main frequency 1/12 and close frequencies, respectively. These figures demonstrate the typical shape of the periodograms of the singular vectors generated by a modulated harmonic.

(d) Grouping for finding a refined decomposition of a series

The problem of finding a refined structure of a series by Basic SSA is equivalent to the identification of the eigentriples of the SVD of the trajectory matrix of this series, which correspond to the trend, various oscillatory components, and noise. The principles of grouping for identification of the trend and oscillatory components have been described above.

As regards noise, we should always bear in mind the intrinsic uncertainty of this concept under the lack of a rigorous mathematical model for noise. From the practical point of view, a natural way of noise extraction is the grouping of the eigentriples, which do not seemingly contain elements of trend and oscillations. In doing that, one should be careful about the following.

1. If the frequency range of the noise contains the frequency of a harmonic component of the signal, then the harmonic component reconstructed from the related SVD eigentriples will also include the part of the noise corresponding to this frequency (compare Marple, 1987, Chapter 13.3). Analogously, the extracted trend 'grasps' the low-frequency parts of the noise, if there are any. If the frequency ranges of the signal and noise do not intersect, then this effect does not appear.

2. If the amplitude of a harmonic component of the signal is small and the noise is large, then the singular values corresponding to the harmonic and the noise may be close. That would imply the impossibility of separating the harmonic from the noise on the basis of the analysis of the eigentriples for the whole series. Speaking more formally, the harmonic and noise would not be strongly separable. This effect disappears asymptotically, when $N \to \infty$.

3. The SVD of the trajectory matrix of the pure noise (that is, of an aperiodic stationary sequence) for large N, L and K should be expected to contain at least some (leading) eigentriples looking like harmonics (see Section 6.4.3). The components of the original series, reconstructed from these eigentriples, will look similar. This necessitates a profound control of the interpretability of the reconstructed components.

Certainly, the above discussion concerns also the problem of noise reduction, which is a particular case of the problem under consideration.

(e) Grouping hints

A number of characteristics of the eigentriples of the SVD of the trajectory matrix of the original series may very much help in making the proper grouping for extraction of the components from the series. Let us discuss two of these characteristics.

1. *Singular values*

As mentioned above, if N, L and K are sufficiently large, then each harmonic different from the saw-tooth one produces two eigentriples with close singular values. Moreover, a similar situation occurs if we have a sum of several different harmonics with (approximately) the same amplitudes; though the corresponding singular vectors do not necessarily correspond to a pure harmonic (the frequencies can be mixed), they can still form pairs with close singular values and similar shapes.

Therefore, explicit plateaux in the eigenvalue spectra prompts the ordinal numbers of the paired eigentriples. Fig. 1.34 depicts the plot of leading singular val-

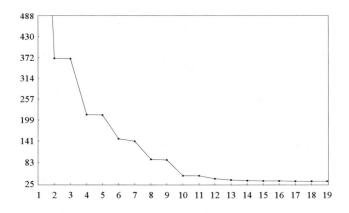

Figure 1.34 *Births: leading singular values.*

ues for the example 'Births' (Section 1.3.4). Five evident pairs with almost equal leading singular values correspond to five (almost) harmonic components of the 'Births' series: eigentriple pairs 2-3, 4-5 and 10-11 are related to a one-week periodicity with frequencies $1/7$, $2/7$ and $3/7$, while pairs 6-7 and 8-9 describe the annual birth cycle (frequencies $\approx 1/365$ and $\approx 2/365$). Note that the first singular value, equal to 4772.5, corresponds to the trend component of the series and is omitted in Fig. 1.34.

Another useful insight is provided by checking breaks in the eigenvalue spectra. As a rule, a pure noise series produces a slowly decreasing sequence of singular values. If such a noise is added to a signal, described by a few eigentriples with

CHOICE OF SSA PARAMETERS

large singular values, then a break in the eigenvalue spectrum can distinguish signal eigentriples from the noise ones.

Note that in general there are no formal procedures enabling one to find such a break. Moreover, for complex signals and large noise, the signal and noise eigentriples can be mixed up with respect to the order of their singular values.

At any rate, singular values give important but supplementary information for grouping; the structure of the singular vectors is more essential.

2. w-*Correlations*

As discussed earlier, a necessary condition for the (approximate) separability of two series is the (approximate) zero w-correlation of the reconstructed components. On the other hand, the eigentriples entering the same group can correspond to highly correlated components of the series.

Thus, a natural hint for grouping is the matrix of the absolute values of the w-correlations, corresponding to the full decomposition (in this decomposition each group corresponds to only one matrix component of the SVD). This matrix for an artificial series F with

$$f_n = e^{n/400} + \sin(2\pi n/17) + 0.5\sin(2\pi n/10) + \varepsilon_n, \quad n = 0,\ldots,339, \quad (1.29)$$

standard Gaussian white noise ε_n, and $L = 85$, is depicted in Fig. 1.35 (w-correlations for the first 30 reconstructed components are shown in 20-colour scale from white to black corresponding to the absolute values of correlations from 0 to 1).

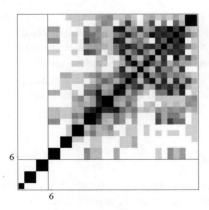

Figure 1.35 *Series* (1.29): *matrix of* w-*correlations*.

The form of this matrix gives an indication of how to make the proper grouping: the leading eigentriple describes the exponential trend, the two pairs of the subsequent eigentriples correspond to the harmonics, and the large sparkling square indicates the white noise components. Note that theoretical results of Section 6.1.3 tell us that such a separation can be indeed (asymptotically) valid.

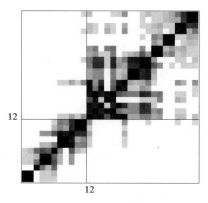

Figure 1.36 *White dwarf: matrix of **w**-correlations.*

For the example 'White dwarf' (Section 1.3.2) with $L = 100$, the matrix of the absolute values of the **w**-correlations of the reconstructed components produced from the leading 30 eigentriples is depicted in Fig. 1.36 in the manner of Fig. 1.35.

It is clearly seen that the splitting of all the eigentriples into two groups — from the first to the 11th and the rest — gives rise to a decomposition of the trajectory matrix into two almost orthogonal blocks, with the first block corresponding to the smoothed version of the original series and the second block corresponding to the residual, see Figs. 1.5 and 1.6 in Section 1.3.2. Note that despite the fact that some **w**-correlations between the eigentriples in different blocks exceed 0.2, the reconstructed components are almost **w**-uncorrelated: $|\rho^{(w)}| = 0.004$.

The similarity of Figs. 1.35 and 1.36 gives us an additional argument in favour of the assertion that in the 'White dwarf' example smoothing leads to noise reduction.

1.6.2 Window length effects

Window length is the main parameter of Basic SSA, in the sense that its improper choice would imply that no grouping activities will help to obtain a good SSA decomposition. Moreover, it is the single parameter of the decomposition.

Selection of the proper window length depends on the problem in hand, and on preliminary information about the time series. In the general case no universal rules and unambiguous recommendations can be given for the selection of the window length. The main difficulty here is caused by the fact that variations in the window length may influence both weak and strong separability features of SSA, i.e., both the orthogonality of the appropriate time series intervals and the closeness of the singular values.

However, there are several general principles for the selection of the window length L that have certain theoretical and practical grounds. Let us discuss these principles.

(a) General effects

1. The SVDs of the trajectory matrices, corresponding to the window lengths L and $K = N - L + 1$, are equivalent (up to the symmetry: left singular vectors $\leftrightarrow$ right singular vectors). Therefore, *for the analysis of structure of time series by Basic SSA it is meaningless to take the window length larger than half of the time series length.*

2. Bearing in mind the previous remark, *the larger the window length is, the more detailed is the decomposition of the time series.* The most detailed decomposition is achieved when the window length is approximately equal to half of time series length, that is when $L \sim N/2$. The exceptions are the so-called series of finite rank, where for any L larger than d and $N > 2d - 1$ (d is the rank of the series) the number of nonzero components in the SVD of the series is equal to d and does not depend on the window length (see Section 1.6.1 for examples and Section 5.1 for general results).

3. The effects of weak separability.

    - Since the results concerning weak separability of time series components are mostly asymptotic (when $L, K \to \infty$), in the majority of examples to achieve a better (weak) separation one has to choose large window lengths. In other words, *a small window length could mix up interpretable components.*
    - If the window length L is relatively large (say, it is equal to several dozen), then the (weak) *separation results are stable with respect to small perturbations in L.*
    - On the other hand, for *specific series and tasks, there are concrete recommendations for the window length selection,* which can work for a relatively small N (see section 'Window length for periodicities' below).

4. The effects of closeness of singular values.
    The negative effects due to the closeness of the singular values related to different components of the original series (that is, the absence of strong separability in the situation where (approximate) weak separability does hold), are not easily formalized in terms of the window length. These effects are often difficult to overcome by means of selection of L alone.

    Let us mention two other issues related to the closeness of the singular values.

    - *For the series with a complex structure, too large window length L can produce an undesirable decomposition of the series components of interest,* which may lead, in particular, to their mixing with other series components.

This is an unpleasant possibility, especially since a significant reduction of L can lead to a poor quality of the (weak) separation.

- Alternatively, sometimes in these circumstances *even a small variation in the value of L can reduce mixing and lead to a better separation of the components*, i.e., provide a transition from weak to strong separability. At any rate, it is always worthwhile trying several window lengths.

(b) Window length for extraction of trends and smoothing

1. Trends

In the problem of trend extraction, a possible contradiction between the requirements for weak and strong separability emerges most frequently. Since trend is a relatively smooth curve, its separability from noise and oscillations requires large values of L.

On the other hand, if trend has a complex structure, then for values of L that are too large, it can be described only by a large number of eigentriples with relatively small singular values. Some of these singular values could be close to those generated by oscillations and/or noise time series components.

This happens in the example 'Births', where the window length of order 1000 and more (the series length is 5113) leads to the situation where the components of the trend are mixed up with the components of the annual and half-year periodicities (a short description of the series 'Births' is provided in Section 1.3.4; other aspects relating to the choice of the window length in this example are discussed below).

The problem is complex; there is a large variety of situations. We briefly consider, on a quantitative level, two extreme cases: when the trend can be extracted relatively easily, and when the selection of the window length for extraction of trend is difficult or even impossible.

(i) *Trends: reliable separation*

Let $F = F^{(1)} + F^{(2)}$ where $F^{(1)}$ is a trend and $F^{(2)}$ is the residual. We assume the following.

1. The series $F^{(1)}$ is 'simple'. The notion of 'simplicity' can be understood as follows:

    - From the theoretical viewpoint, the series $F^{(1)}$ is well approximated by a series with finite and small rank d (for example, if it looks like an exponential, $d = 1$, a linear function, $d = 2$, a quadratic function, $d = 3$, etc.). See Section 5.1 for a description of the series of finite rank.
    - We are interested in the extraction of the general tendency of the series rather than of the refined trend.
    - In terms of frequencies, the periodogram of the series $F^{(1)}$ is concentrated in the domain of rather small frequencies.

- In terms of the SSA decomposition, the few first eigentriples of the decomposition of the trajectory matrix of the series $F^{(1)}$ are enough for a reasonably good approximation of it, even for large L.

2. Assume also that the series $F^{(1)}$ is much 'larger' than the series $F^{(2)}$ (for instance, the inequality $||F^{(1)}|| \gg ||F^{(2)}||$ is valid).

Suppose that these assumptions hold and the window length L provides a certain (weak, approximate) separation of the time series $F^{(1)}$ and $F^{(2)}$. We can expect that in the SVD of the trajectory matrix of the series F, the leading eigentriples will correspond to the trend $F^{(1)}$; i.e., they will have larger singular values than the eigentriples corresponding to $F^{(2)}$. In other words, strong separation occurs. Moreover, the window length L, sufficient for the separation, should not be very large in this case in view of the 'simplicity' of the trend.

This situation is illustrated by the example 'Production' (Section 1.3.1, Figs. 1.1 and 1.2), where both trend versions are described by the leading eigentriples. Analogously, if we are interested only in extracting the main tendency of the series 'Unemployment' (Section 1.3.6), then, according to Fig. 1.37, taking just one leading eigentriple will be a perfectly satisfactory decision for $L = 12$.

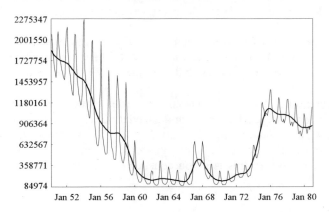

Figure 1.37 *Unemployment: $L = 12$ for extraction of the main tendency.*

(ii) *Trends: difficult case*
Much more difficult situations arise if we want to extract a refined trend $F^{(1)}$, when the residual $F^{(2)}$ has a complex structure (for example, it includes a large noise component) with $||F^{(2)}||$ being large. Then large L can cause not only mixing of the ordinal numbers of the eigentriples corresponding to $F^{(1)}$ and $F^{(2)}$ (this is the case of the 'Unemployment' example), but also closeness of the corresponding singular values, and therefore a lack of strong separability.

Certainly, there are many intermediate situations between these two extremes. Consider, for instance, the 'England temperatures' example (average annual tem-

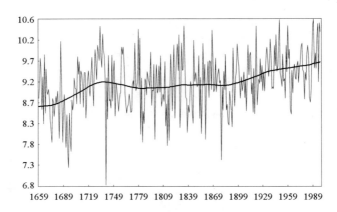

Figure 1.38 *England temperatures: $L = 48$ for extraction of the main tendency.*

peratures, Central England, from 1659 to 1998). Here the problem of extraction of a smooth trend can easily be solved: under the choice of L equal to several dozen, the first eigentriple always describes the general tendency; see Fig. 1.38 for the choice $L = 48$.

This happens because relatively small values of L are enough to provide (weak) separability, the trend has a simple form, and it thus corresponds to only one eigentriple; this eigentriple is leading due to a relatively large mean value of the series.

At the same time, if we wish to centre the series (which may seem a natural operation since in this kind of problem the deviation from the mean is often the main interest), then small values of L, say $L < 30$, do not provide weak separation. Large values of L, say $L > 60$, mix up the trend eigentriple with some other eigentriple of the series; this is a consequence of the complexity of the series structure.

2. Smoothing

Generally, the recommendations concerning the selection of the window length for the problem of smoothing are similar to the case of the trend extraction. This is because these two problems are closely related. Let us describe the effects of the window length in the language of frequencies.

Treating smoothing as removing of the high-frequency part of a series, we have to take the window length L large enough to provide separation of this low-frequency part from the high-frequency one. If the powers of all low frequencies of interest are significantly larger than those of the high ones, then the smoothing problem is not difficult, and the only job is collecting several leading eigentriples. This is the case for the 'Tree rings' and 'White dwarf' examples of Section 1.3.2. Here, the larger we take L, the narrower the interval of low frequencies we can extract.

CHOICE OF SSA PARAMETERS

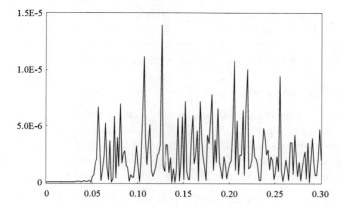

Figure 1.39 *White dwarf: $L = 100$, periodogram of residuals.*

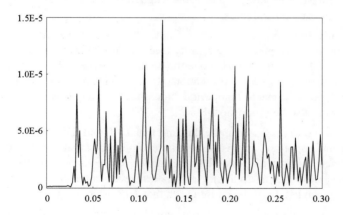

Figure 1.40 *White dwarf: $L = 200$, periodogram of residuals.*

For instance, in Section 1.3.2, the smoothing of the series 'White dwarf' has been done with $L = 100$, with the result of the smoothing being described by the leading 11 eigentriples. In the periodogram of the residuals (see Fig. 1.39) we can see that for this window length the powers of the frequencies in the interval $[0, 0.05]$ are practically zero.

If we take $L = 200$ and 16 leading eigentriples for the smoothing, then this frequency interval is reduced to $[0, 0.03]$ (see Fig. 1.40). At the same time, for $L = 10$ and two leading eigentriples, the result of smoothing contains the frequencies from the interval $[0, 0.09]$.

Visual inspection shows that all smoothing results look similar. Also, their eigenvalue shares are equal to $95.9\% \pm 0.1\%$. Certainly, this effect can be explained by the specificity of the series: its frequency power is highly concentrated in the narrow low-frequency region.

More difficult problems with smoothing occur when powers of low and high frequencies are not separable from each other by their values.

(c) Window length for periodicities

The problem of choosing the window length L for extraction of a periodic component $F^{(1)}$ out of the sum $F = F^{(1)} + F^{(2)}$ has certain peculiarities related to the correspondence between the window length and the period. In general, these peculiarities are the same for the pure harmonics and for complex periodicities, and even for modulated periodicities. Thus, we do not consider these cases separately.

1. For the problem of extraction of a periodic component with period T, it is natural to measure the length of the series in terms of the number of periods. Specifically, if $F^{(1)}$ is asymptotically separable from $F^{(2)}$, then to achieve the separation we must have, as a rule, the length of the series N such that the ratio N/T is at least several units.

2. For relatively short series, it is preferable to take into account the conditions for pure (nonasymptotic) separability (see Section 1.5); if one knows that the time series has a periodic component with an integer period T (for example, if this component is a seasonal component), then it is better to take the window length L proportional to that period. Note that from the theoretical viewpoint, $N-1$ must also be proportional to T.

3. In the case of a long series, the demand that L/T and $(N-1)/T$ be integers is not that important. In this case, it is recommended that the window length be chosen as large as possible (for instance, close to $N/2$, if the computer facilities allow one to do this). Nevertheless, even in the case of long series it is recommended that L be chosen such that L/T is an integer.

4. If the series $F^{(2)}$ contains a periodic component with period $T_1 \approx T$, then to extract $F^{(1)}$ we generally need a larger window length than for the case when such a component is absent (see Section 6.1.2 for the theory).

5. Since two harmonic components with equal amplitudes produce equal singular values, asymptotically, when L and K tend to infinity, a large window length can cause a lack of strong separability and therefore a mixing up of the components.

    If in addition the frequencies of the two harmonics are (almost) equal, then a contradiction between the demands for the weak and strong separability can occur; close frequencies demand large window lengths, but the large window lengths lead to approximately equal singular values.

To demonstrate the effect of divisibility of L by T, let us return to the 'Eggs' example (Section 1.3.3). Figs. 1.41 and 1.42 depict the matrices of w-correlations for the full decomposition of the series with $L = 12$ and $L = 18$. It is

CHOICE OF SSA PARAMETERS

clearly seen that for $L = 12$ the matrix is essentially diagonal, which means that the eigentriples related to the trend and different seasonal harmonics are almost **w**-uncorrelated. This means that the choice $L = 12$ allows us to extract all the harmonic components of the series.

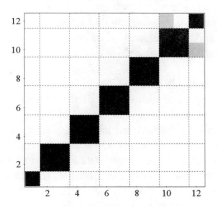

Figure 1.41 *Eggs:* $L = 12$, **w**-*correlations.*

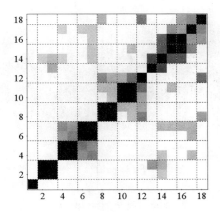

Figure 1.42 *Eggs:* $L = 18$, **w**-*correlations.*

For $L = 18$ (that is, when the period 12 does not divide L), only the leading seasonality harmonics can be extracted in a proper way; the other components have relatively large **w**-correlations.

The choice $L = 13$ would give results that are slightly worse than for $L = 12$, but much better than for $L = 18$. This confirms the robustness of the method with respect to small variations in L.

(d) Refined structure

In doing a simultaneous extraction of different components from the whole series, all the aspects discussed above should be taken into account. Thus, in basically all the examples of Section 1.3, where the periodicities were the main interest, the window length was a multiple of the periods. At the same time, if in addition the trends were to be extracted, L was reasonably large.

For the short series and/or series with a complex structure, these simple recommendations may not suffice and the choice of the window length becomes a more difficult problem.

For instance, in the example 'War' (Section 1.3.7), the choice $L = 18$ is dictated by the specific amplitude modulation of the harmonic components of the series, which is reflected in the shape of the trend. When the window length is reduced to 12, the amplitude-modulated harmonics are mixed up with the trend (the effect of the small window length). On the other hand, the decompositions with $L = 24$ and $L = 36$ lead to a more detailed decomposition of the sum of the trend and the annual periodicity (six eigentriples instead of four for $L = 12$), where the components are again mixed up (the effect of a too large L).

Note that if we wish to solve the problem of extracting the sum of the trend and the annual periodicity, then the choice of $L = 36$ is preferable to $L = 18$.

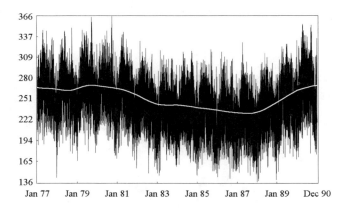

Figure 1.43 *Births: trend.*

To demonstrate the influence of the window length on the result of the decomposition, let us consider another more complex example, namely 'Births' (Section 1.3.4).

In the series 'Births' (daily data for about 14 years, $N = 5113$) there is a one-week periodicity ($T_1 = 7$) and an annual periodicity ($T_2 = 365$). Since $T_2 \gg T_1$, it is natural to take the window length as a multiple of T_2.

The choice $L = T_2$, as was shown in Section 1.3.4, guarantees the simultaneous extraction of both weekly and annual periodicities. Moreover, this window length allows also the extraction of the trend of the series (see Fig. 1.43) using the single leading eigentriple. Note that these results are essentially the same as for the cases $L = 364$ and $L = 366$.

At the same time, if we would choose $L = 3T_2 = 1095$, then the components of the trend will be mixed up with the components of the annual and half-year periodicities; this is a consequence of the complex shape of the trend and the closeness of the corresponding eigenvalues. Thus, large values of the window length lead to violation of strong separability.

If the problem of separation of the trend from the annual periodicity is not important, then values of L larger than 365 work well. If the window length is large, we can separate the global tendency of the series (trend + annual periodicity) from the weekly periodicity + noise even better than for $L = 365$ (for $L = 1095$ this component is described by several dozen eigentriples rather than by 5 eigentriples for $L = 365$). In this case, the weekly periodicity itself is perfectly separable from the noise as well.

Note also that if we were to take a small (in comparison to the annual period) window length (for example, $L = 28$), then the global behaviour of the series would be described by just one leading eigentriple. The weekly periodicity could also be separated, but a little worse than for large L.

In even more complex cases, better results are often achieved by the application of Sequential SSA, which after extraction of a component with a certain L requires a repeated application of SSA to the residuals with different L. An example of sequential SSA is described in Section 1.7.3.

(e) Hints

If we are not using the knowledge about the subject area and the series, then the only source of information helping in the window length selection is the series itself. Often the shape of the graph of the series (leading, for instance, to a visual identification of either a trend or a strong harmonic) is an effective indicator. At the same time, it is impossible to describe all possible manipulations with the series that may help in the selection of L, especially if we bear in mind that the corresponding algorithms should be fast (faster than the numerical calculation of the SVD of a large matrix) and should use the specifics of the problem.

We consider just one way of getting recommendations for selecting the window length, namely, the method based on the periodogram analysis of the original series and parts of the series.

1. If the resolution of the periodogram of the series F is good enough (that is, the series is sufficiently long), then the periodogram can help in determining the periods of the harmonic components of the series and thus in selecting the window length for their separation. Moreover, the presence of distinct powerful frequency ranges in the periodogram indicates possible natural components of the series related to these frequency ranges.

2. One of the sufficient conditions for approximate weak separability of two series is the smallness of all the spectral correlations for all the subseries of length L (and also $K = N - L + 1$) of these two series (see Section 1.5.3). Assume that:

- periodograms of all the subseries of length L and K of the series F have the same structure,
- this structure is characterized by the presence of distinct and distant powerful frequency ranges.

In this case, the choice of a window length equal to L would most probably lead to a splitting of the series F into the components corresponding to these frequency ranges. This suggests that a preliminary periodogram analysis of at least several subseries might be useful.

A control of the correct choice of the window length is made at the grouping stage; the possibility of a successful grouping of the eigentriples means that the window length has been properly selected.

1.7 Supplementary SSA techniques

Supplementary SSA techniques may often improve Basic SSA for many specific classes of time series and for series of a complex structure. In this section, we consider several classes of this kind and describe the corresponding techniques. More precisely, we deal with the following series and problems:

1. The time series is oscillating around a linear function, and we want to extract this linear function.
2. The time series is stationary-like, and we want to extract several harmonic components from the series.
3. The time series has a complex structure (for example, its trend has a complex form, or several of its harmonic components have almost equal amplitudes), and therefore for any window length a mixing of the components of interest occurs.
4. The time series is an amplitude-modulated harmonic and we require its envelope.

The last problem is rather specific. Its solution is based on the simple idea that squared amplitude-modulated harmonic is a sum of low and high-frequency series that can be separated by Basic SSA.

The technique for the first two problems is in a way similar; having information about the time series, we use a certain decomposition of the trajectory matrix, which is different from the straightforward SVD and adapted to the series structure.

A lack of strong separability (in the presence of weak separability) is one of the main difficulties in Basic SSA. One of possible ways to overcome this difficulty is to enlarge the singular value of a series component of interest by adding a series similar to this component.

Alternatively, we can use Sequential SSA. This means that we extract some components of the initial series by the standard Basic SSA and then extract other components of interest from the residuals.

Suppose, for example, that the trend of the series has a complex form. If we choose a large window length L, then certain trend components would be mixed with other components of the series. For small L, we would extract the trend but obtain mixing of the other series components which are to be extracted.

A way to solve the problem, and this is a typical application of Sequential SSA, is to choose a relatively small L to extract the trend or its part, and then use a large window length to separate components of interest in the residual series.

Let us describe these approaches in detail and illustrate them with examples.

1.7.1 Centring in SSA

Consider the following extension of Basic SSA. Assume that we have selected the window length L. For $K = N - L + 1$, consider a matrix $\mathbf{A}$ of dimension $L \times K$ and pass from the trajectory matrix $\mathbf{X}$ of the series F to the matrix $\mathbf{X}^* = \mathbf{X} - \mathbf{A}$. Let $\mathbf{S}^* = \mathbf{X}^*(\mathbf{X}^*)^\mathrm{T}$, and denote by λ_i and U_i ($i = 1, \ldots, d$) the nonzero eigenvalues and the corresponding orthonormal eigenvectors of the matrix $\mathbf{S}^*$. Setting $V_i = (\mathbf{X}^*)^\mathrm{T} U_i / \sqrt{\lambda_i}$ we obtain the decomposition

$$\mathbf{X} = \mathbf{A} + \sum_{i=1}^{d} \mathbf{X}_i^* \qquad (1.30)$$

with $\mathbf{X}_i^* = \sqrt{\lambda_i} U_i V_i^\mathrm{T}$, instead of the standard SVD (1.2). At the grouping stage the matrix $\mathbf{A}$ will enter one of the resultant matrices as an addend. In particular, it can produce a separate time series component after the application of diagonal averaging.

If the matrix $\mathbf{A}$ is orthogonal to all the $\mathbf{X}_i^*$ (see Section 4.4), then the matrix decomposition (1.30) yields the decomposition

$$\|\mathbf{X}\|_{\mathcal{M}}^2 = \|\mathbf{A}\|_{\mathcal{M}}^2 + \sum_{i=1}^{d} \|\mathbf{X}_i^*\|_{\mathcal{M}}^2$$

of the squared norms of the corresponding matrices.

Here we consider two ways of choosing the matrix $\mathbf{A}$, thoroughly investigated in Sections 4.4 and 6.3. We follow the terminology and results from these sections.

(a) Single and Double centring

Single centring is the row centring of the trajectory matrix. Here

$$\mathbf{A} = \mathcal{A}(\mathbf{X}) = [\mathcal{E}_1(\mathbf{X}) : \ldots : \mathcal{E}_1(\mathbf{X})],$$

where each ith component of the vector $\mathcal{E}_1(\mathbf{X})$ ($i = 1,\ldots,L$) is equal to the average of the ith components of the lagged vectors $X_1,\ldots,X_K$.

Thus, under Single centring we consider the space $\mathrm{span}(X_1^{(c)},\ldots,X_K^{(c)})$ with $X_i^{(c)} = X_i - \mathcal{E}_1(\mathbf{X})$ rather than $\mathrm{span}(X_1,\ldots,X_K)$. In other words, we shift the origin to the centre of gravity of the lagged vectors and then use the SVD of the obtained matrix. Of course, Single centring is a standard procedure in the principal component analysis of multidimensional data.

For the *Double centring*, SVD is applied to the matrix, computed from the trajectory matrix by subtracting from each of its elements the corresponding row and column averages and by adding the total matrix average. In other words, in this case we have

$$\mathbf{A} = \mathcal{A}(\mathbf{X}) + \mathcal{B}(\mathbf{X}) \tag{1.31}$$

with $\mathcal{B}(\mathbf{X}) = [\mathcal{E}_{12}(\mathbf{X}) : \ldots : \mathcal{E}_{12}(\mathbf{X})]^{\mathrm{T}}$, where the jth component of the vector $\mathcal{E}_{12}(\mathbf{X})$ ($j = 1,\ldots,K$) is equal to the average of all the components of the vector $X_j^{(c)}$.

Under Single centring the addend $\mathbf{A}$ has the same form as the other components of the decomposition (1.30), provided we have included the normalized vector of averages $U_{0(1)} = \mathcal{E}_1(\mathbf{X})/\|\mathcal{E}_1(\mathbf{X})\|$ in the list of eigenvectors U_i. Indeed, $\mathbf{A} = U_{0(1)} Z_{0(1)}^{\mathrm{T}}$ with $Z_{0(1)} = \|\mathcal{E}_1(\mathbf{X})\| \mathbf{1}_K$. (Each component of the vector $\mathbf{1}_K \in \mathbf{R}^K$ is equal to 1.)

In the Double centring case, we add one more vector to the list of eigenvectors, the vector $U_{0(2)} = \mathbf{1}_L/\sqrt{L}$. Here

$$\mathbf{A} = U_{0(1)} Z_{0(1)}^{\mathrm{T}} + U_{0(2)} Z_{0(2)}^{\mathrm{T}}$$

with $Z_{0(2)} = \sqrt{L} \mathcal{E}_{12}(\mathbf{X})$. We set

$$\lambda_{0(1)} = \|Z_{0(1)}\| = \|\mathcal{E}_1(\mathbf{X})\|\sqrt{K} \text{ and } \lambda_{0(2)} = \|Z_{0(2)}\| = \|\mathcal{E}_{12}(\mathbf{X})\|\sqrt{L}.$$

Moreover, let $V_{0(1)} = Z_{0(1)}/\sqrt{\lambda_{0(1)}}$ and $V_{0(2)} = Z_{0(2)}/\sqrt{\lambda_{0(2)}}$. Then we call $\left(U_{0(i)}, V_{0(i)}, \lambda_{0(i)}\right)$ ($i = 1, 2$) *the first and the second average triples*.

Since $\mathcal{A}(\mathbf{X})$ and $\mathcal{B}(\mathbf{X})$ are orthogonal to each other and to all the other decomposition components (see Section 4.4), we have for the Double centring

$$\|\mathbf{X}\|_{\mathcal{M}}^2 = \lambda_{0(1)} + \lambda_{0(2)} + \sum_{i=1}^d \lambda_i$$

(for the Single centring the term $\lambda_{0(2)}$ is omitted). Therefore, the shares of the average triples and the eigentriples are equal to

$$\lambda_{0(1)}/\|\mathbf{X}\|_{\mathcal{M}}^2, \quad \lambda_{0(2)}/\|\mathbf{X}\|_{\mathcal{M}}^2 \quad \text{and} \quad \lambda_i/\|\mathbf{X}\|_{\mathcal{M}}^2.$$

Note that Basic SSA does not use any centring. Nevertheless, Single centring can have some advantage if the series F can be expressed in the form $F = F^{(1)} + F^{(2)}$, where $F^{(1)}$ is a constant series and $F^{(2)}$ oscillates around zero.

Certainly, if the series length N is large enough, its additive constant component will undoubtedly be extracted by Basic SSA (as well as with the averaging of all the components of the series), but, for the short series, Single centring SSA can work better. Since the analogous, but much brighter, effects are produced by Double centring SSA, we do not consider any Single centring example here.

The effect of Double centring can be explained as follows. If the initial series is a linear one, $\mathbf{X}$ is its trajectory matrix and $\mathbf{A}$ is defined by (1.31), then $\mathbf{A} = \mathbf{X}$. Therefore, for $F = F^{(1)} + F^{(2)}$ with linear $F^{(1)}$, the matrix $\mathbf{A}$ contains the entire linear part of the series F. Theoretically, Double centring leads to the asymptotic extraction of the linear component of the series from rather general oscillatory residuals (see Section 6.3.2).

As usual, nonasymptotic effects occur as well. For fixed N and L, let us consider the series F which is the sum of a linear series $F^{(1)}$ and a pure harmonic $F^{(2)}$ with an integer period T. If the window length L and $K = N - L + 1$ divide T, then the matrix $\mathbf{A}$ defined by (1.31) coincides with the trajectory matrix of the series $F^{(1)}$. The residual matrix $\mathbf{X}^* = \mathbf{X} - \mathbf{A}$ corresponds to the trajectory matrix of the harmonic series $F^{(2)}$. Therefore, here we obtain the theoretically precise linear trend extraction (see Section 6.3.2 for the theory).

(b) Double centring and linear regression

Comparison of the extraction of a linear component of a series by Double centring SSA and by linear regression can be instructive. Note that these two methods have different origins and therefore can produce very different results.

As for linear regression, it is a formal procedure for the linear approximation by the least-squares method and gives a linear function of time for any series, even if the series does not have any linear tendency at all. By contrast, Double centring SSA gives us (usually, approximately) a linear component only if the strong linear tendency is really present. On the other hand, Double centring does not produce a precise linear function but only a pointwise approximation of it. Roughly speaking, linear regression estimates the coefficients of a linear function, while Double centring SSA estimates the values of a linear function at each point.

If the time series has a linear tendency and its length is rather large, then both methods produce similar results. The difference appears for a relatively short time series. For these series, the objective of the linear regression can be in disagreement with the problem of searching for a linear tendency of the series.

Let us illustrate these statements by two examples.

Example 1.5 *'Investment': long time series with a linear tendency*
The theory tells us (see Section 6.3.2) that Double centring SSA extracts (perhaps, approximately) the linear component of the series if the series oscillates near this linear component. Since linear regression automatically approximates any series by a linear function, a correspondence between the results of both methods would indicate that the linear function obtained by the regression method describes the

actual tendency of the series, and thus it is not merely the result of a formal procedure.

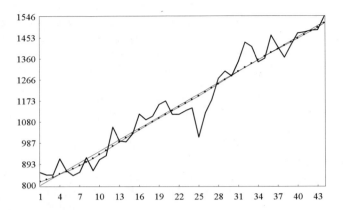

Figure 1.44 *Investment: linear regression and Double centring results.*

The example 'Investment' (the investment series of U.K. economic time series data, quarterly, successive observations, Prothero and Wallis, 1976), illustrates these considerations. The series is presented in Fig. 1.44, thick line.

Let us select the window length $L = 24$ in Double centring SSA, and take both average triples for the reconstruction. Then the reconstructed component (Fig. 1.44, thin line with black points) will resemble a linear function.

For comparison, the result of standard linear regression analysis (thin line) is placed on the same plot. Since both lines (SSA reconstruction curve and linear regression function) are very close, the general linear behaviour of the 'Investment' series can be considered as being confirmed. Note that the 'Investment' series can be regarded as a 'long' series since it oscillates rather rapidly around the regression line.

The second example demonstrates the difference between these two methods and shows the Double centring SSA capabilities for short series.

Example 1.6 *'Hotels': continuation of the extracted tendency*
The 'Hotels' series (hotel occupied room, monthly, from January 1963 to December 1976, O'Donovan, 1983) is a good example for discussing the difference between linear regression and Double centring SSA approaches to the extraction of a linear tendency in a series. Fig. 1.45 depicts the initial series and its linear regression approximation. Despite the fact that the series is not symmetric with respect to the regression line and has an increasing amplitude, the whole linear tendency seems to have been found in a proper way.

The second figure (Fig. 1.46) deals with the first 23 points of the series. Two lines intersect the plot of the series. The thin one is the linear regression line

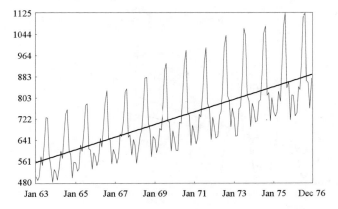

Figure 1.45 *Hotels: time series and its linear regression approximation.*

calculated from this subseries. The thick line is the reconstruction of the series produced by both average triples for the Double centring SSA with window length $L = 12$. This Double centring curve is almost linear but differs from that of the linear regression.

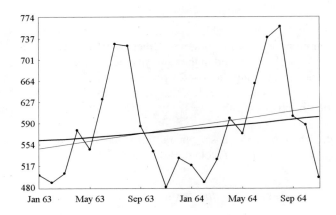

Figure 1.46 *Hotels: short interval. Regression and Double centring lines.*

Fig. 1.47 shows the continuation of both lines for the first 72 points of the 'Hotels' series. The upper linear function ($y = 543.6 + 3.2x$) is a continuation of the linear regression line of the Fig. 1.46. The lowest linear function ($y = 556.8 + 1.9x$) is the linear continuation of the Double centring line (Fig. 1.46). It is very close to the middle linear function ($y = 554.8 + 2.0x$) which is the part of the global linear regression line of Fig. 1.45.

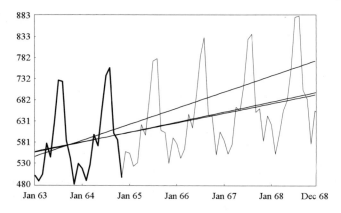

Figure 1.47 *Hotels: continuation of linear functions.*

Comparison of these lines shows that linear regression fails to find a good approximation of the main tendency of the whole series based on the first 23 points. On the other hand, the Double centring line is very close to the global linear regression line, but it uses only the first 23 points of the series rather than the full number 168.

Note that using Single or Double centring SSA, one can extract not only constants or linear components of time series. Other components of interest (such as oscillatory ones) can be extracted in the same manner as in Basic SSA. For example, for the series, containing the first 23 points of the 'Hotels' data and Double centring with $L = 12$ (see Fig. 1.46), the eigentriples 1-2, 3-4, 5-6, 7, 8-9 and 10-11 describe harmonics with $\omega = 1/12, 2/12, 3/12, 6/12, 4/12$ and $5/12$, respectively.

Moreover, if the time series has a general linear-like tendency, then the Double centring approach is often preferable to Basic SSA.

1.7.2 Stationary series and Toeplitz SSA

If the length N of the series F is not sufficiently large and the series is assumed to be stationary, then the usual recommendation is to replace the matrix $\mathbf{S} = \mathbf{XX}^\mathrm{T}$ by some other matrix, which takes into account the stationarity of the series.

Note first that we can consider the *lag-covariance matrix* $\mathbf{C} = \mathbf{S}/K$ instead of $\mathbf{S}$ for obtaining the SVD of the trajectory matrix $\mathbf{X}$. Indeed, the difference between the SVDs of the matrices $\mathbf{S}$ and $\mathbf{C}$ lies only in the magnitude of the corresponding eigenvalues (for $\mathbf{S}$ they are K times larger); the singular vectors of both matrices are the same. Therefore, we can use both $\mathbf{S}$ and $\mathbf{C}$ in Basic SSA with the same effect.

Denote by $c_{ij} = c_{ij}(N)$ the elements of the lag-covariance matrix $\mathbf{C}$. If the time series is stationary, and $K \to \infty$, then $\lim c_{ij} = R_f(|i-j|)$ as $N \to \infty$, where $R_f(k)$ stands for the lag k term of the time series covariance function; see Sections 1.4.1 and 6.4. (Recall that according to our agreement of Section 1.4.1, any infinite stationary series has zero average.)

Therefore, the main idea is to take the Toeplitz version of the lag-covariance matrix, that is to put equal values $\widetilde{c}_{ij}$ in each matrix diagonal $|i - j| = k$. Of course, the convergence $\widetilde{c}_{ij} \to R_f(|i-j|)$ must be kept.

There are several ways of getting the Toeplitz lag-covariance matrices from the series (see Elsner and Tsonis, 1996, Chapter 5.3). The main one is to use the standard estimate of the covariance function of the series and to transform it into an $L \times L$ matrix. More precisely (see Anderson, 1994, Chapter 8.2), for the time series $F = (f_0, \ldots, f_{N-1})$ and a fixed window length L, we take the matrix $\widetilde{\mathbf{C}}$ with the elements

$$\widetilde{c}_{ij} = \frac{1}{N - |i-j|} \sum_{m=0}^{N-|i-j|-1} f_m f_{m+|i-j|}, \quad 1 \le i, j \le L, \qquad (1.32)$$

rather than Basic SSA lag-covariance matrix $\mathbf{C} = \mathbf{S}/K$ with the elements

$$c_{ij} = \frac{1}{K} \sum_{m=0}^{K-1} f_{m+i-1} f_{m+j-1}, \quad 1 \le i, j \le L. \qquad (1.33)$$

Having obtained the *Toeplitz lag-covariance matrix* $\widetilde{\mathbf{C}}$ we calculate its orthonormal eigenvectors $H_1, \ldots, H_L$ and decompose the trajectory matrix:

$$\mathbf{X} = \sum_{i=1}^{L} H_i Z_i^{\mathrm{T}}, \qquad (1.34)$$

where $Z_i = \mathbf{X}^{\mathrm{T}} H_i$. We thus obtain an orthogonal matrix decomposition of the kind discussed in Section 4.2.1. Setting $\lambda_i = ||Z_i||^2$ and $Q_i = Z_i/\sqrt{\lambda_i}$ (here we formally assume that $\widetilde{\mathbf{C}}$ has full rank), we come to the decomposition of the trajectory matrix $\mathbf{X}$ into a sum similar to the usual SVD. The grouping and diagonal averaging can then be made in the standard way. Note that the numbers λ_i (which may be called squared *Toeplitz singular values*) generally do not coincide with the eigenvalues of the matrix $\widetilde{\mathbf{C}}$.

If the initial series is a sum of a constant series with the general term c_0 and a stationary series, then centring seems to be a convenient procedure (since we are dealing with finite time series, the centring can be applied for $c_0 = 0$ as well). One way is to centre the entire series before calculating the matrix (1.32).

The other method is to apply the Single centring. For *Toeplitz SSA* with the lag-covariance matrix (1.32) this means that we extract the product

$$M_{ij} = \left(\frac{1}{n(i,j)} \sum_{m=0}^{n(i,j)-1} f_m \right) \left(\frac{1}{n(i,j)} \sum_{m=0}^{n(i,j)-1} f_{m+|i-j|} \right)$$

(here we used the notation $n(i,j) = N - |i-j|$) from $\widetilde{c}_{ij}$, find the eigenvectors $H_1, \ldots, H_L$ of the above matrix, compute the (single) centred trajectory matrix $\mathbf{X}^*$ as was described in Section 1.7.1, obtain $Z_i = (\mathbf{X}^*)^\mathrm{T} H_i$, and come to the decomposition similar to (1.34) with an additional matrix term $\mathbf{A}$ corresponding to the Single centring. Note that unlike Basic SSA, the Toeplitz SSA is not invariant with respect to the substitution of $K = N - L + 1$ for the window length L, even without centring.

The Toeplitz construction of the lag-covariance matrix seems to have an advantage since the matrix elements (1.32) are generally closer than (1.33) to the terms $R_f(|i-j|)$ of the theoretical covariance function, due to a wider range of averaging. Nevertheless, it is not universally better since we are not dealing with the lag-covariance matrix itself but rather with some specific features of the decompositions of the trajectory matrices, such as separability.

First, the Toeplitz SSA is not aimed at nonstationary series. If the series has a strong nonstationary component, then Basic SSA seems to be preferable. For example, if we are dealing with a pure exponential series, then it is described by a single eigentriple (see Sections 1.6.1 and 5.1 for details) for any window length, while Toeplitz SSA produces L eigentriples for window length L with harmonic-like eigenvectors. The same effect takes place for the linear series, exponential-cosine series, etc. In terms of Section 4.2.1, Toeplitz SSA often produces a decomposition, which is not minimal.

Second, Toeplitz SSA generally produces a nonoptimal decomposition. The decomposition of the trajectory matrix produced by SVD (it is used in Basic SSA and Single and Double centring SSA) is optimal in the the sense that each eigenvalue is the solution of a certain optimization problem; in other words, each eigenvalue is as large as it can be. Therefore, the main series effects are described by the leading SVD eigentriples, but even subsequent eigentriples can be meaningful.

If we have nonoptimal orthogonal decomposition of the trajectory matrix, it is more 'spread' and the problem similar to the problem of small 'almost equal singular values' becomes even more serious.

Moreover, for long stationary series, both methods give practically the same results. Yet, for relatively short stationary and noisy series, Toeplitz SSA can be advantageous.

Example 1.7 *'Tree rings': four modulated harmonics*
Let us consider the 'Tree rings' example (see Section 1.3.2). The periodogram (Fig. 1.48) of the series shows four sharp peaks corresponding approximately to the periods $T_1 = 74$, $T_2 = 52$, $T_3 = 42$ and $T_4 = 12.5$.

If we take the window length $L = 334$, then Basic and Single centring SSA (both using SVD) extract periodicities corresponding to T_1 and T_4 but produce a mixture of the two other periodic components. Standard Toeplitz SSA with Single centring works better (see Fig. 1.49) and extracts all the leading periodicities at once.

SUPPLEMENTARY SSA TECHNIQUES

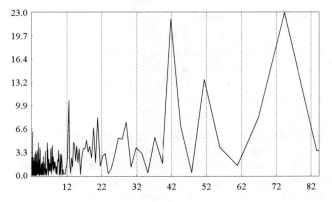

Figure 1.48 *Tree rings: periodogram in periods up to $T = 85$.*

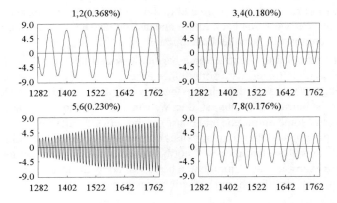

Figure 1.49 *Tree rings: four periodic components. First 500 points.*

Remark 1.1 In Example 1.7 we did not discuss whether the extracted periodicities are the true ones or produced by the aperiodic component of the series (see Section 6.4). Our aim was to demonstrate their extraction.

1.7.3 Close singular values

As was discussed in Section 1.6.2, close singular values of SVD cause difficulties that are difficult to overcome by modifying the window length L. Nevertheless, there are several techniques that can help to solve the problem. Let us discuss two of them.

(a) Series modification

Sometimes one can modify the series in such a manner that this problem disappears. The theoretical base of such effects is the following simple fact: if the time series $F^{(1)}$ and $F^{(2)}$ are weakly separable, then for a wide range of constants $c \neq 0$ the time series $F^{(1)}$ and $cF^{(2)}$ are strongly separable. In practice, we use this fact in some approximate sense.

Let us consider two examples of this kind.

1. If we want to extract a small slowly varying trend whose components are mixed with other series components, then it can be worthwhile to add a constant to the series and use a relatively small window length for the trend extraction. Then the new trend will be described by the leading eigentriple, and there will be no problem in its extraction. The added constant has to be subtracted from the extracted series.

   The example 'England temperatures' (Section 1.6.2) is of this kind if we deal with it in the reverse manner; being centred, the time series is complex for the rough trend extraction, but if we add to the centred series a constant, equal to 9.18 (that is, if we come back to the uncentered data), a rather wide range of window lengths will provide the extraction.

2. Assume that our aim is extracting a harmonic with a known frequency ω and this harmonic is mixed with some other time series components due to their close singular values. If the selected window length L provides a weak separability of the harmonic of interest, then we can add a harmonic of the same frequency (and some amplitude and phase) to the series. Under the proper choice of these parameters, the singular values corresponding to the harmonic will be enlarged enough so that they will not be mixed with any other series components (for example, the harmonic will be described by the leading eigentriples). Therefore, the modified harmonic will be easily extracted.

Example 1.8 *'Rosé wine': adding a harmonic component*

To illustrate the extraction of a harmonic component from the series, let us consider the example 'Rosé wine' described in Section 1.4.1 (see Fig. 1.17 for the time series and Fig. 1.18 for its periodogram). As was mentioned in Section 1.6, the harmonics with frequencies 4/12 and 5/12 are mixed under the choice $L = 84$. Moreover, other window lengths lead to mixing of other harmonics due to a complex nonstationary structure of the series.

However, if we add to the series a harmonic with frequency 4/12 (that is, period 3), zero phase and amplitude 30, then the new quarterly harmonics will be extracted under the choice of the same $L = 84$ and the pair of the second and third eigentriples. The final result is obtained by subtracting the additional harmonic component.

Note that the problem of close singular values can be solved by other modifications of Basic SSA as well. For instance, the Toeplitz SSA helps in extracting harmonic components in the 'Tree rings' example of Section 1.7.2. However, this

SUPPLEMENTARY SSA TECHNIQUES

example seems to be a good illustration of the advantages of a concrete technique related to the problem of close singular values rather than an illustration of the absolute advantage of the Toeplitz SSA.

(b) Sequential SSA

The mixing problem of the time series components (formally, the problem of close singular values for weakly separable series components) may be resolved in one more manner, by the so-called *Sequential SSA*.

The two-step Sequential SSA can be described as follows. First, we extract several time series components by Basic SSA with a certain window length L_1. Then we apply Basic SSA to the residuals and extract several series components once again. The window length L_2 of the second stage is generally different from L_1.

Having extracted two sets of time series components, we can group them in different ways. For instance, if a rough trend has been extracted at the first stage and other trend components at the second stage, then we have to add them together to obtain the accurate trend.

Let us illustrate this by an example.

Example 1.9 *Long 'Unemployment' series: extraction of harmonics*
Consider the 'Unemployment' series starting from January 1948 (note that 'Unemployment' example of Section 1.3.6 has April 1950 as its starting point). The series is depicted in Fig. 1.50 (thin line).

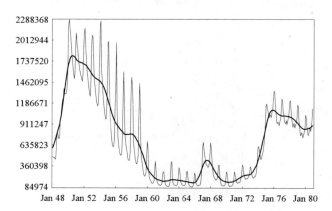

Figure 1.50 *Long 'Unemployment' series: time series from January 1948.*

Comparing Fig. 1.50 with Fig. 1.12, we see that the trend of the long 'Unemployment' series (thick line) has a more complex structure than that of the shorter one. Selection of a large window length would mix the trend and periodic components of the series. For small window lengths the periodic components are not separable from each other, and therefore these lengths are not suitable. Hence, Basic SSA fails to extract (amplitude-modulated) harmonic components of the series.

The two-stage Sequential SSA proves to be a better method in this case. If we apply Basic SSA with $L = 12$ to the initial series, then the first eigentriple will describe the trend, which is extracted rather well: the trend component does not include high frequencies, while the residual component practically does not contain low ones (see Fig. 1.51 for the residual series).

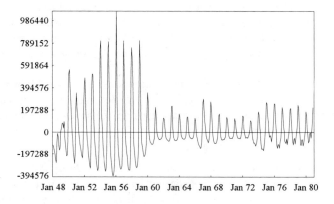

Figure 1.51 *Long 'Unemployment' series: trend residuals.*

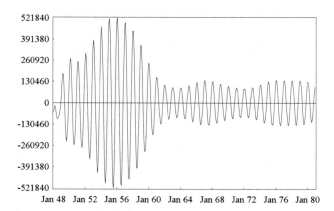

Figure 1.52 *Long 'Unemployment' series: annual periodicity.*

The second Sequential SSA stage is applied to the residual series with $L = 180$. Since the series is amplitude modulated, the main periodogram frequencies (annual $\omega = 1/12$, half-annual $\omega = 1/6$ and 4-months $\omega = 1/4$) are somewhat spread out, and therefore each (amplitude-modulated) harmonic can be described by several (more than 2) eigentriples.

Periodogram analysis of the obtained singular vectors shows that the leading 14 eigentriples with share 91.4% can be related to 3 periodicities: the eigentriples $1, 2, 5 - 8, 13, 14$ describe the annual amplitude-modulated harmonic (Fig. 1.52), the eigentriples $3, 4, 11 - 12$ are related to half-year periodicity, and the eigentriples $9, 10$ describe the 4-months harmonic.

The same technique can be applied to the 'Births' series if we want to obtain better results than those described in Section 1.3.4. (See Section 1.6.2 for a discussion concerning the large window length problem in this example.)

1.7.4 Envelopes of highly oscillating signals

The capabilities of SSA in separating signals with high and low frequencies can be used in a specific problem of enveloping highly oscillating sequences with slowly varying amplitudes. The simple idea of such a technique can be expressed as follows.

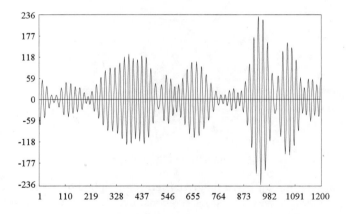

Figure 1.53 *EEG: α-rhythm. First 1200 points.*

Let $f_n = A(n) \cos(2\pi\omega n)$ where ω is large and $A(n)$ is slowly varying. Then

$$g_n \stackrel{\text{def}}{=} 2f_n^2 = A^2(n) + A^2(n) \cos(4\pi\omega n). \quad (1.35)$$

Since $A^2(n)$ is slowly varying and the second term on the right-hand side of (1.35) oscillates rapidly, one can gather the slowly varying terms of the SSA decomposition for g_n, and therefore approximately extract the term $A^2(n)$ from the series (1.35). All we need to do then is to take the square root of the extracted term.

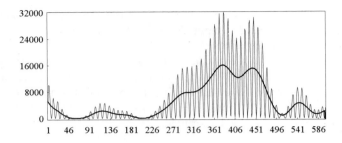

Figure 1.54 *EEG: series G and its slowly varying component. First 600 points.*

Example 1.10 *EEG: envelope of α-rhythm*

This idea is illustrated by the time series F representing an α-rhythm component of an electroencephalogram (EEG). The whole series F consists of approximately 3500 points; its first 1200 points can be seen in Fig. 1.53. The series can be described as an amplitude-modulated harmonic with the main frequency approximately equal to $1/20$.

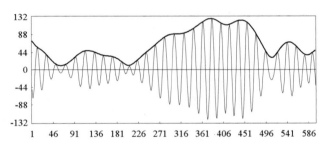

Figure 1.55 *EEG: α-rhythm and its envelope. First 600 points.*

Let us consider the square of the initial series multiplied by 2 and denote it by G. Taking window length $L = 60$ and reconstructing the low-frequency part of the time series G from the eigentriples 1, 4, 7 and 10, we obtain an estimate of $A^2(n)$ (the first 600 points of the reconstructed series are depicted in Fig. 1.54 by the thick line; the thin line corresponds to the series G).

By taking the square root of the estimate we obtain the result. (See Fig. 1.55, where the first 600 points of the initial series with its envelope are depicted.)

It may be interesting to note that the α-rhythm time series under consideration was extracted from the initial EEG signal by 5-stage Sequential SSA with different window lengths (the largest was equal to 600).

Note also that to obtain the resulting envelope we may need some smoothing to remove very small but existing parts of highly oscillating components. As usual, Basic SSA with small window length would do the job.

CHAPTER 2

SSA forecasting

A reasonable forecast of a time series can be performed only if the following conditions are met:

1. The series has a structure.
2. A mechanism (method, algorithm) identifying this structure is found.
3. A method of the time series continuation, based on the identified structure, is available.
4. The structure of the time series is preserved for the future time period over which we are going to forecast (continue) the series.

All these conditions are natural. Of course, condition 4 cannot be validated with the help of the data to be forecasted. Moreover, the structure of the series can hardly be identified uniquely (for example, if the series has a noise component). Therefore, the situation of different (and even 'contradictory') forecasts is not impossible. Thus, it is important not only to realize and indicate the structure under continuation, but also to check its stability.

At any rate, a forecast can be made only if a certain model is built. The model can either be derived from the data or at least checked against the data. In SSA forecasting, these models can be described with the help of the linear recurrent formulae (or equations). Note that in general the dimension (in other words, the order) of the recurrent formulae may be unknown.

The class of series governed by linear recurrent formulae (LRFs) is rather wide and important for practical implications. For instance, an infinite series is governed by some LRF if and only if it can be represented as a linear combination of products of exponential, polynomial and harmonic series. (See Chapter 5 for a review of the entire theory.)

The series governed by LRFs admits natural *recurrent continuation* since each term of such a series is equal to a linear combination of several preceding terms. Of course, the coefficients of this linear combination can be used for the continuation as well.

It is important that we need not necessarily search for an LRF of minimal dimension. Indeed, any other LRF governing the series produces the same continuation.

The theory of Section 5.2, Chapter 5, indicates how to find an LRF, which governs a series, with the help of SSA. The general idea can be described as follows.

Let d be the minimal dimension of all LRFs governing F. (In this case we shall say that the time series F is governed by a minimal LRF of dimension d.) It can be proved that if the window length L is larger than d, and the length of the series is sufficiently large, then the trajectory space of the series F is d-dimensional. Basic SSA provides a natural basis for the trajectory space.

The trajectory space determines (under mild and natural restrictions) an LRF of dimension $L-1$ that governs the series. If we apply this LRF to the last terms of the initial series F, we obtain the continuation of F.

The same idea may work if we want to continue an additive component $F^{(1)}$ of a series F. Here we assume that $F^{(1)}$ is governed by an LRF and is strongly separable from the residual series $F^{(2)} = F - F^{(1)}$ for the selected value of the window length L. It should be mentioned that if two series are strongly separable, then each of them must satisfy some LRF (see Remark 6.1 in Section 6.1.1).

In practice, it is not reasonable to assume that the series of interest is governed by an LRF of relatively small dimension. In this way we come to the concept of *approximate recurrent continuation*, which can and will also be called the *recurrent forecasting*. We thus suppose that the series F under consideration can be expressed as a sum of the series $F^{(1)}$ admitting recurrent continuation and the residual series $F^{(2)}$. If we consider the residuals as a noise, then we have the problem of forecasting the signal $F^{(1)}$ in the presence of the noise $F^{(2)}$. We may also have the problems of forecasting the series $F^{(1)}$ regarded as a trend or a seasonal component of F.

The main assumption is that for a certain window length L, the series components $F^{(1)}$ and $F^{(2)}$ are approximately strongly separable. Then, acting as in Basic SSA, we reconstruct the series $F^{(1)}$ with the help of a selected set of eigentriples and obtain approximations to both the series $F^{(1)}$ and its trajectory space. In other words, we obtain both the LRF, approximately governing $F^{(1)}$, and the initial data for this formula. Hence we obtain a forecast of the series $F^{(1)}$.

The theory of the method can be found in Chapter 5. The contents of the present Chapter are as follow.

Section 2.1 formally describes the general SSA forecasting algorithm. The rest of the chapter is devoted to study of this algorithm and related discussions.

Section 2.2 describes the principles of SSA forecasting and its relations to linear recurrent formulae. Several modifications of the general SSA forecasting algorithm are considered in Section 2.3.

Section 2.4 is devoted to a description of different ways of constructing confidence intervals that can be used for checking the forecast accuracy and stability. After the summarizing Section 2.5, several forecasting examples are presented in Section 2.6.

When dealing with continuation, we always need to bear in mind the length of the series under continuation. Therefore, we usually incorporate this length into the notation of the series and write, for example, F_N rather than simply F.

2.1 SSA recurrent forecasting algorithm

Let us formally describe the forecasting algorithm under consideration.

Algorithm inputs:
 (a) Time series $F_N = (f_0, \ldots, f_{N-1})$, $N > 2$.
 (b) Window length L, $1 < L < N$.
 (c) Linear space $\mathfrak{L}_r \subset \mathbf{R}^L$ of dimension $r < L$. It is assumed that $e_L \notin \mathfrak{L}_r$, where $e_L = (0, 0, \ldots, 0, 1)^T \in \mathbf{R}^L$. In other terms, $\mathfrak{L}_r$ is not a 'vertical' space. In practice, the space $\mathfrak{L}_r$ is defined by its certain orthonormal basis, but the forecasting results do not depend on this concrete basis.
 (d) Number M of points to forecast for.

Notations and Comments:
 (a) $\mathbf{X} = [X_1 : \ldots : X_K]$ (where $K = N - L + 1$) is the trajectory matrix of the time series F_N.
 (b) $P_1, \ldots, P_r$ is an orthonormal basis in $\mathfrak{L}_r$.
 (c) $\widehat{\mathbf{X}} \stackrel{\text{def}}{=} [\widehat{X}_1 : \ldots : \widehat{X}_K] = \sum_{i=1}^{r} P_i P_i^T \mathbf{X}$. The vector $\widehat{X}_i$ is the orthogonal projection of X_i onto the space $\mathfrak{L}_r$.
 (d) $\widetilde{\mathbf{X}} = \mathcal{H}\widehat{\mathbf{X}} = [\widetilde{X}_1 : \ldots : \widetilde{X}_K]$ is the result of the Hankelization of the matrix $\widehat{\mathbf{X}}$. The matrix $\widetilde{\mathbf{X}}$ is the trajectory matrix of some time series $\widetilde{F}_N = (\widetilde{f}_0, \ldots, \widetilde{f}_{N-1})$.
 (e) For any vector $Y \in \mathbf{R}^L$ we denote by $Y_\triangle \in \mathbf{R}^{L-1}$ the vector consisting of the last $L - 1$ components of the vector Y, while $Y^\triangledown \in \mathbf{R}^{L-1}$ is the vector consisting of the first $L - 1$ components of Y.
 (f) We set $\nu^2 = \pi_1^2 + \ldots + \pi_r^2$, where π_i is the last component of the vector P_i ($i = 1, \ldots, L$). Since ν^2 is the squared cosine of the angle between the vector e_L and the linear space $\mathfrak{L}_r$, it can be called the *verticality coefficient* of $\mathfrak{L}_r$.
 (g) Suppose that $e_L \notin \mathfrak{L}_r$. (In other words, we assume that $\mathfrak{L}_r$ is not a vertical space.) Then $\nu^2 < 1$. It can be proved (see Chapter 5, Theorem 5.2) that the last component y_L of any vector $Y = (y_1, \ldots, y_L)^T \in \mathfrak{L}_r$ is a linear combination of the first components $y_1, \ldots, y_{L-1}$:
$$y_L = a_1 y_{L-1} + a_2 y_{L-2} + \ldots + a_{L-1} y_1.$$

Vector $\mathcal{R} = (a_{L-1}, \ldots, a_1)^T$ can be expressed as

$$\mathcal{R} = \frac{1}{1 - \nu^2} \sum_{i=1}^{r} \pi_i P_i^\triangledown \qquad (2.1)$$

and does not depend on the choice of a basis $P_1, \ldots, P_r$ in the linear space $\mathfrak{L}_r$.

SSA recurrent forecasting algorithm:
 In the above notations, define the time series $G_{N+M} = (g_0, \ldots, g_{N+M-1})$ by

the formula

$$g_i = \begin{cases} \widetilde{f}_i & \text{for } i = 0, \ldots, N-1, \\ \sum_{j=1}^{L-1} a_j g_{i-j} & \text{for } i = N, \ldots, N+M-1. \end{cases} \quad (2.2)$$

The numbers $g_N, \ldots, g_{N+M-1}$ form the M terms of the SSA recurrent forecast. For brevity, we call this algorithm *SSA R-forecasting algorithm*.

Remark 2.1 Let us define the linear operator $\mathcal{P}^{(r)} : \mathfrak{L}_r \mapsto \mathbf{R}^L$ by the formula

$$\mathcal{P}^{(r)}Y = \begin{pmatrix} Y_\triangle \\ \mathcal{R}^\mathrm{T} Y_\triangle \end{pmatrix}, \quad Y \in \mathfrak{L}_r. \quad (2.3)$$

If setting

$$Z_i = \begin{cases} \widetilde{X}_i & \text{for } i = 1, \ldots, K, \\ \mathcal{P}^{(r)} Z_{i-1} & \text{for } i = K+1, \ldots, K+M, \end{cases} \quad (2.4)$$

the matrix $\mathbf{Z} = [Z_1 : \ldots : Z_{K+M}]$ is the trajectory matrix of the series G_{N+M}. Therefore, (2.4) can be regarded as the vector form of (2.2).

If $\mathfrak{L}_r$ is spanned by certain eigenvectors corresponding to the SVD of the trajectory matrix of the series F_N, then the corresponding SSA R-forecasting algorithm will be called the *Basic SSA R-forecasting algorithm*.

Remark 2.2 Denote by $\mathfrak{L}^{(L)} = \mathrm{span}(X_1, \ldots, X_K)$ the trajectory space of the series F_N. Suppose that $\dim \mathfrak{L}^{(L)} = r < L$ and $e_L \notin \mathfrak{L}^{(L)}$. If we use the Basic SSA R-forecasting algorithm with $\mathfrak{L}_r = \mathfrak{L}^{(L)}$, then $\mathbf{X} = \widehat{\mathbf{X}} = \widetilde{\mathbf{X}}$ and therefore $\widetilde{F}_N = F_N$. This means that the initial points $g_{N-L+1}, \ldots, g_{N-1}$ of the forecasting recurrent formula (2.2) coincide with the last $L-1$ terms of the series F_N.

2.2 Continuation and approximate continuation

The algorithmic scheme described in the previous section is related to both the series, which are governed by the linear recurrent formulae, and the SSA methodology. Let us describe the ideas that lead to SSA forecasting.

2.2.1 Linear recurrent formulae and their characteristic polynomials

The theory of the linear recurrent formulae and associated characteristic polynomials is well known (for example, Gelfond, 1967, Chapter V, §4). However, we provide here a short survey of the most essential results. A more formal description can be found in Chapter 5.

(a) Series governed by linear recurrent formulae

By definition, a nonzero series $F_N = (f_0, \ldots, f_{N-1})$ is governed by a *linear recurrent formula* (*LRF*) of dimension not exceeding $d \geq 1$ if

$$f_{i+d} = \sum_{k=1}^{d} a_i f_{i+d-k} \tag{2.5}$$

for certain $a_1, \ldots, a_d$ with $a_d \neq 0$ and $0 \leq i \leq N - d + 1$. In the notation of Section 5.2 this is expressed as $\mathrm{fdim}(F_N) \leq d$. If

$$d = \min(k : \mathrm{fdim}(F_N) \leq k),$$

then we write $\mathrm{fdim}(F_N) = d$ and call d the *finite-difference dimension* of the series F_N. In the case when F_N is governed by LRF (2.5) and $d = \mathrm{fdim}(F_N)$, the formula (2.5) is called *minimal*.

If (2.5) holds but we do not require that $a_d \neq 0$, then the time series F_N *satisfies* the LRF (2.5).

The class of series governed by LRFs is rather wide: it contains harmonic, exponential and polynomial series and is closed under term-by-term addition and multiplication. For instance, the exponential series $f_n = e^{\alpha n}$ is governed by the LRF $f_n = a f_{n-1}$ with $a = e^{\alpha}$, the harmonic series $f_n = \cos(2\pi \omega n + \phi)$ satisfies the equation

$$f_n = 2\cos(2\pi\omega) f_{n-1} - f_{n-2},$$

and so on. Other examples, as well as theoretical results, can be found in Section 5.2.

The difference between minimal and arbitrary LRFs governing the same series can be illustrated by the following example. For the exponential series F_N with $f_n = a^n$, $a = e^{\alpha}$ and $N \geq 3$, the LRF $f_n = a f_{n-1}$ is the minimal one and $\mathrm{fdim}(F_N) = 1$. On the other hand, the series $f_n = a^n$ satisfies the equation $f_n = 2a f_{n-1} - a^2 f_{n-2}$ for $2 \leq n \leq N - 1$.

To understand whether the LRF (2.5) is minimal for the series F_N with sufficiently large N, one can apply the following procedure. Consider the window length L ($1 < L < N$) and suppose that $d < \min(L, K)$. In view of (2.5), the L-lagged vectors $X_1, \ldots, X_K$ satisfy the vector recurrent equation

$$X_{i+d} = \sum_{k=1}^{d} a_i X_{i+d-k}, \quad 1 \leq i \leq K - d.$$

Therefore, each X_i is a linear combination of $X_1, \ldots, X_d$. If these vectors are linearly independent, then the LRF (2.5) is minimal and vice versa.

These assertions can be formulated in other terms. Denote by $\mathfrak{L}^{(L)}$ the trajectory space of the series F_N satisfying (2.5). If $d < \min(L, K)$, then the equalities $\mathrm{fdim}(F_N) = d$ and $\dim \mathfrak{L}^{(L)} = d$ are equivalent. Such a reformulation leads to a new concept.

Let $1 \le d \le L$. By definition, an arbitrary series F_N has *L-rank d* (i.e., $\mathrm{rank}_L(F_N) = d$) if $\dim \mathfrak{L}^{(L)} = d$.

If $\mathrm{rank}_L(F_N) = d$ for any L such that $d < \min(L, K)$, then the time series F_N has *rank d* (briefly, $\mathrm{rank}(F_N) = d$).

Roughly speaking, each time series F_N with $\mathrm{fdim}(F_N) = d$ has rank, and this rank is equal to d. The following simple example shows that the opposite assertion is not true: let us take $N = 7$ and $F_N = (1, 1, 1, 1, 1, 1, 2)$; then for each $L = 2, \ldots, 6$ we have $\mathrm{rank}_L(F_N) = 2$, while no LRF of dimension $d < 6$ can govern this series.

However, if $\mathrm{rank}_L(F_N) = d < L$, then the series F_N (with the exception of several first and last terms) is governed by an LRF of dimension $d_0 \le d$.

This LRF can be found by the procedure described in Theorem 5.1 of Chapter 5, but the procedure seems to be difficult for practical computations.

Moreover, let $L > \mathrm{rank}_L(F_N)$ and $e_L \notin \mathfrak{L}^{(L)}$. Let us denote $r = \dim \mathfrak{L}^{(L)}$ and take $\mathfrak{L}_r = \mathfrak{L}^{(L)}$. Then, as shown in Theorem 5.2 in the same chapter, the series F_N satisfies the LRF

$$f_{L+i-1} = a_1 f_{L+i-2} + \ldots + a_{L-1} f_i, \quad 0 \le i \le K-1, \qquad (2.6)$$

where $\mathcal{R} = (a_{L-1}, \ldots, a_1)^{\mathrm{T}}$ is defined in (2.1).

This fact has a purely geometric origin; due to Theorem 5.2, if $\mathfrak{L} \subset \mathbf{R}^L$ is a linear subspace of dimension $r < L$ and $e_L \notin \mathfrak{L}$, then the last component y_L of any vector $Y \in \mathfrak{L}$ is equal to the inner product $\mathcal{R}^{\mathrm{T}} Y^\nabla$, where the vector $Y^\nabla \in \mathbf{R}^{L-1}$ consists of the first $L-1$ components of the vector Y and $P_1, \ldots, P_r$ is an orthonormal basis of $\mathfrak{L}$.

(b) Characteristic polynomials and their roots

Let the series $F_N = (f_0, \ldots, f_{N-1})$ have finite-difference dimension d and is governed by the LRF

$$f_{d+i} = a_1 f_{d+i-1} + a_2 y_{d+i-2} + \ldots + a_d y_i, \qquad a_d \ne 0, \qquad (2.7)$$

for $0 \le i \le N - d$. Consider the *characteristic polynomial* of the LRF (2.7):

$$P_d(\lambda) = \lambda^d - \sum_{k=1}^{d} a_k \lambda^{d-k}.$$

Let $\lambda_1, \ldots, \lambda_p$ be the different (complex) roots of the polynomial $P_d(\lambda)$. Since $a_d \ne 0$, these roots are not equal to zero. We also have $k_1 + \ldots + k_p = d$, where k_m are the multiplicities of the roots λ_m ($m = 1, \ldots, p$).

Denote $f_n(m, j) = n^j \lambda_m^n$ for $1 \le m \le p$ and $0 \le j \le k_m - 1$. Theorem 5.3 of Section 5.2 tells us that the general solution of the equation (2.7) is

$$f_n = \sum_{m=1}^{p} \sum_{j=0}^{k_m - 1} c_{mj} f_n(m, j), \qquad (2.8)$$

with certain complex c_{mj}. The specific values of the c_{mj} are defined by the first d elements of the series F_N: $f_0, \ldots, f_{d-1}$.

Thus, each root λ_m produces a component

$$f_n^{(m)} = \sum_{j=0}^{k_m-1} c_{mj} f_n(m, j) \tag{2.9}$$

of the series f_n.

Let us fix m and consider this component in the case $k_m = 1$, which is the main case in practice. Set $\lambda_m = \rho e^{i2\pi\omega}$, $\omega \in (-1/2, 1/2]$, where $\rho > 0$ is the modulus (absolute value) of the root and $2\pi\omega$ is its polar angle.

If ω is either 0 or $1/2$, then λ_m is a real root of the polynomial $P_d(\lambda)$ and the series component $f_n^{(m)}$ is real and is equal to $c_{m0}\lambda_m^n$. This means that $f_n^{(m)} = A\rho^n$ for positive λ_m and $f_n^{(m)} = A(-1)^n\rho^n = A\rho^n \cos(\pi n)$ for negative λ_m. The latter case corresponds to the exponentially modulated saw-tooth sequence.

All other values of ω lead to complex λ_m. In this case, P_d has a complex conjugate root $\lambda_l = \rho e^{-i2\pi\omega}$ of the same multiplicity $k_l = 1$. We thus can assume that $0 < \omega < 1/2$ and describe a pair of conjugate roots by the pair of real numbers (ρ, ω) with $\rho > 0$ and $\omega \in (0, 1/2)$.

If we add together the components $f_n^{(m)}$ and $f_n^{(l)}$ corresponding to these conjugate roots, then we obtain the real series $A\rho^n \cos(2\pi\omega n + \phi)$ with A and ϕ expressed in terms of c_{m0} and c_{l0}.

The asymptotic behaviour of $f_n^{(m)}$ essentially depends on $\rho = |\lambda_m|$. Let us consider the simplest case $k_m = 1$ as above. If $\rho < 1$, then $f_n^{(m)}$ rapidly tends to zero and asymptotically has no influence on the whole series (2.8). Alternatively, the root with $\rho > 1$ and $|c_{m0}| \neq 0$ leads to a rapid increase of $|f_n|$ (at least for a certain subsequence of n).

For example, if $\lambda_m = \rho = 0.8$ and $|c_{m0}| \neq 0$, then $|f_n^{(m)}|$ becomes smaller by approximately a factor 10 in 10 time steps and by a factor $5 \cdot 10^9$ in 100 steps. If $\lambda_m = \rho = 1.2$ (and $|c_{m0}| \neq 0$), then $|f_n^{(m)}|$ is increased approximately 6-fold in 10 time steps and $8 \cdot 10^7$-fold in 100 steps. Similar effects hold for the series component $A\rho^n \cos(2\pi\omega n + \phi)$ corresponding to a pair of conjugate complex roots: the series amplitude $A\rho^n$ rapidly decreases or increases depending on the inequalities $\rho < 1$ or $\rho > 1$.

The root λ_m with $k_m > 1$ produces k_m terms in the sum (2.9). For example, if $\lambda_m = 1$ and $k_m = 2$, then $f_n^{(m)} = An + B$ for some A and B. In other words, the root 1 of multiplicity 2 generates a linear series. Example 5.10 of Section 5.2 treats the general case $k_m = 2$ in detail.

If the series F_N has finite-difference dimension d, then the characteristic polynomial of its minimal LRF (2.7) has d roots. As was mentioned above, the same series satisfies many other LRFs of certain dimensions $r > d$. Consider such an

LRF
$$f_{r+i} = b_1 f_{r+i-1} + b_2 y_{r+i-2} + \ldots + b_r y_i. \tag{2.10}$$

The characteristic polynomial $P_r(\lambda)$ of the LRF (2.10) has r roots with d roots (we call them the *main roots*) coinciding with the roots of the minimal LRF. The other $r - d$ roots are *extraneous*: in view of the uniqueness of the representation (2.9), the coefficients c_{mj} corresponding to these roots are equal to zero. However, the LRF (2.10) governs a wider class of series than the minimal LRF (2.7).

Since the roots of the characteristic polynomial specify its coefficients uniquely, they also determine the corresponding LRF. Consequently, by removing the extraneous roots of the characteristic polynomial $P_r(\lambda)$, corresponding to the LRF (2.10), we can obtain the polynomial describing the minimal LRF of the series.

Example 2.1 *Annual seasonality*
Let the series F_N have the period 12 (for instance, this series describes a seasonality). Then it can be expressed as a sum of a constant and six harmonics:

$$f_n = c_0 + \sum_{k=1}^{5} c_k \cos(2\pi nk/12 + \phi_k) + c_6 \cos(\pi n). \tag{2.11}$$

Under the condition that $c_k \neq 0$ for $k = 0, \ldots, 6$ the series has finite-difference dimension 12. In other words, the characteristic polynomial of the minimal LRF governing the series (2.11) has 12 roots. All these roots have the modulus 1. Two real roots ($+1$ and -1) correspond to the first and the last terms in (2.11). The harmonic term with frequency $\omega_k = k/12$ generates two complex conjugate roots $\exp(\pm i 2\pi k/12)$, which have polar angles $\pm 2\pi k/12$.

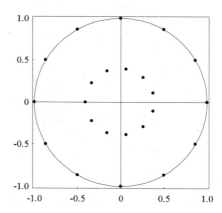

Figure 2.1 *Annual seasonality: main and extraneous roots.*

Let us now consider an LRF that is not minimal. Let N be large enough. If we select certain $L > 13$ and take $r = 12$, $\mathfrak{L}_r = \mathfrak{L}^{(L)}(F_N)$, then the vector

$\mathcal{R} = (a_{L-1}, \ldots, a_1)^\mathrm{T}$ defined in (2.1) produces the LRF

$$f_{i+L-1} = a_1 f_{i+L-2} + \ldots + a_{L-1} f_i, \qquad (2.12)$$

which is not minimal but governs the series (2.11).

Let us take $c_0 = \ldots = c_6 = 1$, $\phi_1 = \ldots = \phi_5 = 0$ and $L = 24$. The roots of the characteristic polynomial of the formula (2.12) are depicted in Fig. 2.1. We can see that the main 12 roots of the polynomial form a regular dodecagon, with the vertices on the unit circle of the complex plane. Eleven extraneous roots can be seen around zero; they have small moduli.

2.2.2 Recurrent continuation of time series

If the time series F_N is governed by an LRF (2.10) of dimension $r < N$, then there exists a natural *recurrent continuation* of such a series produced by the same formula (2.10). Whether LRF is minimal or not is of no importance since the extraneous roots have no influence on the series F_N.

(a) L-continuation

It is important to reformulate the concept of recurrent continuation in purely geometrical terms. Let us start with a definition.

Consider a time series $F_N = (f_0, \ldots, f_{N-1})$ and fix a window length $1 < L < N$. Denote by $X_1, \ldots, X_K$ the corresponding L-lagged vectors, and set $\mathfrak{L}^{(L)} = \mathrm{span}(X_1, \ldots, X_K)$. Let $d = \dim \mathfrak{L}^{(L)}$. (In other terms, the L-rank of the series F_N is equal to d.) Evidently, $d \leq \min(L, K)$.

We say that the series F_N *admits a continuation in* $\mathfrak{L}^{(L)}$ (or, briefly, *admits L-continuation*) if there exists a uniquely defined $\widetilde{f}_N$ such that all L-lagged vectors of the series $\widetilde{F}_{N+1} = (f_0, \ldots, f_{N-1}, \widetilde{f}_N)$ belong to $\mathfrak{L}^{(L)}$. In this case, the series $\widetilde{F}_{N+1}$ (as well as the number $\widetilde{f}_N$) will be called the *one-step L-continuation* of the series F_N.

Theorem 5.4 and Remark 5.9 in Section 5.3 provide the complete description of those series that admit L-continuation. For the moment, the following is important.

1. If $e_L \in \mathfrak{L}^{(L)}$, then F_N does not admit L-continuation. As a consequence, if $d = L$, then the series cannot be L-continued since the uniqueness condition does not hold.

2. If $d < L \leq K$ and $e_L \notin \mathfrak{L}^{(L)}$, then the series F_N admits L-continuation. From now on we assume that these assumptions concerning $\mathfrak{L}^{(L)}$ are satisfied.

3. The one-step L-continuation of the series F_N can be performed by the formula

$$\widetilde{f}_N = \sum_{k=1}^{L-1} a_k f_{N-k}, \qquad (2.13)$$

where the vector $\mathcal{R} = (a_{L-1}, \ldots, a_1)^\mathrm{T}$ is defined in the formula (2.1) applied to the space $\mathfrak{L}_r = \mathfrak{L}^{(L)}$.

4. The series F_N is governed by the same LRF (2.13), that is

$$f_{i+L} = \sum_{k=1}^{L-1} a_k \, f_{i+L-k}, \qquad 0 \le i \le N - L - 1.$$

5. If the series F_N admits a one-step L-continuation, then it can be L-continued for an arbitrary number of steps. Therefore, we can consider an infinite series F which is the L-continuation of F_N.

6. Let the series F_N satisfy an LRF

$$f_{i+d_0} = \sum_{k=1}^{d_0} b_k \, f_{i+d-k}, \qquad 0 \le i \le N - d_0 - 1, \qquad (2.14)$$

and $d_0 \le \min(L-1, K)$. Then $d \le d_0$, $e_L \notin \mathfrak{L}^{(L)}$ and the series will admit L-continuation, which can be produced by the same formula (2.14).

These properties are not surprising in view of the results discussed above concerning the correspondence between the series with $\mathrm{fdim}(F_N) = d$ and $\mathrm{rank}_L(F_N) = d$. Reformulated in terms of continuation, this correspondence means that under the conditions $\mathrm{rank}_L(F_N) < L \le K$ and $e_L \notin \mathfrak{L}^{(L)}$ the concepts of recurrent continuation and L-continuation are equivalent.

(b) Recurrent continuation and Basic SSA forecasting

Let us return to the forecasting algorithm of Section 2.1, considering the case of Basic SSA R-forecasting.

Suppose that $\mathfrak{L}_r = \mathfrak{L}^{(L)}$, $e_L \notin \mathfrak{L}^{(L)}$ and $r < L \le K$. Then

$$r = \mathrm{rank}_L(F_N) = \mathrm{fdim}(F_N)$$

and the series F_N is governed by an LRF of order r. In other words, the series F_N admits L-continuation.

Since the vectors X_i belong to the linear space $\mathfrak{L}_r$, the matrix $\widetilde{\mathbf{X}}$ of the forecasting algorithm coincides with the trajectory matrix $\mathbf{X}$ for the initial series F_N.

Denote by F_{N+M} recurrent continuation of the series F_N for M steps. This continuation can be performed with the help of the LRF (2.6), as the latter governs the series F_N. By the algorithm description, the forecasting formula (2.2) is produced by the same LRF (2.6).

Therefore, the series G_{N+M} defined by the formula (2.2) is equal to F_{N+M} and the SSA R-forecasting algorithm with $\mathfrak{L}_r = \mathfrak{L}^{(L)}$ produces recurrent continuation of the series F_N. The vector form (2.4) of the algorithm corresponds to the L-continuation.

To obtain the vector $\mathcal{R}$, we must have an orthonormal basis of the linear space $\mathfrak{L}^{(L)}$, see formula (2.1). Dealing with SSA, the SVD of the trajectory matrix

X for the series F_N provides us with the eigenvectors (left singular vectors) $U_1, \ldots, U_r$, which form a natural basis of $\mathfrak{L}^{(L)}$. Therefore, if the series F_N admits L-continuation, the latter can be performed with the help of SSA.

Other choices of $\mathfrak{L}_r$ can lead to continuation of the series components. Let $F_N = F_N^{(1)} + F_N^{(2)}$ with nonzero $F_N^{(1)}$ and $F_N^{(2)}$. Denote the L-lagged vectors of the series $F_N^{(1)}$ by $X_1^{(1)}, \ldots, X_K^{(1)}$ and set $\mathfrak{L}^{(L,1)} = \mathrm{span}(X_1^{(1)}, \ldots, X_K^{(1)})$.

Let $r = \dim \mathfrak{L}^{(L,1)}$ and assume that $r < L \leq K$ and $e_L \notin \mathfrak{L}^{(L,1)}$. Then the series $F_N^{(1)}$ admits L-continuation and the SSA R-forecasting algorithm with $\mathfrak{L}_r = \mathfrak{L}^{(L,1)}$ performs this continuation.

Suppose that series $F_N^{(1)}$ and $F_N^{(2)}$ are strongly separable for the window length $L \leq K$ (see Section 1.5) and denote by **X** the trajectory matrix of the series F_N. Then the SVD of the matrix **X** produces both the space $\mathfrak{L}_r$ and the series $F_N^{(1)}$.

Indeed, let U_i ($i = 1, \ldots, L$) be the eigenvectors of the matrix $\mathbf{S} = \mathbf{XX}^\mathrm{T}$ and let $I = \{j_1, \ldots, j_r\} \subset \{1, \ldots, L\}$ be the set of indices corresponding to the time series $F_N^{(1)}$. If we take $P_i = U_{j_i}, i = 1, \ldots, r$, then $\mathfrak{L}_r = \mathrm{span}(P_1, \ldots, P_r)$ and $r < L$. The series $F_N^{(1)}$ can be obtained in terms of the resultant Hankel matrix, which is produced by the grouping of the elementary matrices corresponding to the set of indices I.

Therefore, Basic SSA gives rise to the continuation of the series component which is accomplished by the Basic SSA R-forecasting algorithm. Note that if $F_N^{(1)}$ and $F_N^{(2)}$ are strongly separable, then both dimensions of their trajectory spaces are smaller than L.

2.2.3 Approximate continuation

The problems of exact continuation have mainly a theoretical and methodological sense. In practice, it is not wise to assume that the series obtained by measurements is governed by some LRF of relatively small dimension. Thus, we pass to the concept of approximate continuation, which is of greater importance in practice.

(a) Approximate separability and forecasting errors

Let $F_N = F_N^{(1)} + F_N^{(2)}$ and suppose that the series $F_N^{(1)}$ admits a recurrent continuation. Denote by d the dimension of the minimal recurrent formula governing $F_N^{(1)}$. If $d < \min(L, K)$, then $d = \mathrm{rank}_L(F_N^{(1)})$.

If $F_N^{(1)}$ and $F_N^{(2)}$ are strongly separable for some window length L, then we can perform recurrent continuation of the series $F_N^{(1)}$ by the method described in Section 2.2.2. We now assume that $F_N^{(1)}$ and $F_N^{(2)}$ are approximately strongly separable and discuss the problem of approximate continuation of the series $F_N^{(1)}$.

If $F_N^{(2)}$ is small enough and signifies an error or noise, this continuation can be regarded as a forecast of the signal $F_N^{(1)}$ in the presence of noise $F_N^{(2)}$. In other cases we can describe the problem as that of forecasting an interpretable component $F_N^{(1)}$ of F_N: for example, forecasting its trend or seasonal component.

As above, to do the continuation we use the Basic SSA R-forecasting algorithm described in Section 2.1. Formally, we assume that the following conditions hold.

1. The series of length N and window length L provide approximate strong separability of the series $F_N^{(1)}$ and $F_N^{(2)}$.

2. Let
$$\mathbf{X} = \sum_i \sqrt{\lambda_i} U_i V_i^\mathrm{T}$$
be the SVD of the trajectory matrix $\mathbf{X}$ of the series F_N. Then the choice of the eigentriples $\{(\sqrt{\lambda_i}, U_i, V_i)\}_{i \in I}$, $I = (i_1, \ldots, i_r)$, associated with $F_N^{(1)}$ allows us to achieve (approximate) separability.

3. $d \stackrel{\text{def}}{=} \text{fdim}(F_N^{(1)}) \leq r < L \leq K$.

4. $e_L \notin \text{span}(U_i, i \in I)$. In other terms, $\sum_{i \in I} u_{iL}^2 < 1$, where u_{iL} is the last component of the eigenvector U_i.

If these conditions hold, then we can apply the (Basic) SSA R-forecasting algorithm, taking $\mathfrak{L}_r = \text{span}(U_i, i \in I)$ and $P_j = U_{i_j}$. The result $g_N, \ldots, g_{N+M-1}$ is called the *approximate recurrent continuation* of the series F_N.

Let us discuss the features of this forecasting method. The forecast series g_n ($n \geq N$) defined by (2.2), generally does not coincide with recurrent continuation of the series $F_N^{(1)}$. The errors have two origins. The main one is the difference between the linear space $\mathfrak{L}_r$ and $\mathfrak{L}^{(L,1)}$, the trajectory space of the series $F_N^{(1)}$. Since the LRF (2.2) is produced by the vector $\mathcal{R}$ and the latter is strongly related to the space $\mathfrak{L}_r$ (see Proposition 5.5 of Chapter 5), the discrepancy between $\mathfrak{L}_r$ and $\mathfrak{L}^{(L,1)}$ produces an error in the LRF governing the forecast series. In particular, the finite-difference dimension of the forecast series g_n ($n \geq N$) is generally greater than d.

The other origin of the forecasting errors lies in the initial data for the forecast. For recurrent continuation, the initial data is $f_{N-L+1}^{(1)}, \ldots, f_{N-1}^{(1)}$, where $f_n^{(1)}$ is the nth term of the series $F_N^{(1)}$. In the Basic SSA R-forecasting algorithm, the initial data consists of the last $L-1$ terms $g_{N-L+1}, \ldots, g_{N-1}$ of the reconstructed series. Since generally $f_n^{(1)} \neq g_n$, the initial data produces its own error of forecasting.

On the other hand, if the quality of approximate separability of $F_N^{(1)}$ and $F_N^{(2)}$ is rather good and we select the proper eigentriples associated with $F^{(1)}$, then we can expect that the linear spaces $\mathfrak{L}_r$ and $\mathfrak{L}^{(L,1)}$ are close. Therefore, the coefficients in the LRF (2.2) are expected to be close to those of the LRF governing recurrent continuation of the series $F_N^{(1)}$. Analogously, approximate separability implies that the reconstructed series g_n is close to $f_n^{(1)}$, and therefore the errors of the initial forecasting data are small. As a result, in this case we can expect that the Basic SSA R-forecasting procedure provides a reasonable approximation to recurrent continuation of $F_N^{(1)}$, at least in the first few steps.

The following artificial example illustrates the role of separability in forecasting.

Example 2.2 *Separability and forecasting*

Let us consider the series $F_N = F_N^{(1)} + F_N^{(2)}$ with $N = 100$,

$$f_n = f_n^{(1)} + f_n^{(2)}, \quad f_n^{(1)} = 3a^n, \quad f_n^{(2)} = \sin(2\pi n/10)$$

and $a = 1.01$. Note that the series $F_N^{(2)}$ has finite-difference dimension 2 and $F_N^{(1)}$ is governed by the minimal LRF $f_n^{(1)} = af_{n-1}^{(1)}$.

If we want to forecast the series $F_N^{(1)}$, then we have to choose the window length L and take just one eigenvector of the corresponding SVD as the basis of the linear space $\mathfrak{L}_1$. (In this example, the leading eigenvector is acceptable for a wide range of L.)

Evidently, the forecasting result depends on L. The choice of the window length L can be expressed in terms of separability: a proper L ought to provide good separability characteristics. Let us compare the choice of two window lengths, $L = 50$ and $L = 15$, from the viewpoint of forecasting. Since exponential and harmonic series are asymptotically separable, the window length $L = 50$ seems to provide a good separation, while $L = 15$ should be regarded as too small.

The results for both Basic R-forecasting procedures are depicted in Fig. 2.2, where the top thick line starting at $n = 101$ corresponds to $L = 50$, and the analogous bottom thick line relates to $L = 15$. The thin line indicates the initial series F_N continued up to $n = 190$, which is the last forecasting point.

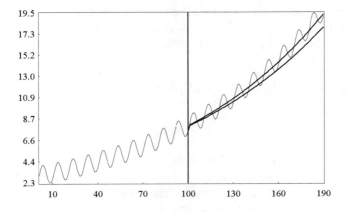

Figure 2.2 *Forecasting accuracy: two variants of window length.*

For $L = 50$, the choice of the first eigentriple in correspondence with $F_N^{(1)}$ leads to the w-correlation $\rho_{12}^{(w)} = 0.0001$ and the maximum cross-correlation $\rho^{(L,K)} = 0.034$. Therefore, the achieved separability should be regarded as rather good. If we take $L = 15$, then we obtain $\rho_{12}^{(w)} = 0.0067$ and $\rho^{(L,K)} = 0.317$, which means that the separation is poorer.

If we compare the forecasting results at the point $n = 190$, then we observe that the window length $L = 50$ provides the relative forecast error of about 2%, while the choice $L = 15$ gives almost 9%. This difference is not surprising since the window length $L = 15$ is too small for achieving good separability.

Note that both forecasts underestimate the true series. This can be explained in terms of the characteristic polynomials. Indeed, the main root of the polynomial $P_{14}(\lambda)$ corresponding to $L = 15$ is equal to 1.0091. The analogous root for $L = 50$ is 1.0098. The (single) root of the polynomial corresponding to the minimal LRF governing $F_N^{(1)}$ is $a = 1.01$. The arrangement of the roots coincides with the arrangement of the two forecasts and the exponential series $f_n^{(1)} = a^n$.

(b) Approximate continuation and the characteristic polynomials

Let us return to the errors of separability and forecasting. The discrepancies between $\mathfrak{L}_r$ and $\mathfrak{L}^{(L,1)}$ can be described in terms of the characteristic polynomials. We have three LRFs: (i) the minimal LRF of dimension d governing the series $F_N^{(1)}$, (ii) the continuation LRF of dimension $L-1$, which also governs $F_N^{(1)}$, but produces $L-d-1$ extraneous roots in its characteristic polynomial P_{L-1}, and (iii) the forecasting LRF governing the forecast series g_n $(n \geq N)$. The characteristic polynomial $P_{L-1}^{(f)}$ of the forecasting LRF also has $L-1$ roots.

If $\mathfrak{L}_r$ and $\mathfrak{L}^{(L,1)}$ are close, then the coefficients of continuation and forecasting recurrent formulae must be close too. Therefore, all simple roots of the forecasting characteristic polynomial $P_{L-1}^{(f)}$ must be close to that of the continuation polynomial P_{L-1}. The roots λ_m with multiplicities $k_m > 1$ could be perturbed in a more complex manner.

Example 2.3 *Perturbation of the multiple roots*
Let us consider the series F_N with

$$f_n = (A + 0.1\,n) + \sin(2\pi n/10), \quad n = 0, \ldots, 199.$$

Evidently, $F_N = F_N^{(1)} + F_N^{(2)}$ with the linear series $F_N^{(1)}$ defined by $f_n^{(1)} = A + 0.1\,n$ and the harmonic series $F_N^{(2)}$ corresponding to $f_n^{(2)} = \sin(2\pi n/10)$.

The series F_N has finite-difference dimension $\mathrm{fdim}(F_N) = 4$. Therefore, any LRF governing F_N produces a characteristic polynomial with four main roots. These main roots do not depend on A; the linear part of the series generates one real root $\lambda = 1$ of multiplicity 2, while the harmonic series corresponds to two complex conjugate roots with modulus $\rho = 1$ and $\omega = 0.1$.

Our aim is to forecast the series $F_N^{(1)}$ for $A=0$ and $A=50$ with the help of the Basic SSA forecasting algorithm. In both cases, we take the window length $L=100$ and choose the eigentriples that correspond to the linear part of the initial time series F_N. (For $A=0$ we take the two leading eigentriples, while for $A=50$ the appropriate eigentriples have the ordinal numbers 1 and 4.) Since the series $F_N^{(1)}$ and $F_N^{(2)}$ are not exactly separable for any choice of L, we deal with approximate separability.

The forecasting polynomials $P_{L-1}^{(f)}$ with $A=0$ and $A=50$ demonstrate different splitting of the double root $\lambda=1$ into two simple ones. For $A=0$ there appear two complex conjugate roots with $\rho=1.002$ and $\omega=0.0008$, while in the case $A=50$ we obtain two real roots equal to 1.001 and 0.997. All extraneous roots are less than 0.986.

This means that for $A=0$ the linear series $F_N^{(1)}$ is approximated by a low-frequency harmonic with a slightly increasing exponential amplitude. In the case $A=50$ the approximating series is the sum of two exponentials, one of them is slightly increasing and another one is slightly decreasing.

These discrepancies lead to quite different long-term forecasting results: oscillating for $A=0$ and exponentially increasing for $A=50$.

In the case of a large discrepancy between $\mathfrak{L}_r$ and $\mathfrak{L}^{(L,1)}$, both the main and the extraneous roots of the continuation polynomial can differ significantly, and the error of the forecasting procedure can be rather large.

Evidently, such an error depends on the order $L-1$ of the characteristic polynomials as well; the bigger the number of the perturbed extraneous roots, the less precise the forecasting procedure may become.

On the other hand, the conditions for approximate separability are usually asymptotic and require relatively large L. In practice, this means that we have to take the smallest window length L providing a sufficient (though approximate) separability.

2.3 Modifications to Basic SSA R-forecasting

The Basic SSA R-forecasting algorithm discussed in Section 2.2 should be regarded as the main forecasting algorithm due to its direct relation to the linear recurrent formulae. Nevertheless, there exist several natural modifications to this algorithm that can give better forecasts in specific situations.

2.3.1 SSA vector forecasting

Let us return to Basic SSA and assume that our aim is to extract a certain additive component $F_N^{(1)}$ from a series F_N. In this algorithm, for an appropriate window length L, we obtain the SVD of the trajectory matrix of the series F_N and select the eigentriples $(\sqrt{\lambda_i}, U_i, V_i)$, $i \in I = (j_1, \ldots, j_r)$, corresponding to $F_N^{(1)}$. Then we obtain the resultant matrix

$$\mathbf{X}_I = \sum_{i \in I} \sqrt{\lambda_i} U_i V_i^\mathrm{T}$$

and, after diagonal averaging, we obtain the reconstructed series $\widetilde{F}_N^{(1)}$ that estimates $F_N^{(1)}$.

Note that the columns $\widehat{X}_1, \ldots, \widehat{X}_K$ of the resultant matrix $\mathbf{X}_I$ belong to the linear space $\mathfrak{L}_r = \mathrm{span}(U_i, i \in I)$. If $F_N^{(1)}$ is strongly separable from $F_N^{(2)} \stackrel{\mathrm{def}}{=} F_N - F_N^{(1)}$, then $\mathfrak{L}_r$ coincides with $\mathfrak{L}^{(L,1)}$ (the trajectory space of the series $F_N^{(1)}$) and $\mathbf{X}_I$ is a Hankel matrix (in this case $\mathbf{X}_I$ is the trajectory matrix of the series $F_N^{(1)}$). If $F_N^{(1)}$ and $F_N^{(2)}$ are approximately strongly separable, then $\mathfrak{L}_r$ is close to $\mathfrak{L}^{(L,1)}$ and $\mathbf{X}_I$ is approximately a Hankel matrix.

Briefly, the idea of 'vector forecasting' can be expressed as follows. Let us imagine that we can continue the sequence of vectors $\widehat{X}_1, \ldots, \widehat{X}_K$ for M steps in such a manner that:

1. The continuation vectors Z_m ($K < m \leq K + M$) belong to the same linear space $\mathfrak{L}_r$.
2. The matrix $\mathbf{X}_M = [\widehat{X}_1 : \ldots : \widehat{X}_K : Z_{K+1} : \ldots : Z_{K+M}]$ is approximately a Hankel matrix.

Having obtained the matrix $\mathbf{X}_M$ we can obtain the series G_{N+M} by diagonal averaging. Since the first elements of the reconstructed series $\widetilde{F}_N^{(1)}$ coincide with the elements of G_{N+M}, the latter can be considered to be a forecast of $F_N^{(1)}$.

Now let us give a formal description of the *SSA vector forecasting algorithm* (briefly, *V-forecasting*) in the same manner as was done in Section 2.1 for the SSA recurrent forecasting algorithm.

Preliminaries:

- The SSA vector forecasting algorithm has the same inputs and conditions as the SSA R-forecasting algorithm.
- The notation in (a)-(g) of Section 2.1 is kept. Let us introduce some more notation.

Consider the matrix

$$\Pi = \mathbf{V}^\nabla (\mathbf{V}^\nabla)^\mathrm{T} + (1 - \nu^2) \mathcal{R} \mathcal{R}^\mathrm{T}, \qquad (2.15)$$

where $\mathbf{V}^\nabla = [P_1^\nabla : \ldots : P_r^\nabla]$. The matrix Π is the matrix of the linear operator that performs the orthogonal projection $\mathbf{R}^{L-1} \mapsto \mathfrak{L}_r^\nabla$ (see Proposition 5.9 in Section 5.3), where $\mathfrak{L}_r^\nabla = \mathrm{span}(P_1^\nabla, \ldots, P_r^\nabla)$.

We define the linear operator $\mathcal{P}^{(v)} : \mathfrak{L}_r \mapsto \mathbf{R}^L$ by the formula

$$\mathcal{P}^{(v)} Y = \begin{pmatrix} \Pi Y_\triangle \\ \mathcal{R}^\mathrm{T} Y_\triangle \end{pmatrix}, \quad Y \in \mathfrak{L}_r. \qquad (2.16)$$

SSA vector forecasting algorithm:

1. In the notation above we define the vectors Z_i as follows:

$$Z_i = \begin{cases} \widehat{X}_i & \text{for } i = 1, \ldots, K \\ \mathcal{P}^{(v)} Z_{i-1} & \text{for } i = K+1, \ldots, K+M+L-1. \end{cases} \quad (2.17)$$

2. By constructing the matrix $\mathbf{Z} = [Z_1 : \ldots : Z_{K+M+L-1}]$ and making its diagonal averaging we obtain a series $g_0, \ldots, g_{N+M+L-1}$.
3. The numbers $g_N, \ldots, g_{N+M-1}$ form the M terms of the SSA vector forecast.

If $\mathfrak{L}_r$ is spanned by certain eigenvectors obtained by Basic SSA, we shall call the corresponding algorithm the *Basic SSA vector forecasting algorithm*. Let us discuss its features.

(a) *Continuation*

If $\mathfrak{L}_r$ is the trajectory space of the series F_N (in other words, if we act under the assumptions of Section 2.2.2), then the result of the vector forecasting coincides with that of the recurrent one. Thus, in this case the V-forecasting algorithm performs recurrent continuation of the series F_N.

More precisely, in this situation the matrix Π is the identity matrix, and (2.16) coincides with (2.3). Furthermore, the matrix $\mathbf{Z}$ has Hankel structure and diagonal averaging is the identical operation.

The same coincidence holds if $F_N = F_N^{(1)} + F_N^{(2)}$, the series $F_N^{(1)}$ and $F_N^{(2)}$ are strongly separable, and $\mathfrak{L}_r$ is the trajectory space of the series $F_N^{(1)}$. The Basic SSA V-forecasting then performs recurrent continuation of $F_N^{(1)}$.

(b) *Forecasting*

Though the results are the same, the essentials of recurrent and vector forecasting are different. Briefly, recurrent forecasting performs recurrent continuation directly (with the help of LRF), while vector forecasting deals with L-continuation. In the case of approximate continuation, the two forecasting algorithms usually give different results.

In a typical situation, there is no time series such that the linear space $\mathfrak{L}_r$ (for $r < L - 1$) is its trajectory space, and therefore (see Proposition 5.6) this space cannot be the trajectory space of the series to be forecasted. The recurrent forecasting method uses $\mathfrak{L}_r$ to obtain the LRF of the forecast series.

The vector forecasting procedure tries to perform the L-continuation of the series in $\mathfrak{L}_r$; any vector $Z_{i+1} = \mathcal{P}^{(v)} Z_i$ belongs to $\mathfrak{L}_r$, and $Z_{i+1}^{\triangledown}$ is as close to $(Z_i)_\triangle$ as it can be. The last component of Z_{i+1} is obtained from $Z_{i+1}^{\triangledown}$ by the LRF applied in the recurrent forecasting. Since the matrix $\mathbf{Z}$ is not a Hankel one, diagonal averaging works in the same manner as in Basic SSA.

(c) *Details*

Both forecasting methods have two general stages: diagonal averaging and continuation. For the recurrent forecasting, diagonal averaging is used to obtain the reconstructed series, and continuation is performed by applying the LRF. In the vector forecasting method, these two stages are used in the reverse order; first, vector continuation in $\mathfrak{L}_r$ is performed and then diagonal averaging gives the forecast values.

Note that in order to get M forecast terms the vector forecasting procedure performs $M+L-1$ steps. The aim is the permanence of the forecast under variations in M: the M-step forecast ought to coincide with the first M values of the forecast for $M+1$ or more steps. In view of the features of diagonal averaging, we have to produce $L-1$ extra steps.

(d) *Comparison*

If the series admits recurrent continuation, then the results for both Basic SSA forecasting methods coincide. In the case of approximate continuation they differ. Typically, a poor approximation implies a large difference between the two forecasts.

In the case of approximate separability it is hard to compare the recurrent and vector forecasting methods theoretically. Generally, the approximate coincidence of the two forecasting results can be used as an argument in favour of the forecasting stability.

Recurrent forecasting is simpler to interpret due to the description of LRFs in terms of the characteristic polynomials. On the other hand, results of data analysis show that the vector forecasting method is usually more 'conservative' (or less 'radical') in those cases when the recurrent forecasting method demonstrates rapid increase or decrease.

2.3.2 Toeplitz SSA forecasting

Using Basic SSA recurrent and vector forecasting, we take $\mathfrak{L}_r$ to be spanned by certain eigenvectors U_k, $k \in I$, of the SVD applied to the trajectory matrix $\mathbf{X}$ of the series F_N. In other words, the basis vectors P_i of $\mathfrak{L}_r$ have the form $P_i = U_{j_i}$ (see Section 2.2.2). Other decompositions of the trajectory matrix lead to another choice of $\mathfrak{L}_r$.

If the original series can be regarded as a stationary one, then as defined in (1.34) the Toeplitz SSA decomposition

$$\mathbf{X} = \sum_{i=1}^{L} H_i Z_i^T$$

can be used in place of the SVD in Basic SSA. Here the H_i stands for the ith eigenvector of the Toeplitz lag-covariance matrix defined in (1.32). (See Section 1.7.2 in Chapter 1 for details.)

Let us consider the SSA R-forecasting algorithm of Section 2.1. If we select a set of indices $I = (j_1, \ldots, j_r)$ and take $P_i = H_{j_i}$ as the basis vectors in $\mathfrak{L}_r$, then we obtain the *Toeplitz SSA R-forecasting algorithm*. Evidently, one can use the vector forecasting variant in Toeplitz forecasting as well.

As was mentioned in Section 1.7.2, for relatively short intervals of stationary-like series, the Toeplitz SSA may give better separability characteristics than Basic SSA. Therefore, if we have a problem of continuation of a sum of several harmonic components of a stationary series, then Toeplitz forecasting may have an advantage.

Moreover, if L is much smaller than $K = N - L + 1$, then the Toeplitz lag-covariance matrix has a more regular structure than the standard lag-covariance matrix used in Basic SSA. The eigenvectors of the Toeplitz lag-covariance matrix are also more regular. Since forecasting is based on the space $\mathfrak{L}_r$ generated by the eigenvectors (and does not use both the factor vectors and the singular values), for stationary time series Toeplitz SSA forecasting may give more stable results.

2.3.3 Centring in SSA forecasting

To elucidate the characteristics of the (single) centring variant of SSA forecasting, we start with a series that admits recurrent continuation.

Consider the series F_N with $\mathrm{fdim}(F_N) = d \geq 1$ and sufficiently large N. As was described in Section 2.2, if we take the window length L such that $d < \min(L, K)$ and suppose that the corresponding trajectory space $\mathfrak{L}^{(L)}$ is not a vertical one, then $\dim \mathfrak{L}^{(L)} = d$ and the choice $\mathfrak{L}_r = \mathfrak{L}^{(L)}$ leads to recurrent continuation of the series F_N, which is performed by SSA recurrent forecasting algorithm of Section 2.1. Let us consider another way of doing such a continuation.

By definition, the space $\mathfrak{L}^{(L)}$ is spanned by the L-lagged vectors $X_1, \ldots, X_K$ of the series F_N. In the same manner as in Section 1.7, we denote by $\mathcal{E} = \mathcal{E}_1(\mathbf{X})$ the vector of the row averages of the trajectory matrix $\mathbf{X}$. In other words, we set

$$\mathcal{E} = (X_1 + \ldots + X_K)/K. \tag{2.18}$$

Evidently, $\mathcal{E} \in \mathfrak{L}^{(L)}$. We set

$$\mathfrak{L}_\mathcal{E}^{(L)} = \mathrm{span}(X_1 - \mathcal{E}, \ldots, X_K - \mathcal{E}) = \mathfrak{L}^{(L)} - \mathcal{E}. \tag{2.19}$$

Then (see Section 4.4) the dimension $r \stackrel{\mathrm{def}}{=} \dim \mathfrak{L}_\mathcal{E}^{(L)}$ is equal to either d or $d-1$. Assume that $r \geq 1$ (the case $r = 0$ corresponds to a constant series F_N).

If $e_L \notin \mathfrak{L}_\mathcal{E}^{(L)}$, then according to the proof of Theorem 5.2, the last component y_L of any vector $Y \in \mathfrak{L}_\mathcal{E}^{(L)}$ is equal to the linear combination of its first $L-1$ components:

$$y_L = \sum_{k=1}^{L-1} a_k y_{L-k}, \tag{2.20}$$

where the vector $\mathcal{R} = (a_{L-1}, \ldots, a_1)^\mathrm{T}$ is obtained from $\mathfrak{L}_\mathcal{E}^{(L)}$ by the formula (2.1), with $P_1, \ldots, P_r$ standing for an orthonormal basis in $\mathfrak{L}_\mathcal{E}^{(L)}$.

Let us now consider the infinite series F, which is recurrent continuation of F_N, and denote by X_i $(i > K)$ the ith L-lagged vector in the series F. Since $X_1, \ldots, X_K$ span the space $\mathfrak{L}^{(L)}$, and $X_i \in \mathfrak{L}^{(L)}$ for any $i > K$, it follows that $X_i - \mathcal{E} \in \mathfrak{L}_{\mathcal{E}}^{(L)}$ for any i.

Let us denote by $z_k^{(i)}$ the kth component of the vector $X_i - \mathcal{E}$. In view of (2.20) we obtain

$$z_L^{(i)} = \sum_{k=1}^{L-1} a_k z_{L-k}^{(i)}. \tag{2.21}$$

Rewriting (2.21) in terms of $X_i = (f_{i-1}, \ldots, f_{i+L-2})^{\mathrm{T}}$ we come to the equalities

$$f_{i+L-2} = \sum_{k=1}^{L-1} a_k f_{i+L-2-k} + \varepsilon_L - \mathcal{R}^{\mathrm{T}} \mathcal{E}^{\nabla}, \quad i \geq 1, \tag{2.22}$$

where ε_L is the last component of the vector $\mathcal{E}$.

Thus, we have arrived at the *heterogeneous linear recurrent formula*, governing the series F_N and performing its recurrent continuation. Evidently, if $\mathcal{E} = \mathbf{0}_L$, then (2.22) coincides with recurrent continuation formula which is obtained in terms of $\mathfrak{L}^{(L)}$, see Section 2.2.

The transition from the trajectory space $\mathfrak{L}^{(L)}$ to the space (2.19) is considered in Sections 4.4 and 1.7, where the features of the centring versions of the SVD and Basic SSA are discussed. In terms of these Sections, $\mathfrak{L}_{\mathcal{E}}^{(L)}$ corresponds to single centring.

Single centring ideas give rise to versions of both recurrent and vector SSA forecasting algorithms for Basic and Toeplitz forecasting. Let us describe these versions in the formal manner of Section 2.1. For brevity, we present only the modified items within the description of the algorithms.

There are two versions of these modifications. If we are reconstructing a component of a time series with the help of the centring variant of the Basic (or Toeplitz) SSA, we can either include the average triple into the list of the eigentriples selected for reconstruction or not. These two possibilities are kept in the *centring variant of SSA forecasting*.

Now let $\mathfrak{L}_r$ be a subspace of $\mathbf{R}^L$ of dimension $r < L$, $e_L \notin \mathfrak{L}_r$, and let $P_1, \ldots, P_r$ be some orthonormal basis of $\mathfrak{L}_r$.

If we do not take average triple for the reconstruction, then:
1. The matrix $\widehat{\mathbf{X}}$ (Section 2.1, *Notation and Comments*, item b) is defined as

$$\widehat{\mathbf{X}} = [\widehat{X}_1 : \ldots : \widehat{X}_K] = \sum_{i=1}^{r} P_i P_i^{\mathrm{T}} (\mathbf{X} - \mathbf{A}), \tag{2.23}$$

where $\mathbf{A} = [\mathcal{E} : \ldots : \mathcal{E}]$ and the vector $\mathcal{E}$ has the form (2.18).

2. Formula (2.2) and its vector version defined by (2.3) and (2.4) are kept for the recurrent variant of SSA centring forecasting. Analogously, for SSA vector forecasting, the formulae (2.16) and (2.17) are kept.

In the case when we take the average triple for the reconstruction, we have
1. Matrix $\widehat{\mathbf{X}}$ is defined as

$$\widehat{\mathbf{X}} = [\widehat{X}_1 : \ldots : \widehat{X}_K] = \sum_{i=1}^{r} P_i P_i^{\mathrm{T}} (\mathbf{X} - \mathbf{A}) + \mathbf{A},$$

in the same notation as (2.23).

2. (i) In recurrent forecasting, the formula (2.2) is modified as

$$g_i = \begin{cases} \widetilde{f}_i & \text{for } i = 0, \ldots, N-1, \\ \sum_{j=1}^{L-1} a_j g_{i-j} + a & \text{for } i = N, \ldots, N+M-1 \end{cases}$$

with $a = \varepsilon_L - \mathcal{R}^{\mathrm{T}} \mathcal{E}^{\nabla}$. To modify its vector form (2.3), (2.4), we keep the latter formula and replace (2.3) by

$$\mathcal{P}^{(rc)} Y = \begin{pmatrix} Y_\Delta - \mathcal{E}^{\nabla} \\ \mathcal{R}^{\mathrm{T}}(Y_\Delta - \mathcal{E}^{\nabla}) \end{pmatrix} + \mathcal{E},$$

where the operator $\mathcal{P}^{(rc)}$ maps $\mathfrak{L}_r + \mathcal{E}$ to $\mathbf{R}^L$.

(ii) In SSA vector forecasting variant, the formula (2.17) is kept and (2.16) is replaced by

$$\mathcal{P}^{(vc)} Y = \begin{pmatrix} \Pi(Y_\Delta - \mathcal{E}^{\nabla}) \\ \mathcal{R}^{\mathrm{T}}(Y_\Delta - \mathcal{E}^{\nabla}) \end{pmatrix} + \mathcal{E}, \quad Y \in \mathfrak{L}_r + \mathcal{E}.$$

If we use *Basic SSA centring forecasting*, then the vectors P_i $(1 \leq i \leq r)$ are selected from the set of the SVD eigenvectors for the matrix $\mathbf{X} - \mathbf{A}$. In the *Toeplitz* variant, the Toeplitz decomposition of $\mathbf{X} - \mathbf{A}$ is used instead.

Note that the double centring variant of SVD (see Section 4.4) can hardly be used for forecasting in the style under consideration. The main reason for this is that the double centring is applied to both the rows and columns of the trajectory matrix, while the SSA forecasting algorithm of Section 2.1 and all its modifications and variants are based on the linear space $\mathfrak{L}_r$, which is associated only with the columns of the trajectory matrix.

2.3.4 Other ways of modification

There exist numerous versions of the forecasting methods based on the SSA ideas. Let us mention several of these versions, stating them as problems to be solved rather than as methods recommended for direct use in practice.

(a) Minimal recurrent formula: Schubert and reduction methods

The linear recurrent formula applied in the recurrent SSA forecasting algorithm has dimension $L-1$ (L is the window length), while the minimal recurrent formula governing the series F_N (if any) can have a much smaller dimension. Therefore, for a window length sufficient for approximate separability, it is natural to look for the LRF of relatively small dimension to perform a reasonable forecast.

Assume that the series F_N admits a recurrent continuation. One way of finding its minimal LRF is described in Theorem 5.1 of Section 5.2, where such an LRF is explained in geometrical terms of the Schubert basis (*Schubert* method). Another possibility arises if we can distinguish the main and extraneous roots of the characteristic polynomial of the LRF. In this case we can remove the extraneous roots and come to the minimal formula (*reduction* method).

Both methods are theoretically exact if $\text{fdim}(F_N) < \min(L, K)$. However, their practical usefullness is not at all obvious since we deal with approximate separability, which produces perturbations of all results.

The stability of the Schubert method under data perturbations has not yet been checked. Therefore, there is a danger that not only the coefficients of the obtained 'minimal' LRF but even its dimension can vary significantly under small variations in the data. Also, the method seems to be much more complicated than Basic SSA R-forecasting.

The modification of the Basic SSA R-forecasting algorithm based on the reduction of the polynomial roots works well if the main roots are properly indicated and the perturbation in the data is not very large. Otherwise the forecasting results can be unpredictable. An example of applying the reduction recurrent forecasting algorithm can be found in Section 2.6.1.

Note that both methods can be used only for recurrent forecasting. Moreover, the problem of the initial data arises again; the errors in the initial data for the minimal LRF can affect the forecast more severely than for an LRF of large dimension.

(b) The nearest subspace

If F_N admits recurrent continuation, then the choice $\mathfrak{L}_r = \mathfrak{L}^{(L)}$ leads to the LRF governing F_N. In the case of approximate separability, the forecasting LRF is calculated through the selected linear space $\mathfrak{L}_r$, which typically cannot be the trajectory space of any time series (see Proposition 5.6 in Section 5.2).

One can try to solve this annoying contradiction in the following manner. Let us state the problem of finding a linear space $\mathfrak{L}'_r$ as follows: (a) the space has the same dimension r as the initial space $\mathfrak{L}_r$, (b) $\mathfrak{L}'_r$ is the trajectory space of a certain time series, and (c) $\mathfrak{L}'_r$ is the closest to $\mathfrak{L}_r$ (the cosine of the angle between these spaces is maximum).

If the errors in data are not very large, then such a space can be regarded as an appropriate 'estimate' of the trajectory space of a series under recurrent continuation. The space $\mathfrak{L}'_r$ being found, the corresponding LRF of dimension $L - 1$ appears, and the specific form of the forecast by this LRF depends on the initial data. Since the vector consisting of the last $L-1$ points of the reconstructed series does not generally belong to $\mathfrak{L}'_r$, we can perform its orthogonal projection onto this linear space and take the result for the initial forecast data.

FORECAST CONFIDENCE BOUNDS

We do not discuss here the general algorithmic problem of finding this nearest subspace. Let us consider the simplest case $r = 1$ when the space $\mathfrak{L}_1$ is spanned by a vector $X = (x_1, \ldots, x_L)^{\mathrm{T}}$, and $\mathfrak{L}'_1$ must be spanned by the vector $Y_a = (1, a, \ldots, a^{L-1})^{\mathrm{T}}$. Then the optimal a gives the maximum value for the expression $|(X, Y_a)|/\|X\|\,\|Y_a\|$ and can be obtained by simple calculations.

The optimal a being obtained, we must find the corresponding LRF, that is the vector $\mathcal{R}$ (see formula (2.1) in Section 2.1). In the one-dimensional case this problem is rather simple as well, since all the components of the formula (2.1) are expressed in terms of the single vector $P_1 = Y_a/\|Y_a\|$.

Omitting the calculations we present the result for the case $|a| \neq 1$:

$$\mathcal{R} = C(a)(1, a, \ldots, a^{L-2})^{\mathrm{T}} \tag{2.24}$$

with

$$C(a) = \frac{a^{L-1}(a^2 - 1)}{a^{2L-2} - 1}.$$

We can now apply the LRF so obtained to the appropriate initial data.

Evidently the one-dimensional case is convenient for the reduction of the extraneous polynomial roots; the LRF defined by (2.24) defines a characteristic polynomial with a single main root $\lambda = a$. Therefore, taking the last term of the reconstructed series as the initial point and applying the recurrent formula $f_n = a f_{n-1}$, we make the forecast based on both ideas: that of the nearest subspace and the minimal LRF.

2.4 Forecast confidence bounds

According to the main SSA forecasting assumptions, the component $F_N^{(1)}$ of the series F_N ought to be governed by an LRF of relatively small dimension, and the residual series $F_N^{(2)} = F_N - F_N^{(1)}$ ought to be approximately strongly separable from $F_N^{(1)}$ for some window length L. In particular, $F_N^{(1)}$ is assumed to be a finite subseries of an infinite series $F^{(1)}$, which is a recurrent continuation of $F_N^{(1)}$. These assumptions cannot be ignored, but fortunately they hold for a wide class of practical problems.

To establish confidence bounds for the forecast, we have to apply even stronger assumptions, related not only to $F_N^{(1)}$, but to $F_N^{(2)}$ as well. First, let us consider $F_N^{(2)}$ as a finite subseries of an infinite random noise series $F^{(2)}$ that perturbs the signal $F^{(1)}$. The other assumptions can hardly be formulated in terms of $F_N^{(2)}$ only; they mainly deal with the residual series $\widetilde{F}_N^{(2)} = F_N - \widetilde{F}_N^{(1)}$, where $\widetilde{F}_N^{(1)}$ is the reconstructed component of F_N. Since $\widetilde{F}_N^{(1)} \approx F_N^{(1)}$, the features of $\widetilde{F}_N^{(2)}$ are strongly related to those of $F_N^{(2)}$. A more precise formulation of the additional assumptions depends on the problem we are solving and the method that we are applying.

Here we consider the following two problems, related to construction of the confidence bounds for the forecast. The first problem is to construct a confidence interval for the entire series $F = F^{(1)} + F^{(2)}$ at some future point in time $N + M$. The second problem can be formulated as a construction of confidence bounds for the signal $F^{(1)}$ at the same future point in time.

These two problems will be solved in different ways. The first uses the information about the forecast errors obtained by processing the series. This variant can be called the empirical one. The second requires additional information about the model governing the series $\widetilde{F}_N^{(2)}$ to accomplish a bootstrap simulation of the series F_N (see Efron and Tibshirani, 1986, Section 5, for general bootstrap concepts).

Let us briefly discuss both problems of constructing the confidence bounds for the Basic SSA R-forecasting method. All other SSA forecasting procedures can be treated analogously.

2.4.1 Empirical confidence intervals for the forecast of the initial series

Assume that we have already obtained the forecast value $\widetilde{f}_{N+M-1}^{(1)}$, that is, we have already performed M steps of the Basic SSA R-forecasting procedure. By definition, we use $\widetilde{f}_{N+M-1}^{(1)}$ as the forecast of the (future) term $f_{N+M-1}^{(1)}$ of the signal $F^{(1)}$. As was already mentioned, our problem is to build up a confidence interval for the (future) term f_{N+M-1} of the series F.

Let us consider the *multistart M-step recurrent continuation* procedure. We take a relatively small integer M and apply M steps of recurrent continuation produced by the forecasting LRF modifying the initial data from $(\widetilde{f}_0^{(1)}, \ldots, \widetilde{f}_{L-2}^{(1)})$ to $(\widetilde{f}_{K-M}^{(1)}, \ldots, \widetilde{f}_{N-M-1}^{(1)})$, $K = N - L + 1$.

The last points $g_{j+M+L-1}$ of these continuations can be compared with the values $f_{j+M+L-1}$ of the initial series F_N. We thus obtain the *multistart M-step residual series* H_{K-M+1} with

$$h_j^{(M)} = f_{j+M+L-2} - g_{j+M+L-2}, \quad j = 0, \ldots, K - M.$$

Suppose for the moment that the reconstructed series $\widetilde{F}_N^{(1)}$ coincides with $F_N^{(1)}$ and the forecasting LRF governs it. Then $g_k = f_k^{(1)}$ and the multistart M-step residual series coincides with the last $K - M + 1$ terms of the stationary noise series $F_N^{(2)}$.

If these suppositions are not valid, then $h_j^{(M)}$ does not coincide with $f_{j+M+L-2}^{(2)}$. Even so, let us assume that the multistart M-step residual series is stationary and ergodic in the sense that its empirical cumulative distribution function (c.d.f.) tends to the theoretical c.d.f. of the series as $N \to \infty$. Then, having the series H_{K-M+1} at hand, we can estimate certain of its quantiles (for example, the upper and lower 2.5% ones).

Note that the terms $g_{j+M+L-2}$ are obtained through the same number of steps with the same LRF as the forecast value $\widetilde{f}_{N+M-1}^{(1)}$, and their initial data is taken

FORECAST CONFIDENCE BOUNDS

from the same reconstructed series. Since forecasting requires the assumption that the series structure is kept in the future, the obtained empirical c.d.f. of the multistart M-step residual series can be used to construct the empirical confidence interval for f_{N+M-1}.

More formally, let us fix a confidence level γ ($0 < \gamma < 1$), and set $\alpha = 1 - \gamma$. If $c_{\alpha/2}^-$ and $c_{\alpha/2}^+$ stand for the lower and upper $\alpha/2$-quantiles, calculated through the empirical c.d.f. of the multistart M-step residual series, then we obtain the *empirical confidence interval*

$$\left(\widetilde{f}_{N+M-1}^{(1)} + c_{\alpha/2}^-, \widetilde{f}_{N+M-1}^{(1)} + c_{\alpha/2}^+ \right),$$

which covers f_{N+M-1} with an approximate confidence level γ. Evidently, the number K has to be sufficiently large for the empirical c.d.f. to be stable.

If the multistart M-step residual series can be regarded as white noise, then the other variant of empirical confidence intervals is meaningful. Assuming the Gaussian white noise hypothesis, the standard symmetrical confidence bounds of f_{N+M-1} can be constructed with the help of the sample average and the sample variance of the multistart M-step residual series. Of course, the white noise hypothesis can be checked with the help of the standard statistical procedures.

2.4.2 Bootstrap confidence bounds for the forecast of a signal

Let us consider a method of constructing confidence bounds for the signal $F^{(1)}$ at the moment of time $N + M - 1$. In the unrealistic situation, when we know both the signal $F^{(1)}$ and the true model of the noise $F_N^{(2)}$, the Monte Carlo simulation can be applied to check the statistical properties of the forecast value $\widetilde{f}_{N+M-1}^{(1)}$ relative to the actual term $f_{N+M-1}^{(1)}$.

Indeed, assuming that the rules for the eigentriple selection are fixed, we can simulate S independent copies $F_{N,i}^{(2)}$ of the process $F_N^{(2)}$ and apply the forecasting procedure to S independent time series $F_{N,i} \stackrel{\text{def}}{=} F_N^{(1)} + F_{N,i}^{(2)}$. Then the forecasting results will form a sample $\widetilde{f}_{N+M-1,i}^{(1)}$ ($1 \leq i \leq S$), which should be compared against $f_{N+M-1}^{(1)}$. In this way the *Monte Carlo confidence bounds* for the forecast can be build up.

Since in practice we do not know the signal $F_N^{(1)}$, we cannot apply this procedure. Let us describe the bootstrap (for example, Efron and Tibshirani, 1986, Section 5) variant of the simulation for constructing the confidence bounds for the forecast.

Under a suitable choice of the window length L and the corresponding eigentriples, we have the representation $F_N = \widetilde{F}_N^{(1)} + \widetilde{F}_N^{(2)}$, where $\widetilde{F}_N^{(1)}$ (the reconstructed series) approximates $F_N^{(1)}$, and $\widetilde{F}_N^{(2)}$ is the residual series. Suppose now that we have a (stochastic) model of the residuals $\widetilde{F}_N^{(2)}$. (For instance, we can postulate some model for $F_N^{(2)}$ and, since $\widetilde{F}_N^{(1)} \approx F_N^{(1)}$, apply the same model for $\widetilde{F}_N^{(2)}$ with the estimated parameters.)

Then, simulating S independent copies $\widetilde{F}_{N,i}^{(2)}$ of the series $F_N^{(2)}$, we obtain S series $F_{N,i} \stackrel{\text{def}}{=} \widetilde{F}_N^{(1)} + \widetilde{F}_{N,i}^{(2)}$ and produce S forecasting results $\widetilde{f}_{N+M-1,i}^{(1)}$ in the same manner as in the Monte Carlo simulation variant.

More precisely, any time series $F_{N,i}$ produces its own reconstructed series $\widetilde{F}_{N,i}^{(1)}$ and its own forecasting linear recurrent formula LRF_i for the same window length L and the same set of the eigentriples. Starting at the last $L-1$ terms of the series $\widetilde{F}_{N,i}^{(1)}$, we perform M steps of forecasting with the help of its LRF_i to obtain $\widetilde{f}_{N+M-1,i}^{(1)}$.

As soon as the sample $\widetilde{f}_{N+M-1,i}^{(1)}$ ($1 \leq i \leq S$) of the forecasting results is obtained, we can calculate its (empirical) lower and upper quantiles of a fixed level γ and obtain the corresponding confidence interval for the forecast. This interval (called the *bootstrap confidence interval*) can be compared with the forecast value $\widetilde{f}_{N+M-1}^{(1)}$ obtained from the initial forecasting procedure. A discrepancy between this value and the obtained confidence interval can be caused by the inaccuracy of the stochastic model for $\widetilde{F}_N^{(2)}$.

The average of the bootstrap forecast sample (*bootstrap average forecast*) estimates the mean value of the forecast, while the mean square deviation of the sample shows the accuracy of the estimate.

The simplest model for $\widetilde{F}_N^{(2)}$ is the model of Gaussian white noise. The corresponding hypothesis can be checked with the help of the standard tests for randomness and normality.

2.4.3 Confidence intervals: comparison of forecasting variants

The aim of this section is to compare different SSA forecasting procedures using several artificial series and Monte Carlo confidence intervals.

Let $F_N = F_N^{(1)} + F_N^{(2)}$, where $F_N^{(2)}$ is Gaussian white noise with standard deviation σ. Assume also that the signal $F_N^{(1)}$ admits a recurrent continuation. We can and shall perform a forecast of the series $F_N^{(1)}$ for M steps using different variants of SSA forecasting and appropriate eigentriples associated with $F_N^{(1)}$.

If the signal $F_N^{(1)}$ and its recurrent continuation are known, then we can apply the Monte Carlo procedure described in the previous section to check the accuracy of the forecasting results and compare different ways of forecasting.

To do that, we simulate a large number of independent copies $F_{N,i}^{(2)}$ of $F_N^{(2)}$, produce the time series $F_{N,i} = F_N^{(1)} + F_{N,i}^{(2)}$, and forecast their signal component $F_N^{(1)}$ using the eigentriples of the same ordinal numbers as that for the initial series F_N. Evidently this procedure is meaningful only if the choice of the eigentriples is stable enough for different realizations of the white noise $F_N^{(2)}$.

FORECAST CONFIDENCE BOUNDS

Monte Carlo forecast of the signal $F_N^{(1)}$ is useful in at least two respects: its average (called the *Monte Carlo average forecast*) shows the bias produced by the corresponding forecasting procedure, while the upper and lower quantiles indicate the role of the random component in the forecasting error.

Several effects will be illustrated with the help of this technique. First, we shall compare some forecasting variants from the viewpoint of their accuracy. The second matter to be demonstrated is the role of the proper window length. Lastly, we compare different variants of the confidence intervals in forecasting.

Throughout all the examples, we use the following notation: N stands for the length of the initial series, M is the number of forecasting steps, and σ denotes the standard deviation of the Gaussian white noise $F_N^{(2)}$. The confidence intervals are obtained in terms of the 2.5% upper and lower quantiles of the corresponding empirical c.d.f. using the sample size $S = 1000$.

(a) Periodic signal: recurrent and vector forecasting

Let $N = 100$, $M = 100$, $\sigma = 0.5$. Let us consider a periodic signal $F_N^{(1)}$ of the form

$$f_n^{(1)} = \sin(2\pi n/17) + 0.5\sin(2\pi n/10).$$

The series $F_N^{(1)}$ has difference dimension 4, and we use four leading eigentriples for its forecasting under the choice $L = 50$. The initial series $F_N = F_N^{(1)} + F_N^{(2)}$ and the signal $F_N^{(1)}$ (the thick line) are depicted in Fig. 2.3.

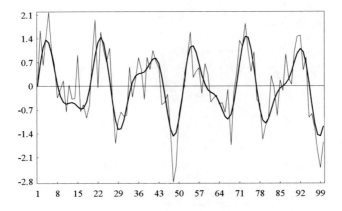

Figure 2.3 *Periodic signal and the initial series.*

Let us apply the Monte Carlo simulation for the Basic SSA recurrent and vector forecasting algorithms.

Fig. 2.4 shows the confidence Monte Carlo intervals for both methods and the true continuation of the signal $F_N^{(1)}$ (thick line). Confidence intervals for R-forecasting are marked by dots, while thin solid lines correspond to vector forecasting. We can see that these intervals practically coincide for relatively small numbers of forecasting steps, while the vector method has an advantage in the long-term forecasting.

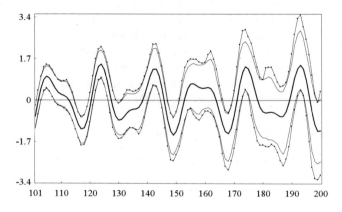

Figure 2.4 *Periodic signal: confidence intervals for the recurrent and vector forecasts.*

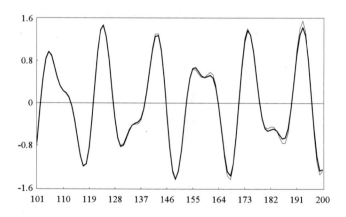

Figure 2.5 *Periodic signal: Basic Monte Carlo average R-forecast.*

The bias in the Basic SSA R-forecast is demonstrated in Fig. 2.5, where the thick line depicts the true continuation of the series $F_N^{(1)}$ and the thin line corresponds to the average of the Monte Carlo average R-forecast. We see that the bias is sufficiently small.

FORECAST CONFIDENCE BOUNDS

Note that the bias in the vector method almost coincides with that in the recurrent one. Therefore, the advantage of vector forecasting can be expressed mainly in terms of its stability rather than in the bias. The bias in both methods is caused by the nonlinear structure of the forecasting procedures.

(b) Periodic signal: Basic and Toeplitz recurrent forecasting

The same series with the same forecasting parameters serves as an example for comparing the Basic and Toeplitz R-forecasting methods. As usual, we apply the centring variant of the Toeplitz forecasting algorithm, though the results of the comparison do not depend on this choice.

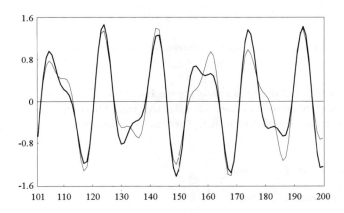

Figure 2.6 *Periodic signal: Toeplitz Monte Carlo average R-forecast.*

Fig. 2.6 is analogous to Fig. 2.5 and shows the bias in Toeplitz R-forecasting. In comparison with the Basic R-forecast, we see that the bias is rather large. The explanation lies in the fact that in contrast to Basic SSA, the four leading eigentriples in the Toeplitz SSA decomposition of the signal $F_N^{(1)}$ do not describe the entire signal; their share is approximately 99.8%. From the formal viewpoint, the Toeplitz decomposition of the trajectory matrix is not the minimal one (see Sections 4.2.1 and 1.7.2).

Indeed, if we consider the signal $F_N^{(1)}$ as the initial series and produce its Toeplitz forecast with $L = 50$ and 4 leading eigentriples, then the result will be very close to the Monte Carlo average forecast, presented in Fig. 2.6 (thin line, the thick line depicts the continuation of the series $F_N^{(1)}$).

The situation with the confidence intervals is different, see Fig. 2.7. The Monte Carlo confidence intervals for the Toeplitz forecast (depicted by thick lines) are typically inside that for the Basic forecast (thin lines). This is not surprising since the Toeplitz SSA gives more stable harmonic-like eigenvectors for stationary time series.

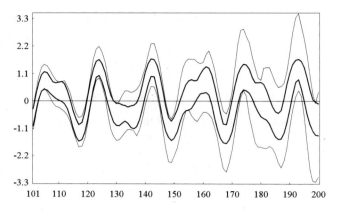

Figure 2.7 *Periodic signal: confidence intervals for the Basic and Toeplitz R-forecasts.*

Note that the confidence intervals for the Basic and Toeplitz forecasting algorithms are shifted relative to each other due to a large bias in the Toeplitz method. We conclude that Toeplitz forecasting proves to be less precise (on average), but more stable.

(c) Separability and forecasting

Consider the series $F_N^{(1)}$ with

$$f_n^{(1)} = 3a^n + \sin(2\pi n/10), \quad a = 1.01,$$

and $N = 100$. This series is governed by an LRF of dimension 3.

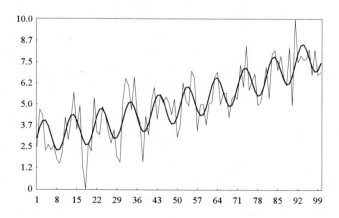

Figure 2.8 *Separability and forecasting: the signal and the initial series.*

FORECAST CONFIDENCE BOUNDS

Taking $\sigma = 1$ and two window lengths $L = 15$ and $L = 50$, we consider Basic SSA R-forecasting of the series $F_N = F_N^{(1)} + F_N^{(2)}$ for 90 steps. Our aim is to compare the accuracy of these two variants of forecasting of the signal $F_N^{(1)}$ with the help of the Monte Carlo simulation. The first three eigentriples are chosen for the reconstruction in both variants. The series F_N and the signal $F_N^{(1)}$ (thick line) are depicted in Fig. 2.8.

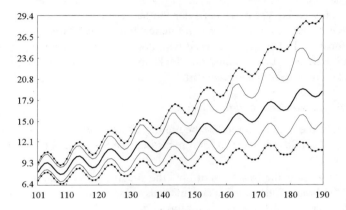

Figure 2.9 *Separability and forecasting: two confidence intervals.*

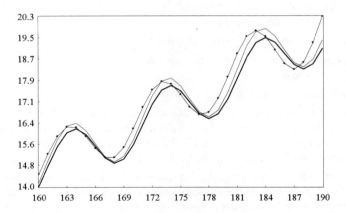

Figure 2.10 *Separability and forecasting: comparison of biases.*

The influence of separability on forecasting in the absence of noise has already been discussed (see Example 2.2 in Section 2.2). We now explain this influence in statistical terms of bias and confidence intervals.

Fig. 2.9 shows that the Monte Carlo forecasting confidence intervals for $L = 15$ (thin line marked with dots) are much wider than that for $L = 50$. This is not surprising since the separability characteristics for $L = 15$ are: $\rho_{12}^{(w)} = 0.0083$ and $\rho^{(L,K)} = 0.26$, while for $L = 50$ we have $\rho_{12}^{(w)} = 0.0016$ and $\rho^{(L,K)} = 0.08$.

Note that both confidence intervals are almost symmetric with respect to the true continuation of the signal (the thick line in Fig. 2.9). This means that in this example the choice of the window length does not have a big influence on the bias of the forecasts. Yet if we consider the last forecast points (Fig. 2.10), we can see that the choice $L = 50$ is again better. Indeed, the Monte Carlo average forecast for $L = 15$ (thin line, marked with dots) has a small but apparent phase shift relative to the true continuation (thick line), while for the choice $L = 50$ (thin line) there is almost no phase shift.

(d) Confidence intervals of different kinds

According to the discussion at the beginning of this section, we can construct three kinds of confidence interval for forecasting (see Section 2.4 for their detailed description).

First, as we know the true form of both the signal $F_N^{(1)}$ and the noise $F_N^{(2)}$, we can build the Monte Carlo confidence intervals, which can be considered to be the true confidence intervals for the signal forecast.

Second, we can apply the bootstrap simulation for the same purpose. Here we use the same Gaussian white noise assumption but calculate its variance in terms of the residuals of the reconstruction.

Third, the empirical confidence bounds for the forecast of the entire series $F_N = F_N^{(1)} + F_N^{(2)}$ can be built as well.

The last two methods are more important in practice since neither $F_N^{(1)}$ nor $F_N^{(2)}$ is usually known. Our aim is to compare three kinds of confidence bounds by a simple example.

Consider the exponential series $F_N^{(1)}$ with $f_n^{(1)} = 3a^n$, $a = 1.01$ and $N = 190$. As above, we assume that $F_N^{(2)}$ is a realization of the Gaussian white noise and take $\sigma = 1$. Since we want to deal with the empirical confidence intervals, we truncate the series at $n = 160$ and use the truncated series as the initial one. A comparison of the confidence intervals is performed for 30 Basic SSA R-forecasting steps with $L = 50$. Since $F_N^{(1)}$ is governed by an LRF of dimension 1, we take one leading eigentriple for reconstruction and forecasting in all cases.

The series F_N (thin oscillating line) is depicted in Fig. 2.11 together with its reconstruction, the Basic SSA R-forecast (thick lines) and the corresponding empirical intervals. The vertical line corresponds to the truncation point.

Figs. 2.12-2.14 show three variants of the confidence intervals on the background of the series F_N. Fig 2.12 represents the empirical intervals around the forecast of the signal $F_N^{(1)}$ (thick line). Since the empirical intervals are built for the entire series F_N, it is not surprising that they cover the series values. Note

FORECAST CONFIDENCE BOUNDS

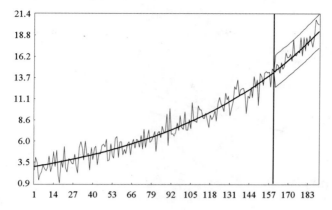

Figure 2.11 *Exponential signal: the initial series and forecast.*

that the length of the empirical confidence intervals is almost constant due to the homogeneity of the residuals used for their construction (see Section 2.4).

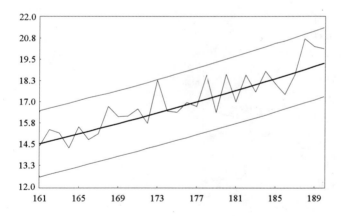

Figure 2.12 *Exponential signal: empirical confidence intervals.*

The bootstrap confidence intervals are shown in Fig. 2.13, where the thick line corresponds to the exponential signal $F_N^{(1)}$. The intervals are shifted relative to the signal (and they are symmetric relative to its forecast) because the bootstrap simulation uses the reconstructed series, which differs from the signal itself. Note that the empirical confidence intervals in Fig. 2.12 are also shifted relative to the signal $F_N^{(1)}$.

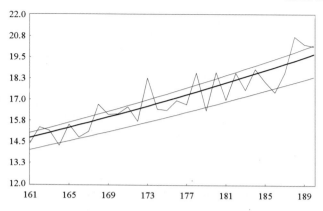

Figure 2.13 *Exponential signal: bootstrap confidence intervals.*

Lastly, the Monte Carlo confidence intervals are depicted in Fig. 2.14 together with the signal $F_N^{(1)}$ (thick line). In this case the intervals appear to be symmetric around the signal.

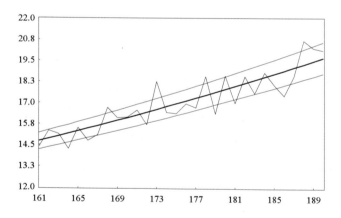

Figure 2.14 *Exponential signal: Monte Carlo confidence intervals.*

Comparing the intervals, we can see that the lengths of the bootstrap and Monte Carlo intervals are very similar and are smaller than those of the empirical intervals. The latter is natural since the first two bound the signal and the third one bounds the entire series.

One more difference is that the intervals obtained by simulation are enlarging in time, while the empirical ones are rather stable. Thus, we can use the empirical confidence intervals only for relatively short-term forecasting.

2.5 Summary and recommendations

Let us summarize the material of the previous sections, taking as an example the Basic SSA R-forecasting method. Other versions of SSA forecasting can be described and commented on similarly.

1. *Statement of the problem*

    We have a series $F_N = F_N^{(1)} + F_N^{(2)}$ and have the problem of forecasting its component $F_N^{(1)}$. If $F_N^{(2)}$ can be regarded as noise, then the problem is that of forecasting the signal $F_N^{(1)}$ in the presence of a noise $F_N^{(2)}$.

2. *The main assumptions*

    - The series $F_N^{(1)}$ admits a recurrent continuation with the help of an LRF of relatively small dimension d.
    - There exists a number L such that the series $F_N^{(1)}$ and $F_N^{(2)}$ are approximately strongly separable for the window length L. This is an important assumption since any time series $F_N^{(1)}$ is an additive component of F_N in the sense that $F_N = F_N^{(1)} + F_N^{(2)}$ with $F_N^{(2)} = F_N - F_N^{(1)}$. The assumption of (approximate) separability means that $F_N^{(1)}$ is a natural additive component of F_N from the viewpoint of the SSA method.

3. *Proper choice of parameters*

    Since we have to select the window length L providing a sufficient quality of separability and to find the eigentriples corresponding to $F_N^{(1)}$, all the major rules of Basic SSA are applicable here. Note that in this case we must separate $F_N^{(1)}$ from $F_N^{(2)}$, but we do not need the decomposition of the entire series $F_N = F_N^{(1)} + F_N^{(2)}$.

4. *Specifics and dangers*

    The SSA forecasting problem has some specifics in comparison with the Basic SSA reconstruction problem:

    - Since the chosen window length L produces an LRF of dimension $L-1$, which is applied as a recurrent continuation formula, the problem of extraneous roots for its characteristic polynomial becomes important. The choice $L = d + 1$ with d standing for the dimension of the minimal LRF, must be optimal. Unfortunately, in practice, small values of L do not usually provide sufficient separability. As a result, one has to try to select the minimal window length that is greater than d and provides reasonable separability.
    - The linear space $\mathcal{L}_r$ of dimension r determining the forecasting LRF is spanned by the eigenvectors of the chosen eigentriples. Since the condition $r \geq d$ has to be fulfilled, the number of eigentriples selected as corresponding to $F_N^{(1)}$ has to be at least d.
    - In Basic SSA, if we enlarge the set of proper eigentriples by some extra eigentriples with small singular values, then the result of reconstruction will

essentially be the same. When dealing with forecasting, such an operation can produce large perturbations since the space $\mathcal{L}_r$ will be perturbed a lot; its dimension will be enlarged, and therefore the LRF governing the forecast will be modified. (Note that the magnitude of the extra singular values is not important in this case.) Hence, the eigentriples describing $F_N^{(1)}$ have to be determined very carefully.

5. *Characteristics of forecasting*

Let us mention several characteristics that might be helpful in judging the forecasting quality.

- *Separability characteristics.* All the separability characteristics considered in detail in Section 1.5 are of importance for forecasting.
- *Polynomial roots.* The roots of the characteristic polynomial of the forecasting LRF can give insight into the behaviour of the forecast. These polynomial roots can be useful in answering the following two questions:

  (a) We expect that the forecast has some particular form (for example, we expect it to be increasing). Do the polynomial roots describe such a possibility? For instance, an exponential growth has to be indicated by a single real root (slightly) greater than 1; if we try to forecast the annual seasonality, then pairs of complex roots with frequencies $\approx k/12$ have to exist, and so on.

  (b) Is it possible to obtain a hazard inconsistent forecast? In terms of the polynomial roots, each extraneous root increases such a possibility. Even so, if the modulus of the root is essentially less than 1, then a slight perturbation of the proper initial data should not produce large long-term errors. Since the polynomial roots with moduli greater than 1 correspond to the series components with increasing envelopes (see Section 2.2.1), large extraneous roots may cause problems even in short-term forecasting.

- *Verticality coefficient.* The verticality coefficient ν^2 is the squared cosine of the angle between the space $\mathcal{L}_r$ and the vector e_L. The condition $\nu^2 < 1$ is necessary for forecasting. If ν^2 is close to 1, then, in view of (2.1), the coefficients of the forecasting LRF will be large and therefore some roots of the characteristic polynomial will have large moduli too. If the expected behaviour of the forecast does not suggest a rapid increase or decrease, then a large value of the verticality coefficient indicates a possible difficulty with the forecast. This typically means that extra eigentriples are taken to describe $F_N^{(1)}$ (alternatively, the approach in general is inappropriate).

6. *The role of the initial data*

Apart from the number M of forecast steps, the formal parameters of the Basic SSA R-forecasting algorithm are the window length L and the set I of eigentriples describing $F_N^{(1)}$. These parameters determine both the forecasting

LRF (2.1) and the initial data for the forecast. Evidently, the forecasting result essentially depends on this data, especially when the forecasting LRF has extraneous roots.

The SSA R-forecasting method uses the last terms $\widetilde{f}^{(1)}_{N-L+1}, \ldots, \widetilde{f}^{(1)}_{N-1}$ of the reconstructed series $\widetilde{F}^{(1)}_N$ as the initial forecasting data. Due to the properties of diagonal averaging, the last (and the first) terms of the series $F^{(1)}_N$ are usually reconstructed with a poorer precision than the middle ones. This effect may cause essential forecast errors.

For example, any linear (and nonconstant) series $f_n = an + b$ is governed by the minimal LRF $f_n = 2f_{n-1} - f_{n-2}$, which does not depend on a and b. The parameters a and b used in the forecast are completely determined by the initial data f_0 and f_1. Evidently, errors in this data may essentially modify the behaviour of the forecast (for example, change a tendency to increase into a tendency to decrease).

Thus, it is important to check the last points of the reconstructed series (for example, to compare them with the expected future behaviour of the series $F^{(1)}_N$).

7. *Reconstructed series and LRFs*
In the situation of strong separability of $F^{(1)}_N$ and $F^{(2)}_N$ and proper eigentriple selection, the reconstructed series is governed by the LRF which completely corresponds to the series $F^{(1)}_N$. Discrepancies in such a correspondence indicate possible errors: insufficient separability (which can be caused by the bad quality of the forecasting parameters) or general inefficiency of the model. Two characteristics of the correspondence may be useful here.

- *Global discrepancies.* Rather than using an LRF for forecasting, we can use it for approximation of either the whole reconstructed series or its subseries. For instance, if we take the first terms of the reconstructed series as the initial data (instead of the last ones) and make $N - L + 1$ steps of the procedure, we can check whether the reconstructed series can be globally approximated with the help of the LRF.

  Evidently, we can use another part of the reconstructed series as the initial data while taking into consideration the poor quality of its first terms or possible heterogeneity of the dynamics of the series $F^{(1)}_N$.

- *Local discrepancies.* The procedure above corresponds to long-term forecasting. To check the short-term correspondence of the reconstructed series and the forecasting LRF, one can apply a slightly different method.

  This method is used in Section 2.4.1 to construct empirical confidence intervals and is called the multistart recurrent continuation. According to it, for a relatively small Q we perform Q steps of the multistart recurrent continuation procedure, modifying the initial data from $(\widetilde{f}^{(1)}_0, \ldots, \widetilde{f}^{(1)}_{L-1})$ to $(\widetilde{f}^{(1)}_{K-Q}, \ldots, \widetilde{f}^{(1)}_{N-Q})$, $K = N - L + 1$. The continuation is computed with the help of the forecasting LRF. The results are to be compared with $\widetilde{f}_L^{(1)}, \ldots, \widetilde{f}_{N-1}^{(1)}$.

Since both the LRF and the initial data have errors, the local discrepancies for small Q are usually more informative than the global ones. Moreover, by checking different Q we can estimate the maximal number M of steps for a reasonable forecast.

8. *Forecasting stability and reliability*

   While the correctness of the forecast cannot be checked via the intrinsic properties of the data, its reliability can be examined. Let us mention several methods for carrying out such an examination.

   - *Different algorithms.* We can try different forecasting algorithms (for example, recurrent and vector) with the same parameters. If their results approximately coincide, we have an argument in favour of the stability of forecasting.

   - *Different initial data.* Since the last terms of the reconstructed series can have significant errors, forecasting can start at one of the previous points. Then we would have several forecasts and, of course, can compare them and get an opinion about the forecasting stability.

   - *Different window lengths.* If the separability characteristics are stable under a small variation in the window length L, we can compare the forecasts for different L.

   - *Forecasting of truncated series.* Let us truncate the initial series F_N by removing the last few terms from it. If the separability conditions are stable under such an operation, then we can forecast the truncated terms and compare the result with the initial series F_N and the reconstructed series $\widetilde{F}_N^{(1)}$ obtained without truncation. If the forecast is regarded as adequate, then its continuation by the same LRF can be regarded as reliable. Of course, this forecast can be compared with the one obtained without truncation, and, if they are similar, we can approve the forecasting stability.

9. *Confidence intervals*

   Though both empirical and bootstrap variants of the confidence intervals are not absolute, they give important additional information about forecasting. Let us summarize their features and peculiarities.

   - *General assumptions.* The model of the initial series is $F_N = F_N^{(1)} + F_N^{(2)}$, where $F_N^{(1)}$ is a signal and $F_N^{(2)}$ is assumed to be 'noise'. The series $F_N^{(1)}$ is supposed to admit a recurrent continuation, and the problem of its forecasting is under solution.

   - *Goals.* The empirical confidence intervals are constructed for the entire series F_N, which is assumed to have the same structure in the future. Bootstrap confidence intervals are built for the continuation of the signal $F_N^{(1)}$.

- *Data sources and additional assumptions.* The data source for construction of the bootstrap confidence intervals is the residual series $\widetilde{F}_N^{(2)} = F_N - \widetilde{F}_N^{(1)}$, where $\widetilde{F}_N^{(1)}$ is the reconstructed series. It is assumed that the statistical model for the residual series is being built and the corresponding parameters have been estimated through the data $\widetilde{F}_N^{(2)}$.

  The data used to construct the empirical confidence intervals has the form of the residual series as well, but now we have the multistart residual series defined in Section 2.4.1.

  Note that in order to build the empirical confidence interval for the forecast at time $N + Q$, we use the multistart Q-step residual series as the sample and calculate the corresponding intervals through its c.d.f., which is shifted by the value $\widetilde{f}_{N+Q-1}^{(1)}$ of the forecast.

  For fixed Q, the multistart Q-step residual series consists of $N - L - Q + 2$ terms. It is assumed that the corresponding empirical distribution is stable if N would increase and therefore this distribution can be used up to time $N + Q$.

  Evidently, the method is meaningful only if $N - L - Q + 2$ is sufficiently large (say, if it exceeds several dozen).

- *Checking the assumptions.* The simplest situation when the assumptions for both variants of the confidence intervals are valid is when the corresponding residual series are Gaussian white noises. This hypothesis can be checked by standard statistical methods.

  Nevertheless, confidence bounds make sense even if such a hypothesis is formally rejected. Then, at any rate, we would have a scale for the precision of the forecast and can compare different forecasting methods from the viewpoint of their stability.

- *Bootstrap average forecast.* Dealing with the bootstrap simulation, we can obtain additional information concerning forecasting. For instance, having obtained the forecast for the series $F_{N,i}$ ($1 \leq i \leq S$), we can take the average of the corresponding (random) forecasts and obtain the *bootstrap average forecast* of the series $F_N^{(1)}$. Therefore, we can use one more forecasting variant, which can be compared to other variants of forecasts.

2.6 Examples and effects

2.6.1 'Wages': Forecast of the exponential tendency

The series 'Wages' (annual wages, U.S., from 1900 to 1970, Hipel and McLeod, 1994) demonstrates a tendency that can be approximately considered as an exponential one. Therefore, it is natural to use a one-dimensional linear space $\mathcal{L}_1$ for its forecasting. Here we compare several ways of performing such a forecast for the period from 1971 to 1980.

Let us truncate the series at 1959 and forecast this truncated series up to 1980, that is for 21 years. The truncated series has length $N = 60$ and under the selection of the window length $L = 30$ we take the leading eigenvector U_1 as the basis of $\mathfrak{L}_1$. The w-correlation of the reconstructed series and the residuals is 0.025.

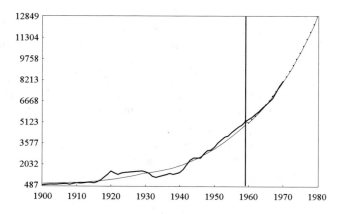

Figure 2.15 *Wages: truncation and forecast.*

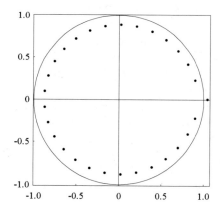

Figure 2.16 *Wages: polynomial roots.*

The result of the Basic SSA recurrent forecasting procedure is shown in Fig. 2.15. Here the thick line corresponds to the 'Wages' series, the vertical line shows the point of truncation, the thin line depicts the reconstructed series and the thin line with dots is the result of forecasting.

Since the first 11 forecasted values are close to real wages for the period from

EXAMPLES AND EFFECTS

1960 to 1970, it is natural to suppose that the subsequent 10 values of forecast would describe the future with good precision.

The linear recurrent formula applied in this forecasting has order $d = 29$. The roots of the corresponding characteristic polynomial are depicted in the complex-plane representation in Fig. 2.16.

Let us compare the result of the Basic SSA recurrent forecasting algorithm with the results of its various modifications.

Firstly, we apply the reduction method, described in Section 2.3.4, for finding the minimal recurrent forecasting formula. Evidently, the dimension of such a formula ought to be 1. Fig. 2.16 shows that the characteristic polynomial has single real root $\lambda_1 \approx 1.0474$, while all the other roots are complex ones with moduli significantly smaller than 1. If we would decide to have a purely exponential forecast, we can keep the root λ_1 and remove all the others since they are extraneous. Then the new (minimal) LRF will have the form $f_n = af_{n-1}, a = \lambda_1$.

Secondly, since we have obtained the one-dimensional space $\mathfrak{L}_1$, the nearest-subspace method (Section 2.3.4) for finding the minimal forecasting recurrent formula can be applied as well. For both minimal formulae the initial data for forecasting consists of the last term of the reconstructed series.

Table 2.1 *Wages: forecasting results.*

Method	Start Time	Forecast 1980	Accuracy
recurrent	1960	12849	0.021
	1967	13099	0.028
	1971	12958	—
MinLRF(1)	1960	13091	0.023
	1967	13074	0.027
	1971	12916	—
MinLRF(2)	1960	12074	0.018
	1967	13188	0.029
	1971	13015	—
vector	1960	12339	0.017
	1967	12694	0.015
	1971	12676	—

Lastly, we use the vector variant of Basic SSA forecasting (Section 2.3.1). Thus, we have four Basic SSA forecasting modifications. Moreover, for each modification we take three variants of the initial forecast point: 1960 (then we take $L = 30$), 1967 ($L = 33$) and 1971 ($L = 36$). The leading eigenvector serves as a basis of $\mathfrak{L}_1$ in all cases. For comparison, we perform forecasting up to 1980.

The forecasting results are gathered in Tables 2.1 and 2.2. The abbreviations 'recurrent' and 'vector' denote the recurrent and vector Basic SSA forecasting methods, respectively. Two other forecasting variants use minimal (one-dimensional) LRFs, achieved by the reduction of the extraneous polynomial roots: the first, 'MinLRF(1),' is obtained by the reduction of the LRF (2.1) of the Basic SSA forecasting procedure, and the second, 'MinLRF(2),' uses the main root of the polynomial defined by the nearest-subspace LRF (2.24).

Table 2.1 shows the forecasting results at 1980 and their accuracy relative to the 'Wages' data. The column 'Accuracy' contains the results of comparison of the forecasts with the initial series: if we consider the forecast results and 'Wages' data under comparison as vectors, then the accuracy is calculated as the distance between these vectors divided by the norm of the 'Wages' vector. Evidently, the comparison can be performed for 1960-1971 and 1967-1971, depending on the initial forecasting point.

Table 2.2 *Wages: coefficients of the minimal LRFs.*

Method	Start Time	Coefficient
MinLRF(1)	1960	1.0474
	1967	1.0473
	1971	1.0470
MinLRF(2)	1960	1.0434
	1967	1.0479
	1971	1.0478

Table 2.2 shows the (single) parameter a of the minimal recurrent formula $f_n = a f_{n-1}$ for different starting times and two methods.

Let us briefly discuss the forecasting results. Table 2.1 shows that the Basic SSA recurrent forecasting algorithm (as well as its reduction modifications) gives similar results for all starting times. Moreover, the accuracy is similar too. This confirms the validity of the choice $r = 1$ for the space $\mathfrak{L}_r$ and demonstrates the stability of forecasting.

The vector forecasting procedure gives more conservative results. Since the accuracy of vector forecasting is slightly better, the conservative forecast ought to be taken into consideration as well.

The nearest-subspace method seems to be less stable with respect to variation of the initial forecast point, though the forecasting results of the forecasts from 1967 and 1971 correspond to that of the Basic SSA recurrent algorithm.

Table 2.2 explains this effect. The minimal recurrent formula calculated via the reduction of the polynomial roots is almost the same for all initial forecast points, while the coefficient of the minimal LRF produced by the nearest subspace is far less stable.

2.6.2 'Eggs': Minimal LRF

The series 'Eggs' (monthly, from January 1938 to December 1940, see Section 1.3.3 for a detailed description) has an evident structure: it can be described as a sum of a slowly decreasing almost constant trend which can be approximated by an exponential sequence, an annual seasonal component, and a small noise. Therefore, we can express the 'Eggs' series F_N as $F_N = F_N^{(1)} + F_N^{(2)}$ with $N = 36$,

$$f_n^{(1)} = c_0 a_0^n + \sum_{k=1}^{5} c_k \cos(2\pi n k/12 + \phi_k) + c_6 \cos(\pi n), \qquad (2.25)$$

$a_0 \lesssim 1$, and a noise series $F_N^{(2)}$. If we would prefer, we can change each pure harmonic component $\cos(2\pi n k/12 + \phi_k)$ of (2.25) for the exponential-modulated harmonic $a_k^n \cos(2\pi n k/12 + \phi_k)$ with $a_k \approx 1$ ($k = 1, \ldots, 6$).

At any rate, if $c_k \neq 0$ for $k = 0, \ldots, 6$, the series $F_N^{(1)}$ is governed by an LRF of dimension 12, and, therefore, the minimal window length for proper forecasting is equal to $L = 13$. In view of the general concepts of Section 1.5, this window length can be expected to provide good separability of $F_N^{(1)}$ and $F_N^{(2)}$ and, consequently, a reasonable forecast.

Fig. 2.17 depicts two variants of such forecasts up to January 1942. (They are represented by the thin lines, with the thick one corresponding to the initial 'Eggs' series.) Both forecasts are performed with the help of the Basic SSA R-forecasting procedure, with $L = 13$ and the eigentriples 1-12, but have different initial forecast points.

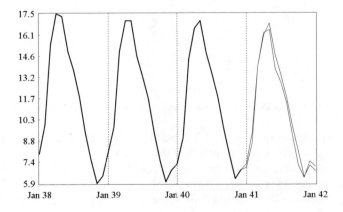

Figure 2.17 *Eggs: forecast by the minimal LRF.*

The first starts at January 1941 and produces 13 steps. The second is based on the first 25 points of the series only. It starts at February 1940 and performs 25 steps. Therefore, we deal with two LRFs of the (minimal) dimension 12 and two initial data sets.

Fig. 2.17 shows that the results of both forecasts during the period from January 1941 to January 1942 are very similar, though the first of them gives slightly larger values due to a small phase shift.

Therefore, for the 'Eggs' series, the minimal LRF produces stable and reasonable results.

2.6.3 'Precipitation': Toeplitz forecasting

The series 'Precipitation' (in *mm*, monthly, Eastport, U.S., from January 1896 to December 1950, Hipel and McLeod, 1994) has a slowly decreasing trend and an oscillatory component with a complex structure (Fig. 2.19, thin oscillating line). Its periodogram, which is depicted in Fig. 2.18 in the frequency scale with marked periods, shows that in addition to the usual annual harmonics and other high-frequency components, there are two periodogram peaks corresponding to approximately 62-months and 104-months periodicities.

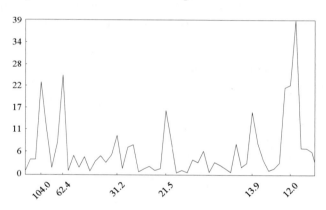

Figure 2.18 *Precipitation: main periodogram peaks and the associated periods.*

As was mentioned in Sections 1.7.2 and 2.3.2, for such data the Toeplitz SSA with centring can give better results than Basic SSA.

Let us truncate the series 'Precipitation' at December 1947 and perform the forecast of its trend for 84 steps (that is from January 1948 to December 1962), taking $L = 312$ for the Toeplitz SSA R-forecasting variant with centring (Section 2.3.3). To reconstruct the trend, we take the average triple and the third Toeplitz eigentriple.

Fig. 2.19 shows the results of forecasting as well as the empirical confidence intervals corresponding to the 2.5% upper and lower quantiles and relating to the entire series 'Precipitation'. The trend of the truncated series (as well as its

EXAMPLES AND EFFECTS

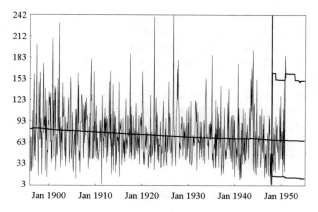

Figure 2.19 *Precipitation: trend forecast*

forecast) is represented by the thick line. The thick vertical line corresponds to the truncation point. Evidently, an almost linear behaviour of the reconstructed trend is continued into the future.

If we want to forecast not only the general tendency of the series, but also its low-frequency (5 and 8-9 years) oscillation components, we select the average triple and the eigentriples 3, 6-7, 10-11 for the reconstruction. (The two leading eigentriples correspond to the annual periodicity.) The resulting reconstruction and forecast can be found at Fig. 2.20.

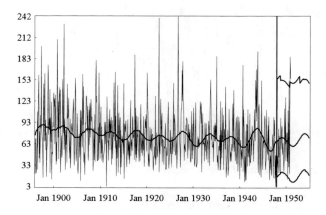

Figure 2.20 *Precipitation: low-frequency oscillations.*

Note that for both forecasting variants, the empirical confidence intervals are not symmetric with respect to the forecast values. The reason lies in the asymmetric behaviour of the initial series: large precipitation values look like 'outliers' on the background of the average series behaviour. Moreover, the hazard-like appearance of these 'outliers' produce abrupt changes in the upper confidence line. If we were to take 80% confidence intervals instead of 95% ones, these irregularities would be removed.

Comparison of the confidence intervals for the forecasting variants of Fig. 2.19 and Fig. 2.20 shows that the latter ones are slightly narrower. It is interesting that the Basic SSA R-forecasting produces very similar forecast values (and confidence intervals as well) for the trend and can hardly make a forecast of the low-frequency components.

2.6.4 'Fortified wine': Vector and recurrent forecasting

The series 'Fortified wine' (fortified wine sales, Australia, monthly, from January 1980 till June 1994) has the same origin as the 'Rosé wine' series, discussed in Example 1.1 of the Section 1.4.1. Therefore, it is not surprising that these two series have basically the same structure: in addition to a smooth trend, there exists an annual seasonality of a complex form, and a noise. The main difference is in the proportion of the amplitudes for different components in the 12-months periodicity.

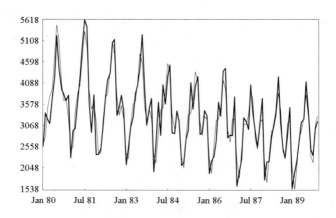

Figure 2.21 *Fortified wine: truncated series and its reconstruction.*

We use this data to illustrate the difference between the recurrent and vector SSA procedures for long-term forecasting. We truncate the series at December 1989 and perform forecasting from January 1990 to December 1994, that is for 60 steps.

Taking the window length $L = 60$, we select the eigentriples 1-7, which correspond to the trend and the main annual harmonics: the 12-months, 6-months and

4-months ones. The truncated series (thick line) and its reconstruction (thin solid line) can be seen in Fig. 2.21.

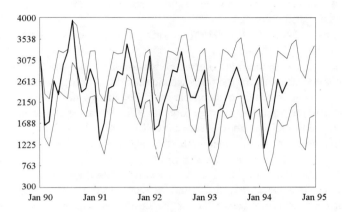

Figure 2.22 *Fortified wine vector forecasting: bootstrap confidence intervals.*

We can see that the approximation is rather good, despite the fact that we did not include high-frequency seasonal harmonics in the reconstruction. This kind of decision might be reasonable if we are interested only in the main seasonal effects. Note, however, that the residual series of such a reconstruction can hardly be regarded as a Gaussian white noise: it contains obvious high-frequency harmonics.

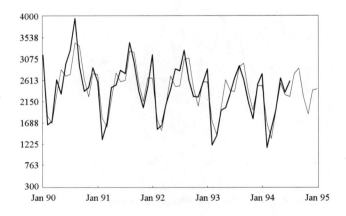

Figure 2.23 *Fortified wine vector forecasting: bootstrap average forecast.*

Nevertheless, while performing forecasting we will apply the Gaussian white noise model for construction of the bootstrap confidence intervals and the bootstrap average forecast. These intervals are not absolute, but still not meaningless. Additional reasons for applying the Gaussian white noise model is that the residual series does not have a visible trend and its empirical c.d.f. is close to the c.d.f. of the normal distribution.

The results of Basic SSA vector forecasting and the bootstrap simulation are depicted in Figs. 2.22-2.24.

The first figure demonstrates a good correspondence between the 95% bootstrap confidence intervals for the forecast and the 54 last points of the initial series (thick line). The next two figures compare the 'Fortified wine' series with its vector forecast and the bootstrap average forecast. (The latter depends not only on the data, but on the model for the residuals as well.) If the forecasting parameters are well chosen and the model is reasonable, then all three curves ought to be similar.

Figs. 2.23 and 2.24 demonstrate this similarity. Both figures represent the average of $S = 1000$ bootstrap forecasts (thin lines). The thick lines correspond to the 'Fortified wine' series (Fig. 2.23) and its vector forecast (Fig. 2.24).

If we apply recurrent forecasting for the same series in the same conditions, then the results of long term forecasting will be much worse. We illustrate these results with figures of the same kind as the previous ones.

Fig. 2.25 shows that the bootstrap confidence intervals for recurrent forecasting become rapidly increasing and almost meaningless for the last third of the forecasting period. Note that for the first year of the forecast, these intervals are similar to that of vector forecasting, although they are a little wider.

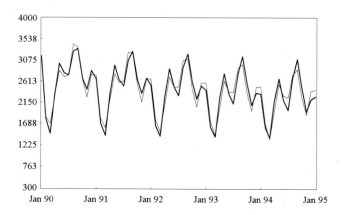

Figure 2.24 *Fortified wine vector forecasting: two forecasts.*

The other pair of figures elucidates this phenomenon. Fig. 2.26 shows that the bootstrap average forecast still corresponds to the 'Fortified wine' series. This correspondence, by the way, can be used as an argument that the Gaussian white noise model for the reconstruction residuals is reasonable.

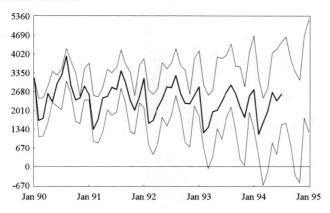

Figure 2.25 *Fortified wine recurrent forecasting: bootstrap confidence intervals.*

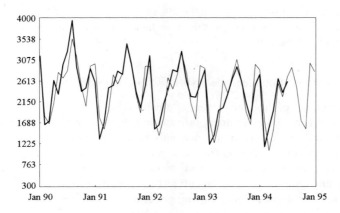

Figure 2.26 *Fortified wine recurrent forecasting: bootstrap average forecast.*

The discrepancies appear when we compare the recurrent forecast with the bootstrap average one (Fig. 2.27). We see that the amplitude of oscillations of the recurrent forecast (the thick line) increases, while the bootstrap average and the initial 'Fortified wine' series have an approximately constant (or even decreasing) amplitude of oscillations (the thin line in Fig. 2.27). The polynomial roots give the explanation: there exists a pair of complex conjugate roots with moduli significantly larger than 1 (they are equal to 1.018). They correspond to the 6-months periodicity and the latter can be explicitly seen on the plot of the recurrent forecast in Fig. 2.27.

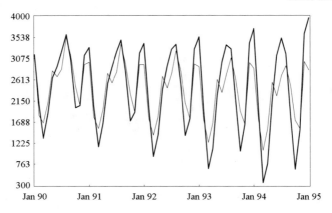

Figure 2.27 *Fortified wine recurrent forecasting: two forecasts.*

Thus, vector forecasting gives reasonable long-term results, while the recurrent one appears to be less stable. On the other hand, the bootstrap average forecast can correct this instability.

2.6.5 'Gold price': Confidence intervals and forecast stability

Previous examples of time series were relatively simple for SSA forecasting since we had in mind enough preliminary information about their behaviour. Moreover, this information was in correspondence with the SSA requirements: we were able to give a reasonable prediction for the number of eigentriples that describe the components under forecasting, as well as the form of the related eigenvectors. Therefore, unpredictable forecasting effects were hardly possible there.

The series 'Gold price' (gold closing price, daily, 97 successive trading days, Hipel and McLeod, 1994), depicted in Fig. 2.28 by the thin line, has a more complex structure. The main difficulties appear if we pay attention to the last points of the series. To perform the forecast, we ought to decide whether the abrupt decrease of 'Gold price' can be regarded as a 'random' effect or as important information for forecasting. Depending on this decision, we select different eigentriples for the reconstruction (and the forecast).

Figs. 2.28 and 2.29 demonstrate the effect of such a decision on the reconstruction (thick lines). Using Basic SSA with $L = 48$ we see that the 6th eigentriple is responsible for the last points of the series.

Indeed, the five leading eigentriples describe the series in a proper way, with the exception of its end (Fig. 2.28). On the other hand, the reconstruction via the six leading eigentriples demonstrates (see Fig. 2.29, thick line) a good approximation of the entire series.

Our aim is to study the recurrent forecasts performed by these variants of the

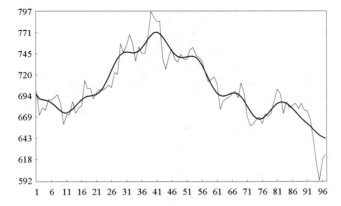

Figure 2.28 *Gold price: initial series and the rough reconstruction.*

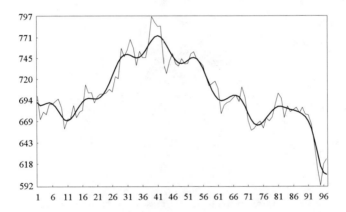

Figure 2.29 *Gold price: initial series and an accurate reconstruction.*

eigentriple selection with the help of the bootstrap confidence intervals under the Gaussian white noise model for the residuals. Yet we start with the *bootstrap confidence bounds for reconstruction*, which are achieved in the same manner and under the same assumptions as the forecasting ones. In the simulation we use the sample size $S = 1000$ and construct the lower and upper 2.5% quantiles for the confidence intervals.

Figs. 2.30 and 2.31 show the confidence intervals for the reconstruction performed by the leading 5 (Fig. 2.30) and 6 (Fig. 2.31) eigentriples, respectively.

In both figures, three thick lines intersect the plot of the 'Gold price' series. The middle ones are the averages of the bootstrap reconstructions (*bootstrap average reconstruction*); they are quite similar to the reconstruction lines of Figs. 2.28 and

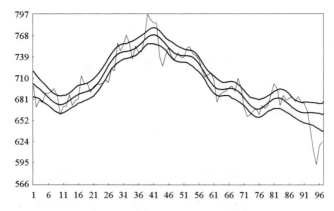

Figure 2.30 *Gold price: confidence intervals for the rough reconstruction.*

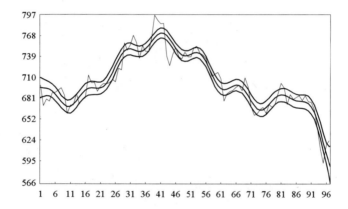

Figure 2.31 *Gold price: confidence intervals for the accurate reconstruction.*

2.29. The other two pairs of thick lines depict the bootstrap confidence intervals. Though the share of the 6th eigentriple is only 0.46%, the confidence intervals of Fig. 2.31 are significantly less than that of Fig. 2.30. Since the initial data for the forecasting LRF corresponds to the last series points, it is important that the confidence intervals become wider towards both edges of the series.

Let us turn to the results of Basic SSA recurrent forecasting depicted in Figs. 2.32 and 2.33. Both figures have the same form; they describe the last 10 points of the 'Gold price' series (thick lines) and 10 forecast points. Four lines intersect the 'Gold price' plots. All of them have a similar sense for the first and last 10 points.

EXAMPLES AND EFFECTS

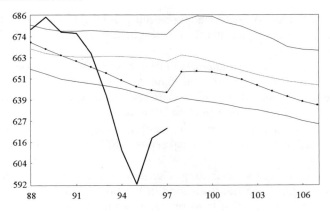

Figure 2.32 *Gold price: forecast based on the five leading eigentriples.*

The thin lines, marked by dots, describe the reconstructed series (up to point 97) and the Basic recurrent forecasts (points 98-107). The other lines demonstrate the results of the bootstrap simulation: the thin middle lines depict the bootstrap average reconstruction (up to the time point 97) and the forecasts; two extreme pairs of the thin solid lines indicate the corresponding confidence intervals.

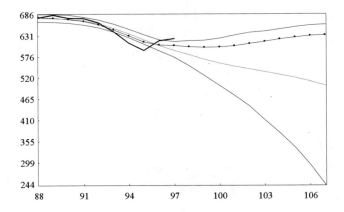

Figure 2.33 *Gold price: forecast based on the six leading eigentriples.*

Figs. 2.32 and 2.33 show that the choice of 5 or 6 leading eigentriples produce very different forecasts.

The forecast of Fig. 2.32 seems to be rather stable since both forecast lines are close to each other and the forecast confidence intervals are not much wider than the reconstructed ones.

In other words, the choice of 5 leading eigentriples leads to stable results which, however, do not take into account the atypical behaviour of the last 'Gold price' points.

Fig. 2.33 demonstrates that attempts to take these points into account for forecasting fail: the confidence intervals become extremely wide and therefore meaningless. The forecast lines differ much. Thus, if we include the 6th eigentriple into the list for forecasting, we come to great instability.

In terms of the bootstrap simulation, the 6th eigentriple is very unstable under random perturbations of the residual series, while the first 5 eigentriples are relatively stable.

The abrupt change of the forecasting behaviour caused by only one weak eigentriple has its origin in the transformation of the linear space $\mathfrak{L}_r$ governing the forecast (see Section 2.1). This transformation can hardly be expressed in terms of the leading characteristic polynomial roots; indeed, the two roots with largest moduli are similar in both cases – they both are real and approximately equal to 0.998. The other roots are significantly smaller.

Even so, we can indicate the characteristic that captures the transformation of the linear space $\mathfrak{L}_r$. This characteristic is related to the verticality of the space. Fig. 2.34 depicts the *verticality function* produced by the SVD of the trajectory matrix of the series 'Gold price'.

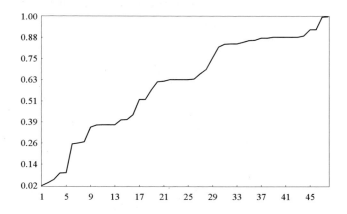

Figure 2.34 *Gold price: verticality function.*

If we set $\mathfrak{L}_i = \text{span}\,(U_1,\ldots,U_i)$ with U_i standing for the ith eigenvector of the SVD, then the verticality function $v(i)$ ($1 \leq i \leq L$) is equal to the squared cosine of the angle between $\mathfrak{L}_i$ and $e_L = (0,\ldots,1)^{\text{T}}$. The verticality coefficient ν^2 (see Section 2.1) can be easily calculated in terms of the verticality function.

The verticality function of Fig. 2.34 shows several relatively big jumps, which indicate the transformation of the linear spaces $\mathfrak{L}_i$ after adding a single eigenvector to its basis. The first jump corresponds to the transition from $\mathfrak{L}_5$ to $\mathfrak{L}_6$, that is to the difference between Figs. 2.32 and 2.33.

Of course, the jump in the verticality function does not necessarily imply abrupt changes in the forecast behaviour. Yet it signals the possibility of such a change.

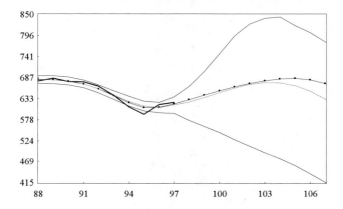

Figure 2.35 *Gold price: forecast based on the nine leading eigentriples.*

The second significant jump of the 'Gold price' verticality function is between the 8th and the 9th eigentriples. It can be checked that if we take the leading seven or eight eigentriples for the reconstruction, then the forecast behaviour will be very similar to that of Fig. 2.33. The adjunction of the 9th eigentriple changes the situation a lot.

The forecasting result is represented in Fig. 2.35 in the manner of Figs. 2.32 and 2.33. It is also unstable and differs from the previous ones.

Note that here the polynomial roots are also meaningful. The two largest root moduli (1.019 and 1.001) correspond to two pairs of complex conjugate roots with frequencies $\omega \approx 1/17$ and $\omega \approx 2/15$. The third root is real and is slightly greater than 0.998.

Evidently, the emergence of new polynomial roots with large moduli may significantly affect even a short-term forecast.

CHAPTER 3

SSA detection of structural changes

3.1 Main definitions and concepts

Let us start with a description of the main model of the time series, whose structural changes we are attempting to detect.

We shall call a time series F_N *homogeneous* if it is governed by some linear recurrent formula (LRF) whose dimension is small relative to N. (See Chapter 2 concerning continuation of such series and series close to them and Chapter 5 concerning the related theory.)

Assume that due to an outside action (or for another reason) a homogeneous time series is being exposed to an instant perturbation; that is, it stops following the original LRF. However, after a certain time period it again becomes governed by an LRF which may be different from the original one. Even if the new LRF coincides with the original one, the behaviour of the time series after the perturbation is generally different from the behaviour of the unperturbed series: this behaviour is determined by the new (perturbed) initial conditions for the LRF.

Thus, we assume that the time series is being exposed to an instant (local) perturbation which sends it from one homogeneous state to another within a relatively short transition time. As a result, the series as a whole stops being homogeneous and the problem of studying this heterogeneity arises.

We shall mostly study the problem of the posterior detection of the heterogeneities without restricting ourselves to the detection of the transition (perturbed) interval of the series, which is the *change-point detection problem*. In addition to this standard detection problem, we shall be interested in comparison of the homogeneous parts of the series before and after the change. The case of two or more local perturbations will be considered as well.

The main idea of the method for solving the above problem of *detection of structural changes* can be described as follows. In view of Section 5.2, the time series F_N governed by the LRF is characterized by the fact that for sufficiently large values of the window length L (this value must be larger than the dimension of the LRF) the L-lagged vectors of this series span the same linear space $\mathfrak{L}^{(L)}$ independently of N (as soon as N is sufficiently large). Moreover, if the LRF is minimal, then the space $\mathfrak{L}^{(L)}$ uniquely defines the LRF and vice versa.

Thus, the violations in the homogeneity of the series can be described in terms of the corresponding lagged vectors: the perturbations force the lagged vectors to leave the space $\mathfrak{L}^{(L)}$. The corresponding discrepancies are defined in terms of the distances between the lagged vectors and the space $\mathfrak{L}^{(L)}$, which can be determined for different parts of the series (for example, before and after the perturbation).

Of course, such an ideal situation is possible only in artificial examples. In practical cases the series are described by LRFs only approximately, and the problem of approximate construction of the spaces $\mathfrak{L}^{(L)}$ becomes important. Here, as in the problems of SSA forecasting (see Chapter 2), the SVD of the trajectory matrices does help. However, unlike the forecasting problems, for studying structural changes in time series, the properties of the SVD of subseries of the initial series are of considerable importance.

As in Chapters 1 and 2, the study in the present chapter is made at a qualitative level; although in some simple cases (which are, as a rule, artificial examples) some quantitative procedures can be suggested.

3.1.1 Heterogeneity matrix and heterogeneity functions

(a) Heterogeneity matrix

Consider two time series $F^{(1)} = F_{N_1}^{(1)}$ and $F^{(2)} = F_{N_2}^{(2)}$ and take an integer L with $2 \leq L \leq \min(N_1 - 1, N_2)$. Denote by $\mathfrak{L}^{(L,1)}$ the linear space spanned by the L-lagged vectors of the series $F^{(1)}$.

Let $U_l^{(1)}$ ($l = 1, \ldots, L$) be the eigenvectors of the SVD of the trajectory matrix of the series $F^{(1)}$. For $l > d \stackrel{\text{def}}{=} \dim \mathfrak{L}^{(L,1)}$, we take vectors from any orthonormal basis of the space orthogonal to $\mathfrak{L}^{(L,1)}$ as the eigenvectors $U_l^{(1)}$.

Let $I = \{i_1, \ldots, i_r\}$ be a subset of $\{1, \ldots, L\}$ and $\mathfrak{L}_r^{(1)} \stackrel{\text{def}}{=} \text{span}(U_l^{(1)}, l \in I)$. Denote by $X_1^{(2)}, \ldots, X_{K_2}^{(2)}$ ($K_2 = N_2 - L + 1$) the L-lagged vectors of the time series $F^{(2)}$.

We introduce a measure called the *heterogeneity index*, which characterizes the discrepancy between the series $F^{(2)}$ and the structure of the series $F^{(1)}$ (described by the subspace $\mathfrak{L}_r^{(1)}$):

$$g(F^{(1)}; F^{(2)}) = \frac{\sum_{l=1}^{K_2} \text{dist}^2 \left(X_l^{(2)}, \mathfrak{L}_r^{(1)} \right)}{\sum_{l=1}^{K_2} \|X_l^{(2)}\|^2}, \qquad (3.1)$$

where $\text{dist}(X, \mathfrak{L})$ is the Euclidean distance between the vector $X \in \mathbf{R}^L$ and the linear space $\mathfrak{L} \subset \mathbf{R}^L$. The heterogeneity index g is the relative error of the optimal approximation of the L-lagged vectors of the time series $F^{(2)}$ by vectors from the space $\mathfrak{L}_r^{(1)}$.

The values of g belong to the interval $[0, 1]$. If all the L-lagged vectors of the series $F^{(2)}$ lie in the subspace $\mathfrak{L}_r^{(1)}$, then $g(F^{(1)}; F^{(2)}) = 0$. Alternatively, if all the L-lagged vectors of $F^{(2)}$ are orthogonal to the subspace $\mathfrak{L}_r^{(1)}$, then $g(F^{(1)}; F^{(2)}) = 1$.

MAIN DEFINITIONS AND CONCEPTS

The heterogeneity index g can be interpreted as a normalized squared error of the multiple regression of vectors $X_1^{(2)}, \ldots, X_{K_2}^{(2)}$ relative to the vectors $U_l^{(1)}, l \in I$.

We now define the *heterogeneity matrix* (**H**-*matrix*) of a time series F_N. The elements of this matrix are values of the heterogeneity index g for different pairs of subseries of the series F_N.

To make the definition we introduce the following objects:

(a) the initial time series F_N: $F_N = (f_0, \ldots, f_{N-1})$, $N > 2$;
(b) the subseries (intervals) $F_{i,j}$ of the time series F_N: $F_{i,j} = (f_{i-1}, \ldots, f_{j-1})$ for $1 \leq i < j \leq N$;
(c) the window length L: $1 < L < N$;
(d) the length B of the base subseries of the series F_N: $B > L$;
(e) the length T of the test subseries of the series F_N: $T \geq L$;
(f) the collection I of different positive integers: $I = \{j_1, \ldots, j_r\}$; we assume that I is such that $j < \min(L, B - L + 1)$ for each $j \in I$;
(g) the *base spaces* ($i = 1, \ldots, N - B + 1$) are spanned by the eigenvectors with the indices in I, obtained by the SVD of the trajectory matrices $\mathbf{X}^{(i,B)}$ of the series $F_{i,i+B-1}$ with window length L. The corresponding set of eigentriples is called the *base set of eigentriples*.

In these terms, the elements g_{ij} of the heterogeneity matrix $\mathbf{G} = \mathbf{G}_{B,T}$ are

$$g_{ij} = g(F_{i,i+B-1}; F_{j,j+T-1}), \tag{3.2}$$

where $i = 1, \ldots, N - B + 1$ and $j = 1, \ldots, N - T + 1$. Thus, in (3.2) the space $\mathfrak{L}_{I,B}^{(L,i)}$ plays the role of $\mathfrak{L}_r^{(1)}$. The series $F_{i,i+B-1}$ are called the *base subseries* (or *base intervals*) of the series F_N, while $F_{j,j+T-1}$ are *test subseries (intervals)*.

By definition, the quantity g_{ij} is the normalized sum of distances between the L-lagged vectors of the series $F_{j,j+T-1}$ and the linear space $\mathfrak{L}_{I,B}^{(L,i)}$. Note that the matrix $\mathbf{G}$ is generally not symmetric even for $T = B$, since the base and the test subseries of the series play different roles in the construction of the heterogeneity index g_{ij}.

If the base spaces $\mathfrak{L}_{I,B}^{(L,i)}$ do not depend on i for a certain range of i, then for any T all ith rows of the matrix $\mathbf{G}$ are equal to each other for the same range of i. In turn, if the set of L-lagged vectors of the series $F_{j,j+T-1}$ does not depend on j for $j_1 \leq j \leq j_2$, then for any B all jth columns of the matrix $\mathbf{G}$ are equal for $j_1 \leq j \leq j_2$.

(b) Heterogeneity functions

On the basis of the heterogeneity matrix $\mathbf{G}$ let us introduce various *heterogeneity functions*.

1. *Row heterogeneity functions*

For fixed $i \in [1, N-B+1]$ the row heterogeneity function is a series $H_{N-T+1}^{(\mathrm{r},i)}$, so that its general term is

$$h_{n-1}^{(r,i)} \stackrel{\text{def}}{=} g_{in} = g(F_{i,i+B-1}; F_{n,n+T-1}), \quad n = 1, \ldots, N - T + 1.$$

Thus, the series $H_{N-T+1}^{(r,i)}$ corresponds to the ith row of the matrix $\mathbf{G}$.

The row heterogeneity function $H_{N-T+1}^{(r,i)}$ reflects the homogeneity of the series F_N (more precisely, of its test subseries $F_{n,n+T-1}$) relative to the fixed base subseries $F_{i,i+B-1}$ (more precisely, relative to the base space $\mathfrak{L}_{I,B}^{(L,i)}$). This is due to the fact that in the construction of the row heterogeneity function the base subseries of the series F_N is fixed, but the test subseries varies.

2. *Column heterogeneity functions*
In addition to the row heterogeneity function we can consider the column heterogeneity functions, which correspond to the columns of the H-matrix. Formally, for fixed $j \in [1, \ldots, N - T + 1]$ the column heterogeneity function $H_{N-B+1}^{(r,j)}$ of length $N - B + 1$ is defined as the time series with general term

$$h_{n-1}^{(c,j)} \stackrel{\text{def}}{=} g_{nj} = g(F_{n,n+B-1}; F_{j,j+T-1}).$$

In this case, the test subseries of the series is fixed, but the base subseries varies.

The column heterogeneity function also reflects the homogeneity of the series F_N (more precisely, the base spaces $\mathfrak{L}_{I,B}^{(L,n)}$) relative to its fixed test subseries $F_{j,j+T-1}$ (more precisely, relative to the L-lagged vectors of this subseries).

When using the column heterogeneity functions in the interpretation of the results we have to bear in mind that in comparison to the row heterogeneity functions, the base and the test subseries of F_N are being interchanged.

3. *Diagonal heterogeneity functions*
The diagonal heterogeneity function is a time series $H_{N-T-\delta+1}^{(d,\delta)}$ with parameter $0 \leq \delta \leq N - T$, such that

$$h_{n-1}^{(d,\delta)} \stackrel{\text{def}}{=} g_{n,n+\delta} = g(F_{n,n+B-1}; F_{n+\delta,n+\delta+T-1})$$

for $n = 1, \ldots, N - T + \delta + 1$. Thus, the series $H_{N-T-\delta+1}^{(d,\delta)}$ corresponds to the 'diagonal' $j = i + \delta$ of the matrix $\mathbf{G}$.

The series $H_{N-T-\delta+1}^{(d,\delta)}$ reflects the local heterogeneity of the series, since both the base and the test subseries of the series F_N vary at the same time. For $T + \delta > B$ the test intervals are ahead of the base intervals. In particular, for $\delta = B$ the test intervals immediately precede the base intervals.

4. *Symmetric heterogeneity function*
When $\delta = 0$ and $T = B$, the base subseries of the series coincides with the test subseries. The heterogeneity matrix $\mathbf{G}$ becomes a square matrix, and the series $H_{N-B+1}^{(s)} \stackrel{\text{def}}{=} H_{N-B+1}^{(d,0)}$ corresponds to its principal diagonal. The general term

$$h_{n-1}^{(s)} \stackrel{\text{def}}{=} g_{n-1}^{(d,0)} = g(F_{n,n+B-1}; F_{n,n+B-1})$$

MAIN DEFINITIONS AND CONCEPTS

of the series $H^{(s)}_{N-B+1}$ is equal to the eigenvalue share:

$$h^{(s)}_{n-1} = 1 - \sum_{l \in I} \lambda^{(n)}_l \bigg/ \sum_{l} \lambda^{(n)}_l, \qquad (3.3)$$

where $\lambda^{(n)}_l$ are the eigenvalues of the SVD of the trajectory matrix of the series $F_{n,n+B-1}$ with window length L. We shall call the series $H^{(s)}_{N-B+1}$ the symmetric heterogeneity function.

The equality (3.3) implies that for $B = T$ the principal diagonal of the heterogeneity matrix $\mathbf{G}$ (namely, the series $H^{(s)}_{N-B+1}$) characterizes the quality of the local description of the series F_N by the eigentriples with indices in I.

Remark 3.1 A single heterogeneity interval in the series F_N can generate one or more radical changes in the behaviour of the heterogeneity functions.

Consider, for instance, the row heterogeneity function (that is, assume that the test interval varies, but the base interval is unchanged). Then an abrupt change in the values of the row heterogeneity function may happen in two regions: when the test interval 'enters' the heterogeneity interval of the original series and when it 'departs' from this heterogeneity interval.

For the column and symmetric heterogeneity functions, a single local heterogeneity in the original series also generates two potential jumps. In the same situation the diagonal heterogeneity functions may have three (for $\delta = B$) and even four jumps (for $\delta \neq B$).

3.1.2 Detection functions

As was already mentioned, posterior change detection problems can be regarded as specific cases of the problem of studying heterogeneities in time series. As a rule, change detection problems are problems of testing homogeneity of the structure of the series with respect to the structure of the initial part of the series ('forward' change). Sometimes the problems of testing changes with respect to the structure of the terminal part of the series are also of interest ('backward' changes). In some cases estimation of the change-point, which is the time of the violation of the homogeneity, is also important.

We shall mostly consider the 'forward' change detection problem, although the 'backward' change detection is of significant importance in the forecasting problems, in finding the homogeneous parts of the original series that can be used for forecasting. The 'forward' change detection problem can easily be transformed into the 'backward' problem by inverting the time, that is, by considering the series $f'_i \stackrel{\text{def}}{=} f_{N-i-1}$.

(a) Structural changes and heterogeneity functions

The specifics of the change detection problem raises the question of the correspondence between the heterogeneity functions and the original series F_N. Let us discuss different cases of this correspondence for the problem of 'forward' change detection.

If we consider the row heterogeneity functions, then the 'forward' change assumes that the test interval is ahead of the base one, and therefore we can formally deal with the *starting base subseries* $F_{1,B}$ which corresponds to the first point of the initial series.

As was already mentioned, a single change in the series F_N must generally correspond to one or more radical changes in the values of heterogeneity functions; for the row heterogeneity function there may be at most two such changes.

Assume that we are interested in the first (after the starting base interval) change in the series. This change corresponds to the 'entrance' of the test interval into the region of heterogeneity, and therefore it is natural to index the values of the row heterogeneity function by the last point of the test interval rather than by the first point of this interval (as is made in the homogeneity matrix). Thus, we obtain the new indexation: the ith term of the row heterogeneity function is indexed by $T - 1 + i$.

In the case when we are trying to find several changes in the original series, it is worthwhile keeping both indexation systems for the terms of the row heterogeneity function (the standard and the new one). This is in order to prevent assigning two structural changes to the original series while observing two radical changes in the values of the row heterogeneity function that are caused by 'passing' the region of heterogeneity by the test interval.

Of course, all the above also relates to the column heterogeneity functions (and to the symmetric heterogeneity functions, where the base and test intervals coincide).

The general case of the diagonal heterogeneity function requires a larger number of indexation systems, since one local heterogeneity can give rise to three or four jumps in the values of this function.

In what follows, we shall not use the entire variety of heterogeneity functions and their indexation systems. In addition to the full information which is contained in the heterogeneity matrix, we shall use only some particular cases of the heterogeneity functions in the form of the so-called detection functions.

(b) Types of detection function

The detection functions considered below will differ from the heterogeneity functions in several aspects.

First, since we are interested only in the 'forward' changes, we shall use only the series $F_{1,B}$ as the base part of the series for both the row and column heterogeneity functions.

Second, for the diagonal (but not symmetric) heterogeneity functions we shall always assume that $\delta = B$. This means that there is no gap between the base and test intervals. We thus always compare neighbouring parts of the time series.

MAIN DEFINITIONS AND CONCEPTS

Finally, the detection and heterogeneity functions have different domains. For example, the indexation of the terms of the heterogeneity functions always starts at 0, but the terms of the row detection function are in correspondence with the end-points of the test intervals. (For the column detection function the indexation corresponds to the last points of the base intervals.) That means that we are mostly interested in the first 'forward' change in the series.

We introduce formally the detection functions, which are related to the heterogeneity functions of the previous section.

1. *Row detection function*

   The row detection function is the series $D_{T,N}^{(r)}$ with terms

   $$d_{n-1}^{(r)} \stackrel{\text{def}}{=} h_{n-T}^{(r,1)} = g(F_{1,B}; F_{n-T+1,n}),\qquad(3.4)$$

   $T \leq n \leq N$. This corresponds to the detection of the change with respect to the initial part of the series (more precisely, with respect to its first B terms, which are represented by the space $\mathfrak{L}_{I,B}^{(L,1)}$).

2. *Column detection function*

   The column detection function is the series $D_{B,N}^{(c)}$ with terms

   $$d_{n-1}^{(c)} \stackrel{\text{def}}{=} h_{n-B}^{(1,c)} = g(F_{n-B+1,n}; F_{1,T}),\qquad(3.5)$$

   $B \leq n \leq N$.

3. *Diagonal detection function*

   This is the series $D_{T+B,N}^{(d)}$ with terms

   $$d_{n-1}^{(d)} \stackrel{\text{def}}{=} h_{n-T-B}^{(d,B)} = g(F_{n-T-B+1,n-T+1}; F_{n-T+1,n}),\qquad(3.6)$$

   $T + B \leq n \leq N$. Since there is no gap between the base and test intervals, this detection function can be used for detection of abrupt structural changes against the background of slow structural changes.

4. *Symmetric detection function*

   Let $T = B$. Then the terms of the series $D_{B,N}^{(s)}$, which is called the symmetric detection function, are defined by

   $$d_{n-1}^{(s)} \stackrel{\text{def}}{=} h_{n-B}^{(s)} = g(F_{n-B+1,n}; F_{n-B+1,n}),\qquad(3.7)$$

   $B \leq n \leq N$. This detection function measures the quality of approximation of the base series by the chosen eigentriples.

The relation between the above heterogeneity/detection functions and the heterogeneity matrix in the case $B = T$ is shown in Fig. 3.1. The rows in this figure are numbered from bottom to top, and the columns are numbered in the standard manner, from left to right. Thus, the bottom ith row corresponds to the base subseries

$$F_{i,i+B-1} = (f_{i-1}, \ldots, f_{i+B-2})$$

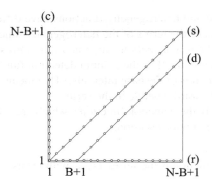

Figure 3.1 *Heterogeneity matrix and four detection functions.*

of the series F_N, and the elements of this row constitute the row heterogeneity function $H_{N-T+1}^{(r,i)}$. Analogously, the jth column from the right of the heterogeneity matrix corresponds to the test subseries

$$F_{j,j+T-1} = (f_{j-1}, \ldots, f_{j+T-2}),$$

and the elements of this column constitute the column heterogeneity function $H_{N-B+1}^{(c,j)}$.

The detection functions are depicted in Fig. 3.1. The row and column detection functions correspond to the first (from below) row and the first column of the matrix. The symmetric detection function is depicted as the principal diagonal of the matrix, while the diagonal function is parallel to the symmetric function, but starts at the point $B+1$ of the first row of a matrix. The indexation of terms of the detection functions is not shown in the figure.

3.2 Homogeneity and heterogeneity

Let us come back to the main model of perturbations of homogeneous (that is, governed by the LRF) time series.

Let F_N be a homogeneous series governed by the minimal LRF of dimension d (in terms of Chapter 5, $\mathrm{fdim}(F_N) = d$). Let us choose integers L and r so that $L \geq d$ and $d \leq r \leq \min(L, N-L+1)$.

If we choose $I = \{1, 2, \ldots, r\}$, then the heterogeneity matrix (3.1) is the zero matrix. Indeed, since $B \geq L$, then for any i we have $\mathfrak{L}^{(L)}(F_{i,i+B-1}) = \mathfrak{L}^{(L)}(F_N)$, and therefore all the L-lagged vectors of the series $F_{j,j+T-1}$ belong to the space $\mathfrak{L}^{(L)}(F_{i,i+B-1})$ for all i,j. This implies that any homogeneous series F_N gives rise to a zero heterogeneity matrix, and the presence of nonzero elements g_{ij} in this matrix is an indication of a violation of homogeneity.

We consider several types of violation of homogeneity and the corresponding heterogeneity matrices.

3.2.1 Types of single heterogeneity

In accordance with the main model of a single local perturbation of homogeneity, the time series F_N is governed by an LRF until a certain time Q. Then an instant perturbation takes place, although in a short time the series again becomes homogeneous and controlled by an LRF. This LRF may differ from the initial one.

If the same LRF is restored, then we have a *temporary* violation of the time series structure; otherwise (when the new LRF is different from the original one) the violation of the structure is *permanent*.

We denote by Q the maximal moment of time such that the series $F_{1,Q-1}$ is homogeneous. This moment of time Q will be called the moment of perturbation or *change-point*. Let $d = \text{fdim}(F_{1,Q-1})$.

Assume that some time $S \geq 0$ after the perturbation, the time series becomes homogeneous again, which means that the series $F_{Q+S,N}$ is homogeneous. We set $d_1 = \text{fdim}(F_{Q+S,N})$. The time interval $[Q, Q+S]$ is called the *transition interval*. The behaviour of the series within the transition interval is of no interest to us.

Let $L \geq \max(d, d_1)$. Assume in addition that $L \leq Q-1$ and $L \leq N-Q-S+1$. If the L-lagged vectors of the series F_N span the original subspace $\mathfrak{L}^{(L)}(F_{1,Q-1})$ after they left the transition interval $[Q, Q+S]$ (that is, if $\mathfrak{L}^{(L)}(F_{1,Q-1}) = \mathfrak{L}^{(L)}(F_{Q+S,N})$), then both homogeneous parts of the time series are governed by the same minimal LRF. This is the case of a temporary heterogeneity. By contrast, if $\mathfrak{L}^{(L)}(F_{1,Q-1}) \neq \mathfrak{L}^{(L)}(F_{Q+S,N})$, then the minimal LRFs that govern the two parts of the series are different and the heterogeneity is permanent.

For instance, a change in the period in one of the harmonic components of the series and a change in the number of harmonic components mean permanent heterogeneity. Alternatively, a change in the phase of one of the harmonic components, a change in a constant component of the series, and change in the slope of a linear additive component of the series mean temporary heterogeneity.

Let us now describe the general form of the heterogeneity matrix (H-matrix) of a locally perturbed homogeneous series. We shall assume that the lengths of the base and test intervals satisfy the condition $\max(B, T) < Q$.

Analogous to the case of a homogeneous series, consider $I = \{1, 2, \ldots, r\}$ and assume that $r = d \leq \min(L, B - L + 1)$. Then all the elements g_{ij} of the heterogeneity matrix $\mathbf{G}_{B,T}$ are zero for $i + B \leq Q$ and $j + T \leq Q$. This is due to the fact that for $i + B \leq Q$ and $j + T \leq Q$, both the base and test subseries of the series F_N are also subseries of the homogeneous series $F_{1,Q-1}$. The values of the other elements of the H-matrix depend on the type of the heterogeneity and the values of parameters.

Schematically, the general form of the H-matrix is depicted in Fig. 3.2. The region $\mathcal{A}$ corresponds to the elements g_{ij} of the H-matrix where the series $F_{i,i+B-1}$

and $F_{j,j+T-1}$ are subseries of the homogeneous series $F_{1,Q-1}$. Thus, in the region $\mathcal{A}$, we have $g_{ij} = 0$. In the region $\mathcal{D}$, both series $F_{i,i+B-1}$ and $F_{j,j+T-1}$ are intervals of the series $F_{Q+S,N}$. Thus, if the dimension d_1 of the series $F_{Q+S,N}$ is not larger than the dimension d of the series $F_{1,Q-1}$, then the region $\mathcal{D}$ is also the zero region.

In case of temporary heterogeneity, all four regions $\mathcal{A}$, $\mathcal{B}$, $\mathcal{C}$ and $\mathcal{D}$ are zero regions.

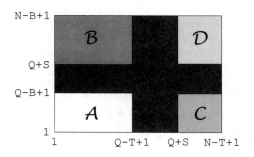

Figure 3.2 *General form of the H-matrix.*

'The heterogeneity cross', that is the region of the elements g_{ij} of the H-matrix with indices (i, j) such that

$$Q - B + 1 \leq i \leq Q + S - 1, \quad Q - T + 1 \leq j \leq Q + S - 1,$$

is also an essential part of this matrix (it is coloured dark in Fig. 3.2). The heterogeneity cross corresponds to those (i, j) where either the base or the test interval has a nonempty intersection with the transition interval. The width of the vertical strip of the cross is equal to $T + S - 1$, and the height of its horizontal strip is $B + S - 1$.

Let us give examples of series and their H-matrices for different types of heterogeneity. Depicting the matrices we shall use the black-and-white scale, where the smaller the value the whiter the colour. White corresponds to zero.

In the majority of examples, the transition interval consists of one point only (that is, $S = 0$). The case $S > 0$ differs only in the width of the heterogeneity cross. An instructive example with $S = 1$ is the case of a homogeneous series with an outlier (Example 3.7).

(a) Permanent violation ('tile-structure' matrices)

As discussed above, permanent violation is characterized by the fact that the two minimal LRFs, which govern the series $F_{1,Q-1}$ and $F_{Q+S,N}$, do not coincide. Since the dimension of the LRF reflects the complexity of the related series, the

HOMOGENEITY AND HETEROGENEITY

main classification will be made in terms of the correspondence between the dimensions $d = \text{fdim}(F_{1,Q-1})$ and $d_1 = \text{fdim}(F_{Q+S,N})$. Moreover, the relation between the spaces spanned by the L-lagged vectors of these subseries of the series F_N will also be taken into account.

For simplicity, we take $S = 0$ in all the examples relating to permanent heterogeneity.

Example 3.1 *Conservation of dimension*
Assume that $\text{fdim}(F_{1,Q-1}) = \text{fdim}(F_{Q,N})$. For instance, let the series F_N have the form

$$f_n = \begin{cases} C_1 \sin(2\pi\omega_1 n + \phi_1) & \text{for } n < Q - 1, \\ C_2 \sin(2\pi\omega_2 n + \phi_2) & \text{for } n \geq Q - 1, \end{cases} \quad (3.8)$$

with $C_1, C_2 \neq 0$, $0 < \omega_1, \omega_2 < 0.5$ and $\omega_1 \neq \omega_2$. The last relation signifies the permanency of the structural change. It is clear that in this case $d = d_1 = 2$.

Under the choice $r = d$, the heterogeneity matrix is a square matrix with elements

$$g_{ij} = 0 \text{ if } \begin{cases} i \leq Q - B \\ j \leq Q - T \end{cases}, \text{ or } \begin{cases} i \geq Q + S \\ j \geq Q + S \end{cases}$$

and, in general, $g_{ij} > 0$ otherwise. Thus, the blocks $\mathcal{A}$ and $\mathcal{D}$ (Fig. 3.2) are zero blocks and the blocks $\mathcal{B}$ and $\mathcal{C}$ are generally not.

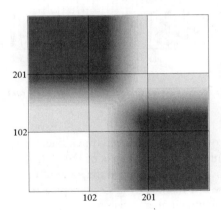

Figure 3.3 *Conservation of dimension: H-matrix.*

This matrix is depicted in Fig. 3.3 for the series (3.8) with $N = 400$, $\omega_1 = 1/10$, $\omega_2 = 1/10.5$, $C_1 = C_2 = 1$, $\phi_1 = \phi_2 = 0$ and $Q = 201$. Other parameters are $B = T = 100$, $L = 50$, $r = 2$ and $I = \{1, 2\}$. Since the values of the parameters (the lengths of the test and base intervals and the window length) are rather large, the matrix has a regular structure with constant values of g_{ij} in the blocks $\mathcal{B}$ and $\mathcal{C}$. The values of g_{ij} vary between the limits $[0, 0.17]$.

In Fig. 3.3, as well as the subsequent figures, the middle horizontal labels indicate the edges of the vertical strip of the heterogeneity cross (i.e., $Q - B + 1$ and Q). The vertical labels ($Q - T + 1$ and Q) correspond to the horizontal strip.

Since we are considering the heterogeneity of the series with respect to its initial part, it is worthwhile using detection functions. These functions are depicted in Fig. 3.4 (three detection functions are shown in the top graph; the series itself is in the bottom one). Note that for pictorial convenience the first 99 terms of the series are omitted.

We can see that all the detection functions start rising at the point 201 (this is the first point with nonzero values of all the detection functions). This is a reflection of the structural change in the series F_N at this moment of time.

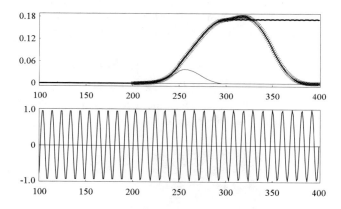

Figure 3.4 *Conservation of dimension: detection functions and the initial series.*

We recall that the row detection function (thick line, top graph of Fig. 3.4) corresponds to the bottom row of the H-matrix of Fig. 3.3. The symmetric detection function (thin line) indicates the principal diagonal of this matrix. The diagonal detection function (the line marked by crosses) depicts the secondary diagonal of the matrix with the column number exceeding the row number by 99.

Also, according to the definition of the detection functions, the indexation has been changed so that the values of g_{ij} correspond to the last points of the test intervals.

In this case, since the length of the test interval is 100, the first term of the row and diagonal detection functions is indexed by 100, and the indexation of the diagonal detection function starts at 200. This indexation system provides a correspondence between the change-points in the behaviour of the detection functions and the time series F_N.

Example 3.2 *Reduction of dimension: general case*
Let us assume that $\text{fdim}(F_{1,Q-1}) > \text{fdim}(F_{Q,N})$, but

$$\mathfrak{L}^{(L)}(F_{1,Q-1}) \not\supset \mathfrak{L}^{(L)}(F_{Q,N}).$$

An example of such a series is the series F_N with

$$f_n = \begin{cases} C_1 \sin(2\pi\omega_1 n) + C_2 \sin(2\pi\omega_2 n) & \text{for } n < Q-1, \\ C_3 \sin(2\pi\omega_3 n) & \text{for } n \geq Q-1, \end{cases} \quad (3.9)$$

where $C_1, C_2, C_3 \neq 0$, $0 < \omega_i < 0.5$, $i = 1, 2, 3$, $\omega_1 \neq \omega_2$, and $\omega_3 \neq \omega_1, \omega_2$. In this case, under the choice $r = 4$ and $I = \{1, 2, 3, 4\}$ (and suitable parameters L, B and T), the heterogeneity matrix has a form similar to the case of the 'conservation of dimension' (Fig. 3.3): the blocks $\mathcal{A}$ and $\mathcal{D}$ are zero blocks, but the blocks $\mathcal{C}$ and $\mathcal{B}$ are not.

Example 3.3 *Reduction of dimension: inheritance of structure*
The heterogeneity matrix has a specific form when the reduction of dimension is caused by the disappearance of one of the series components. In this case we have $\mathfrak{L}^{(L)}(F_{1,Q-1}) \supset \mathfrak{L}^{(L)}(F_{Q,N})$. We called this the inheritance of the series structure after the perturbation. Let us give an example.

Assume that the series F_N has the form (3.9) with $N = 50$, $\omega_1 = \omega_3 = 1/10$, $\omega_2 = 1/5$, $C_1 = 1$, $C_3 = 0.8$, $C_2 = 0.2$, $\phi_1 = \phi_2 = \phi_3 = 0$, and $Q = 32$. We thus have $d = 4$ and $d_1 = 2$.

Since $\omega_1 = \omega_3$, we have $\mathfrak{L}^{(L)}(F_{1,Q-1}) \supset \mathfrak{L}^{(L)}(F_{Q,N})$. Choose $B = 15$, $T = L = 8$. In accordance with our convention, $r = 4$ and $I = \{1, 2, 3, 4\}$. The corresponding H-matrix is depicted in Fig. 3.5 (the values vary from 0 to 0.072).

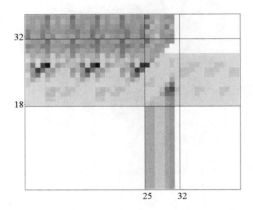

Figure 3.5 *Reduction of dimension: inheritance of structure.*

The blocks $\mathcal{A}, \mathcal{C}$, and $\mathcal{D}$ of this matrix are zero blocks. Only the nonzero block $\mathcal{B}$ deserves a commentary. By definition, an element of the block $\mathcal{B}$ is the normalized sum of distances between certain L-lagged vectors of the series $F_{1,Q-1}$ and the r-dimensional subspace $\mathfrak{L}_r^{(1)}$ spanned by the r leading eigenvectors corresponding to some subseries of the series $F_{Q,N}$.

Since in the present case $r = d > d_1$, and the subspace $\mathfrak{L}^{(L)}(F_{Q,N})$ has dimension d_1, there is an uncertainty in the selection of $r - d_1$ missing eigenvectors (in our case $r - d_1 = 2$), which correspond to the zero eigenvalue. In accordance with our convention, these eigenvectors are chosen as 'arbitrary' orthonormal vectors in the 'zero' eigenspace. (In practice, they are chosen by a concrete computational procedure.)

Thus, despite the fact that the spaces $\mathfrak{L}(F_{1,Q-1})$ and $\mathfrak{L}_r^{(L)}$ intersect, they do not coincide, and the distance from the vectors in the space $\mathfrak{L}(F_{1,Q-1})$ to the space $\mathfrak{L}_r^{(L)}$ is not necessarily zero.

Example 3.4 *Increase of dimension: general case*

Consider the general case of the increase of dimension of the series after the perturbation. We thus assume that $\text{fdim}(F_{1,Q-1}) < \text{fdim}(F_{Q,N})$ and $\mathfrak{L}^{(L)}(F_{1,Q-1}) \not\subset \mathfrak{L}^{(L)}(F_{Q,N})$. This means that the perturbation increases the complexity of the series, which implies structural change. In the simplest case

$$\mathfrak{L}^{(L)}(F_{1,Q-1}) \cap \mathfrak{L}^{(L)}(F_{Q,N}) = \mathbf{0}.$$

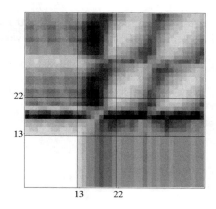

Figure 3.6 *Increase of dimension: the general case.*

Assume, for example, that

$$f_n = \begin{cases} C_1 \sin(2\pi\omega_1 n) & \text{for } n < Q - 1, \\ C_2 \sin(2\pi\omega_2 n) + C_3 \sin(2\pi\omega_3 n) & \text{for } n \geq Q - 1, \end{cases} \quad (3.10)$$

where $C_1, C_2, C_3 \neq 0$, $0 < \omega_1, \omega_2, \omega_3 < 0.5$ and all the ω_i are different. In this

HOMOGENEITY AND HETEROGENEITY

case, the elements of the heterogeneity matrix are such that

$$g_{ij} = 0 \text{ if } \begin{cases} i \leq Q - B \\ j \leq Q - T \end{cases}$$

and (in general) $g_{ij} > 0$ otherwise. Thus, in the general case, only the block $\mathcal{A}$ is the zero block. This matrix is depicted in Fig. 3.6 for the series (3.10) with $N = 50$, $\omega_1 = 1/10$, $\omega_2 = 1/8$, $\omega_3 = 1/5$, $C_1 = 1$, $C_2 = 0.8$, $C_3 = 0.5$ and $Q = 22$. Thus, $r = 2$ and $I = \{1, 2\}$. The other parameters are $B = T = 10$ and $L = 5$. The range of values of the H-matrix is $[0, 0.65]$.

Example 3.5 *Increase of dimension: inheritance of structure*
The more typical case is perhaps when the new structure inherits the structure of the unperturbed series, that is when $\mathcal{L}^{(L)}(F_{1,Q-1}) \subset \mathcal{L}^{(L)}(F_{Q,N})$.

This happens, for instance, when a harmonic is added to the already existing one, which is the case $\omega_1 = \omega_2$ in the series (3.10). In this situation, for appropriate parameters the H-matrix achieves the specific form shown below.

Let the series F_N be (3.10) with $N = 55$, $\omega_1 = \omega_2 = 1/10$, $\omega_3 = 1/5$, $C_1 = 1$, $C_2 = 0.8$, $C_3 = 0.2$, $\phi_1 = \phi_2 = \phi_3 = 0$ and $Q = 32$. Thus, $r = 2$ and $I = \{1, 2\}$. Let the other parameters be $B = 19$ and $T = L = 10$. Then the H-matrix is as depicted in Fig. 3.7. The range of values of its elements is $[0, 0.1]$.

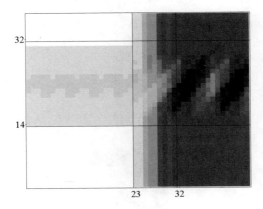

Figure 3.7 *Increase of dimension: inheritance.*

The blocks $\mathcal{A}$ and $\mathcal{B}$ of the matrix of Fig. 3.7 are zero blocks. The equality $\mathcal{B} = \mathbf{0}$ ($\mathbf{0}$ is the zero matrix) is due to the choice of parameters. Specifically, the values for B and L are chosen in such a way that the components of the series $F_{Q,N}$ (harmonics with frequencies $1/10$ and $1/5$) are separable. In this case, for all j such that $Q + S \leq j \leq N - T + 1$, the lagged vectors of the series $F_{1,Q-1}$ belong to the two-dimensional subspace $\mathcal{L}_{I,B}^{(L,j)}$ (where $I = \{1, 2\}$) of the four-dimensional space $\mathcal{L}_{J,B}^{(L,j)}$ (where $J = \{1, 2, 3, 4\}$) spanned by the L-lagged vectors of the series $F_{j,j+B-1}$. Other choices of the parameters L and B may lead to a lack of separability and therefore to a nonzero block $\mathcal{B}$.

(b) Temporary violation ('cross-structure' matrices)

In the case of temporary violation, $\mathfrak{L}^{(L)}(F_{1,Q-1}) = \mathfrak{L}^{(L)}(F_{Q,N})$ and the elements of the H-matrix satisfy the following relation:

if $g_{ij} \neq 0$ then $Q - B < i < Q + S$ or $Q - T < j < Q + S$.

Thus, in the case of temporary violation, all four blocks of the H-matrix (see Fig. 3.2) are zero blocks. Hence the pictorial representation of this matrix has the form of a cross. The horizontal strip reflects how the transition interval distorts the space $\mathfrak{L}^{(L)}(F_{1,Q-1})$ with respect to which the heterogeneity has arisen. The vertical strip shows what kind of influence the heterogeneity has on the lagged vectors of the series.

Consider two cases corresponding to two values of S of particular interest: $S = 0$ and $S = 1$.

Example 3.6 $S = 0$: *change of initial data in the LRF*
In the case of temporary violation and $S = 0$ we have

$$\mathfrak{L}^{(L)}(F_{Q,N}) \neq \mathfrak{L}^{(L)}(F_{Q-1,N}).$$

In particular, this implies that $\mathfrak{L}^{(L)}(F_{1,Q-1}) = \mathfrak{L}^{(L)}(F_{Q,N}) \neq \mathfrak{L}^{(L)}(F_N)$.

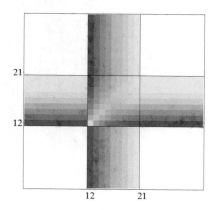

Figure 3.8 *Temporary violation ($S = 0$): H-matrix.*

As an example of this situation we may consider the series

$$f_n = \begin{cases} C_1 \sin(2\pi\omega n + \phi_1) & \text{for } n < Q - 1, \\ C_2 \sin(2\pi\omega n + \phi_2) & \text{for } n \geq Q - 1 \end{cases} \quad (3.11)$$

with $0 < \omega < 0.5$ and $\phi_1 \neq \phi_2$.

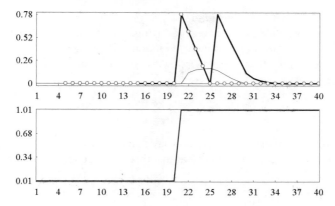

Figure 3.9 *Temporary violation ($S = 0$): detection functions and the time series.*

As another example, take piecewise linear (or piecewise constant) series:

$$f_n = \begin{cases} a_1 n + b_1 & \text{for } n < Q - 1, \\ a_2 n + b_2 & \text{for } n \geq Q - 1 \end{cases} \quad (3.12)$$

with $a_1 \neq a_2$ and/or $b_1 \neq b_2$.

The H-matrix for the piecewise constant series F_N of the form (3.12) with $N = 40$, $a_1 = a_2 = 0$, $b_1 = 0.01$, $b_2 = 1.01$ and $Q = 21$ is represented in Fig. 3.8. Both homogeneous subseries are governed by the same LRF: $f_{n+1} = f_n$. The parameters of the matrix are $r = 1$ and $I = \{1\}$, and also $B = T = 10$ and $L = 5$. The range of values for the elements of the matrix is $[0, 0.78]$.

Consider the detection functions. From their representation in Fig. 3.9 (the detection functions are given in the top graph; the original series is in the bottom one), we observe the start of the growth of the row and diagonal functions at the point $Q = 21$. This is in agreement with the fact that the structure of the series F_N has been perturbed at this time.

Fig. 3.9 depicts the row detection function $D^{(r)}_{10,40}$ (thin line marked with points), the symmetric detection function $D^{(s)}_{10,40}$ (thin line) and the diagonal detection function $D^{(d)}_{20,40}$ (thick line).

Note that the diagonal detection function $D^{(r)}_{20,40}$ has two jumps, when the test and base intervals contact the change-point for the first time. It again starts achieving zero values only when the base interval passes the change-point. Of course, this feature of the diagonal detection function may be undesirable in the case of several changes in the series.

Example 3.7 $S = 1$: *a single outlier*

The presence of a single outlier is one of the most common cases of series heterogeneity. This is the case of a temporary violation with $S = 1$.

Consider a homogeneous series F_N^* with general term f_n^*, and for some Q $(1 < Q < N)$ set

$$f_n = \begin{cases} f_n^* & \text{for } n \neq Q - 1, \\ a & \text{for } n = Q - 1, \end{cases} \quad (3.13)$$

with some $a \neq f_Q^*$. In this case, we shall say that the series F_N has heterogeneity in the form of a single outlier. It is clear that this heterogeneity is temporary.

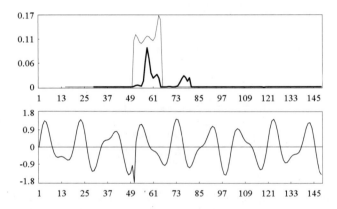

Figure 3.10 *Temporary violation: detection functions and the series with an outlier.*

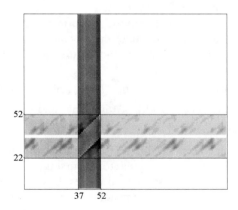

Figure 3.11 *Series with an outlier: H-matrix.*

As an example, consider the series F_N^* with $N = 149$ and

$$f_n^* = \sin(2\pi n/17) + 0.5\sin(2\pi n/10).$$

Take $Q = 51$ and define the series F_N by (3.13) with $a = f_{Q-1}^* - 1.5$. The series F_N is depicted in Fig. 3.10 (bottom graph). For this series $d = d_1 = 4$. Hence, $r = 4$ and $I = \{1, 2, 3, 4\}$.

The H-matrix of the series F_N with $B = 30$ and $T = L = 15$ is represented in Fig. 3.11 (the range is $[0, 0.34]$). From this figure, we can see that the elements of the H-matrix relating to the vertical strip of the heterogeneity cross are on average significantly larger than the corresponding values in its horizontal strip.

This can be explained by the fact that for large B and L ($r < L < B/2$), the spaces $\mathfrak{L}_{I,B}^{(L,i)}$ are stable with respect to a single outlier. Thus, the heterogeneity at point Q has very little effect on the spaces $\mathfrak{L}_{I,B}^{(L,i)}$, but it significantly changes the the lagged vectors of the test subseries $F_{j,j+T-1}$.

The corresponding row and column detection functions are depicted in Fig. 3.10 (top graph). Note that the row function (thin line) detects the outlier more explicitly than the column one. Note also that if B and L were to be reduced, then the vertical and horizontal strips of the cross would be closer in terms of the values of the heterogeneity index.

3.2.2 Multiple heterogeneity

If there are several local regions of heterogeneity in the time series, then its heterogeneity matrix contains submatrices corresponding to each single heterogeneity. Let us give an example with two local heterogeneities.

Example 3.8 *Example of multiple heterogeneity*
Consider the series F_N with

$$f_n = \begin{cases} C_0 + C_2 \sin(2\pi\omega_1 n) & \text{for } 1 < n < Q_1 - 1, \\ C_1 + C_2 \sin(2\pi\omega_1 n) & \text{for } Q_1 - 1 \leq n < Q_2 - 1, \\ C_1 + C_2 \sin(2\pi\omega_2 n) & \text{for } Q_2 - 1 \leq n < N, \end{cases}$$

where $N = 120$, $Q_1 = 41$, $Q_2 = 81$, $C_0 = 0.02$, $C_1 = 0.1$, $C_2 = 1$, $\omega_1 = 0.1$, and $\omega_2 = 10/98$. The series F_{1,Q_1-1}, F_{Q_1,Q_2-1}, and $F_{Q_2,N}$ are homogeneous with governing LRFs of equal dimension 3. Moreover, the LRFs of the first two subseries coincide, but they differ from the LRF of the third subseries. Thus, the first change-point Q_1 corresponds to a temporary violation, while the second change point Q_2 relates to a permanent one.

The series F_N is depicted in Fig. 3.12. In accordance with our convention, $r = 3$ and $I = \{1, 2, 3\}$. Set $B = T = 10$ and $L = 5$. Fig. 3.13 shows the H-matrix of the series F_N (the range for its elements is $[0, 0.09]$).

The elements g_{ij} with $1 \leq i, j \leq 71 = Q_2 - B$ of this matrix correspond to the subseries F_{1,Q_2-1}, which has only one heterogeneity. The representation of

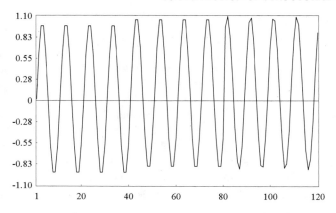

Figure 3.12 *Double heterogeneity: initial series.*

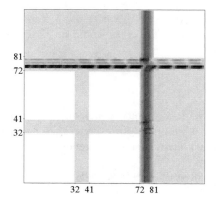

Figure 3.13 *Double heterogeneity: H-matrix.*

the corresponding part of the matrix is similar to Fig. 3.8 and has the form of a cross (temporary violation with $S = 0$).

The elements g_{ij} with $Q_1 = 41 \leq i, j \leq 100 = N - B + 1$ correspond to the subseries $F_{Q_1,N}$, which also has only one heterogeneity with conservation of dimension (this time the heterogeneity is permanent). The representation of the corresponding part of the matrix is similar to Fig. 3.3.

Remark 3.2 All the detection functions above relate to the study of heterogeneities with respect to the initial part of the series. However, in the forecasting problems, the study of heterogeneities with respect to the end part of the series is more important (backward change detection). Then the moment of heterogeneity (change-point) is defined as the minimal P such that the series $F_{P+1,N}$ is homogeneous. If there is only one local heterogeneity in the series, then $P = Q + S$.

The backward detection problem can be reduced to the forward one by the 'inversion' of the series: $f'_i \stackrel{\text{def}}{=} f_{N-i-1}$. Formally, this means the substitution of $r = d_1 = \text{fdim}(F_{P+1,N})$ for $r = d = \text{fdim}(F_{1,Q-1})$ and an appropriate choice of the new parameters B', L' and T'. If $d = d_1$, then we can use 'forward' parameters B, L and T for the backward detection. Then the 'backward' H-matrix is obtained by rotating the 'forward' H-matrix through 180 degrees with respect to the centre of the matrix.

In the latter case, the backward detection functions coincide with the heterogeneity functions for the original series. The shift in the indexation is not needed since the values of the backward detection functions have to refer to the first point of the test interval, which is the case for the heterogeneity functions. Thus, the first (from the right) significant change in values of the heterogeneity functions indicates the first (from the right) heterogeneity in the series.

3.3 Heterogeneity and separability

In the previous sections, the series F_N containing heterogeneities have been considered as single objects. This implied, in particular, that the number r of the eigentriples determining the base spaces $\mathfrak{L}_{B,I}^{(L,i)}$ was equal to the dimension $d = \text{fdim}(F_{1,Q-1})$ of the LRF governing the subseries $F_{1,Q-1}$ of the series F_N.

More realistic is the situation when $F_N = F_N^{(1)} + F_N^{(2)}$, where the additive component $F_N^{(1)}$ is subjected to a perturbation and the series $F_N^{(2)}$ has a sense of a nuisance (for example, $F_N^{(2)}$ is noise).

In this setup, the requirement of (approximate) separability of the corresponding subseries of the series $F_N^{(1)}$ and $F_N^{(2)}$ naturally arises. The concept of separability, being central in the SSA, is discussed in Section 1.5 in detail (see also Section 6.1 for the corresponding theoretical aspects).

The main object describing violations of homogeneity of time series is again the heterogeneity matrix. For a homogeneous series, this is the zero matrix (under a suitable choice of parameters). It was this zero matrix that served as the background in classification of different types of heterogeneity.

In order to describe a variety of forms of the 'background' H-matrices for the problem of detection of structural changes in series components, we first need to study the case of a homogeneous series $F_N^{(1)}$ whose subseries are (approximately) separable from the corresponding subseries of the series $F_N^{(2)}$.

This problem is studied in Section 3.3.1. Section 3.3.2 describes several cases of heterogeneity.

3.3.1 Heterogeneity

Let us start with the 'ideal' case of stable separability.

(a) Stable separability

Let $F_N = F_N^{(1)} + F_N^{(2)}$ and $F_N^{(1)}$ be homogeneous. Set $d = \mathrm{fdim}(F_N^{(1)})$. Assume that for all $i = 1, \ldots, N-B+1$ the subseries $F_{i,i+B-1}^{(1)}$ and $F_{i,i+B-1}^{(2)}$ are strongly L-separable for some window length L such that $d < L < B$. Since $B > d$, the subseries $F_{i,i+B-1}^{(1)}$ are governed by the same LRF that governs the series $F_N^{(1)}$, and therefore $\mathrm{fdim}(F_{i,i+B-1}^{(1)}) = d$. Set $r = d$.

We shall need one more assumption. Specifically, we assume that for any i, the subseries $F_{i,i+B-1}^{(1)}$ is described in the SVD of the L-trajectory matrix of the series $F_{i,i+B-1}$ by the eigentriples indexed by the numbers in $I = \{i_1, \ldots, i_r\}$, which are the same as for the series $F_{1,B}^{(1)}$. If these assumptions hold, then we shall call the series $F_N^{(1)}$ and $F_N^{(2)}$ *stably separable* (more precisely, (B,L)-*stably separable*). If the series are stably separable, then for all $1 \leq i \leq N-B+1$ we have

$$\mathfrak{L}_{I,B}^{(L,i)} = \mathfrak{L}^{(L)}(F_{i,i+B-1}^{(1)}) = \mathfrak{L}^{(L)}(F_N^{(1)}). \qquad (3.14)$$

In view of the equality (3.14), the heterogeneity index g_{ij} defined in (3.2) does not depend on i for any T.

If the series are stably separable, we can express g_{ij} explicitly for all i, j. Indeed, since all the L-lagged vectors of the time series $F_N^{(2)}$ are orthogonal to the subspace $\mathfrak{L}^{(L)}(F_N^{(1)})$, it follows that

$$g_{ij} = \frac{\sum_{l=1}^{T-L+1} \|X_{l,j}^{(2)}\|^2}{\sum_{l=1}^{T-L+1} \|X_{l,j}\|^2}, \qquad (3.15)$$

where $X_{l,j}$ and $X_{l,j}^{(2)}$ are the L-lagged vectors of the time series $F_{j,j+T-1}$ and $F_{j,j+T-1}^{(2)}$, respectively.

Let us comment on the condition of stable separability implying (3.15). The condition of strong separability of the subseries of the series $F_N^{(1)}$ and $F_N^{(2)}$ is, of course, very restrictive. It consists in the requirement of weak separability of the series $F_{i,i+B-1}^{(1)}$ and $F_{i,i+B-1}^{(2)}$ for all i and the condition that the singular values corresponding to these series in the SVD of the trajectory matrices of their sums are different.

Even the demand of weak separability is a strong condition. For a homogeneous series of dimension not exceeding 2, the unique example of two weakly separated series is provided by the pair of exponential-cosine sequences whose parameters satisfy certain relationships (see Sections 5.2 and 6.1.1).

HETEROGENEITY AND SEPARABILITY

However, the class of the series satisfying these conditions significantly widens if we replace the requirement of weak separability by the one of approximate (and, especially, asymptotic) weak separability.

The demand that the eigenvalues be different is necessary for the uniqueness of identification of the weakly separable series in the SVD of the trajectory matrix of their sum.

There is another condition relating to the eigenvalues, namely the constancy of the set I_i indexing the eigentriples which describes the subseries $F_{i,i+B-1}^{(1)}$ of the series $F_N^{(1)}$. As will be seen in examples, strong separability does not imply this constancy (however, for the harmonic series this is true). A more serious matter is that the constancy of I_i is formulated in terms of the ordinal numbers of the eigentriples, but their order can be unstable under even small noise fluctuations.

Let us give two examples of stably separable time series and the corresponding H-matrices.

Example 3.9 *Stable separation: nonperiodic series*
Fig. 3.14 depicts the H-matrix of the series F_N with $N = 100$ and general term $f_n = f_n^{(1)} + f_n^{(2)}$, where

$$f_n^{(1)} = a^n \cos(2\pi n/40), \qquad f_n^{(2)} = a^{-n}, \qquad (3.16)$$

$a = 1.025$. The parameters (guaranteeing stable separation) were chosen as follows: $B = 79$, $L = 40$, $r = 2$ and $I = \{1, 2\}$; this corresponds to selection of the eigentriples related to the series $F_N^{(1)}$. The parameter T was chosen as 79.

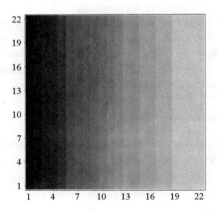

Figure 3.14 *Stable separation: H-matrix for nonperiodic series.*

Note that each column of the H-matrix consists of equal elements but the matrix is not constant. In terms of the heterogeneity functions, all the row heterogeneity functions coincide and are decreasing, while the column ones are different but constant.

The next example deals with purely harmonic series.

Example 3.10 *Stable separation: periodic series*
Let the series $F_N^{(1)}$ and $F_N^{(2)}$ be harmonic with different integer periods $p_1, p_2 > 2$ and amplitudes C_1 and C_2, $C_1 \neq C_2$. Denote by p the greatest common divisor of p_1 and p_2.

Choose the window length L and the parameter B so that p divides both L and $B - L + 1$. Then the series $F_{i,i+B-1}^{(1)}$ and $F_{i,i+B-1}^{(2)}$ are weakly L-separable for all $i = 1, \ldots, N - B + 1$. Also, the inequality $C_1 \neq C_2$ guarantees their strong separability. Moreover, in the SVD of the L-trajectory matrix of the series $F_{i,i+B-1}$, the harmonic with the larger amplitude is related to the larger eigenvalue (of multiplicity two).

Thus, the series $F_N^{(1)}$ and $F_N^{(2)}$ are stably separable. This implies (3.15). Let us show that in the present case all the g_{ij} are equal.

Indeed, since for a periodic harmonic with integer period p and amplitude 1 the sum of its p squares within one period is $p/2$, the norms of the vectors $\|X_{l,j}^{(2)}\|$ and $\|X_{l,j}\|$ do not depend on l and j and their squares are equal to $C_2^2 L/2$ and $(C_1^2 + C_2^2) L/2$, respectively. Thus,

$$g_{ij} = \frac{C_2^2}{C_1^2 + C_2^2},$$

which does not depend on T, the length of the test interval.

The differences in the heterogeneity matrices of Examples 3.9 and 3.10 are related to the fact that the series themselves are essentially different: periodicity of the harmonic series in the latter example guarantees the constancy of the g_{ij}; at the same time, stable separability of the components of a nonperiodic homogeneous series leads only to the equality of all row heterogeneity functions.

Let us now consider the deviations from stable separability of series.

(b) Deviations from stable separability

In this section we consider different deviations from the condition of stable separability of the series $F_N^{(1)}$ and $F_N^{(2)}$, still assuming the homogeneity of the series $F_N^{(1)}$. We have to distinguish two different kinds of deviation: deviations from weak separability, not related to the ordering of the eigenvalues, and the effects of coincidence and rearrangement of the eigenvalues, which have an influence on both strong separability and the constancy of the sets I_i. Of course, these deviations may occur at the same time.

(i) *Deviations from weak separability.* Assume that the series $F_N^{(1)}$ and $F_N^{(2)}$ are approximately weakly separable. This means that either the values of the parameters B and L are close to those guaranteeing (weak) separability or these values are sufficiently large and asymptotic (weak) separability does take place.

HETEROGENEITY AND SEPARABILITY

Assume also that in the SVD of the trajectory matrix of the series $F_{i,i+B-1}$ for all i, the series $F^{(1)}_{i,i+B-1}$ is related to the eigentriples with indices from the same set I. (This requirement corresponds to arrangement stability for the eigentriples in the case of strong separability.)

To start with, we consider two examples when the values of the parameters B and L are close to those guaranteeing exact separability. These examples correspond precisely to the situation described in Examples 3.9 and 3.10. In the former example, under the choice of parameters providing exact separability, the H-matrix consists of equal rows (the series is not periodic), while in the latter example all the elements of the H-matrix are equal (the series is periodic).

Example 3.11 *Approximate weak separability: nonperiodic series*
Fig. 3.15 depicts the H-matrix for the homogeneous series F_N of Example 3.9 with general term (3.16) (compare this figure with Fig. 3.14, where the same scale is used to encode the H-matrix).

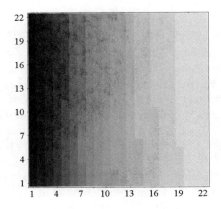

Figure 3.15 *Deviations in parameters: H-matrix for nonperiodic series.*

All the parameters are as in Example 3.9 except for the window length L, which is now $L = 35$ instead of $L = 40$. These parameters guarantee only approximate weak separability rather than exact separability.

Fig. 3.15 shows that the heterogeneity matrix is essentially the same; for instance, the range of its values is (0.005, 0.04). However, the row heterogeneity functions are no longer equal.

Consider now the case when the series $F^{(1)}_N$ and $F^{(2)}_N$ are periodic.

Example 3.12 *Approximate weak separability: periodic series*
Fig. 3.16 depicts the H-matrix for a homogeneous series F_N with $N = 50$ and general term $f_n = f^{(1)}_n + f^{(2)}_n$,

$$f^{(1)}_n = \sin(2\pi n/3), \qquad f^{(2)}_n = 0.5\sin(2\pi n/5).$$

The parameters are $B = 29$, $T = L = 14$, $r = 2$, and $I = \{1, 2\}$. The range of values of the matrix elements is approximately $[0.1, 0.3]$.

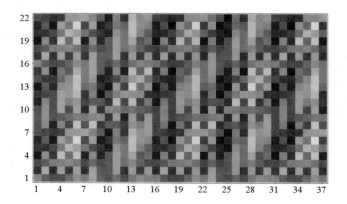

Figure 3.16 *Deviations in parameters: H-matrix for periodic series.*

Note that if we select $L = 15$ (this corresponds to exact separability), then all the elements of the H-matrix will be equal to 0.2 (see Example 3.10). In the case $L = 14$ we have approximate separability.

Even though the value of the window length L is almost the same, the matrix depicted in Fig. 3.16 is very different from the 'ideal' constant matrix corresponding to exact separability. Let us comment on the reasons for this difference.

The lengths B and T of two intervals, the base and the test ones, enter the definition of the heterogeneity matrix. Separability characteristics of the series are strongly influenced by the choice of the length B of the base interval (as well as by the choice of the window length). At the same time, in the case of stable separability and harmonic series, the H-matrix does not depend on T.

In the case of deviations from exact separability, there appears a dependence of the H-matrix on T. The number of terms in the sums in both the numerator and denominator of (3.1) is $T - L + 1$. The smaller T is (and therefore $T - L + 1$), the larger fluctuations the elements of the H-matrix may have.

Large fluctuations in the H-matrix depicted in Fig. 3.16 can be explained by the fact that the number of lagged vectors in the numerator and denominator of (3.1) is only 1 (that is, $T = L$). If we increase the number of these lagged vectors by increasing T, then the variation in values of the elements of the H-matrix will decrease. For instance, for $T = 33$ (in this case $T - L + 1 = 20$) the values of these elements lie in the interval $(1.9, 2.0)$.

Since both series are periodic, there exists a specific choice of the parameter T, which leads to a very small range for the heterogeneity indices g_{ij}.

If we take $T = 29$, then the variation in values across the rows of the H-matrix practically disappears (it becomes 0.005) and all the columns become equal. This is caused by the fact that the number $T - L + 1$ of lagged vectors entering the sums in (3.1) is equal to 15 and is therefore proportional to both periods, 3 and 5.

Note that the choice $T = L$ makes the mosaic structure of the H-matrix more apparent also due to the total disagreement of $T - L + 1 = 1$ with the periods of the series.

Example 3.13 *Asymptotic separability: harmonic series corrupted by noise*

For long series, asymptotic weak separability (see Section 6.1.2) is a more natural concept than the version of approximate separability related to a bad choice of parameters. In the problems under consideration, regular asymptotic separability assumes that the values of L and $B - L + 1$ are large. Of course, this is possible only when N is large.

As a rule, regular asymptotic separability implies that there are small fluctuations around the limiting H-matrices, which are either constant or have the form presented in Fig. 3.14. Natural examples of asymptotic separability arise in the cases of noisy series.

Consider an infinite series F with general term

$$f_n = f_n^{(1)} + f_n^{(2)} = C\sin(2\pi n\omega + \phi) + \varepsilon_n,$$

where ε_n is some (deterministic) stationary chaotic series with covariance function R_ε and spectral density p_ε.

The results of Sections 6.4.4 and 6.4.3 imply that the noise ε_n is asymptotically strongly separable from the signal $F^{(1)}$, and, moreover, the signal $F^{(1)}$ asymptotically corresponds to the first two eigentriples. (In the case of stochastic Gaussian white noise, weak separability is proved in Section 6.1.3.)

Thus, if we choose $r = 2$ and $I = \{1, 2\}$, then asymptotically the elements g_{ij} of the heterogeneity matrix have the constant limit

$$\lim g_{ij} = \frac{2R_\varepsilon(0)}{C^2 + 2R_\varepsilon(0)}. \qquad (3.17)$$

In practice, the closeness of the elements of the H-matrix to the constant value (3.17) is achieved by virtue of the large value of the series length N and the small (relative to C^2) value of the variance $R_\varepsilon(0)$.

Example 3.16 can be regarded as another example demonstrating the effect of asymptotic separability, if the original series is taken only until the moment of rearrangement of the eigentriples. In that example, where the components are not separable for any choice of parameters, the base interval length $B = 100$ and the window length $L = 50$ appear to be large enough, so that the property of asymptotic separability of the series $F^{(1)}_{i,i+B-1}$ and $F^{(2)}_{i,i+B-1}$ provides almost exact separation (until the moment of rearrangement of the eigentriples). We can observe the result of this in the bottom part of the H-matrix in Fig. 3.21.

Let us discuss the deviations from stable separability relating to the changes in the order of the eigenvalues.

(ii) *Rearrangement of the eigentriples.* The following examples are related to the case when there is weak L-separability of the subseries of the series $F_N^{(1)}$ and $F_N^{(2)}$, but the conditions relating to the eigenvalues of the trajectory matrices do not hold for all possible locations of the base interval.

More formally, we assume that

1. For some B and L and all $i = 1, \ldots, N - B + 1$ the series $F_{i,i+B-1}^{(1)}$ and $F_{i,i+B-1}^{(2)}$ are weakly separable;
2. There exists some M (maximal number with this property) such that $B < M$ and for $m = B, B+1, \ldots, M-1$
    - the series $F_{m-B+1,m}^{(1)}$ and $F_{m-B+1,m}^{(2)}$ are strongly separable;
    - in the SVD of the L-trajectory matrix of the series $F_{m-B+1,m}$ the series $F_{m-B+1,m}^{(1)}$ is represented by the eigentriples with the same numbers $I = \{i_1, \ldots, i_r\}$, which hold for the series $F_{1,B}^{(1)}$.

We shall call the moment of time M the moment of *the eigentriple rearrangement* of the series F_N. At this moment we have either termination of strong separability or a rearrangement in the set of the eigentriples representing the subseries of the series $F_N^{(1)}$ (without violation of strong separability). Since the former effect (if it is indeed present) is usually accompanied by the second one, we can consider these two effects together.

Let us give corresponding examples.

Example 3.14 *Eigentriple rearrangement: increase of dimension*
Consider the time series F_N with $N = 400$ and general term $f_n = f_n^{(1)} + f_n^{(2)}$, where

$$f_n^{(1)} = 10\, a^{-n}, \quad f_n^{(2)} = a^n \cos(2\pi n/20),$$

$a = 1.005$. The series F_N is depicted in the bottom graph of Fig. 3.18.

Under the choice $B = 39$ and $L = 20$ (see Section 6.1.1) we have weak separability of the series $F_{i,i+B-1}^{(1)}$ and $F_{i,i+B-1}^{(2)}$ for all i. It is clear that the series $F_{i,i+B-1}^{(1)}$ is described by the single eigentriple. It is straightforward to check that the series $F_{1,B}^{(1)}$ and $F_{1,B}^{(2)}$ are strongly separable and that the series $F_{1,B}^{(1)}$ is represented by the leading eigentriple.

Thus, we have $r = 1$ and $I = \{1\}$. Fig. 3.17 shows the H-matrix for $T = 20$; the plots of the row (thin line) and column (thick line) detection functions are given at the top graph of Fig. 3.18.

We can see that the behaviour of the row functions is totally different from that of the column functions. While the row detection function varies smoothly (this corresponds to its canonical behaviour in the case of stable separability for nonperiodic series), the column function has a jump from almost zero at $n = 320$ to almost 1 at $n = 321$.

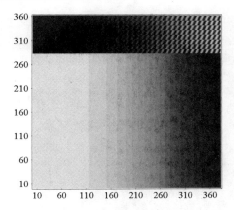

Figure 3.17 *Eigentriple rearrangement and dimension increase: H-matrix.*

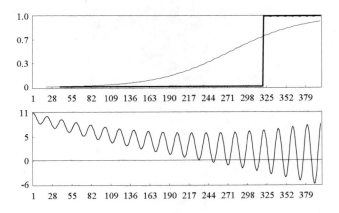

Figure 3.18 *Eigentriple rearrangement: detection functions and the time series.*

An additional analysis shows that this jump identifies the moment of the rearrangement of the order of the eigentriples which describe the series $F^{(1)}_{m-B+1,m}$: for $m \leq 321$ these series were described by the leading eigentriple, but starting at $m = 322$ the two leading eigentriples describe the series $F^{(2)}_{m-B+1,m}$ with only the third one corresponding to $F^{(1)}_{m-B+1,m}$. Thus, the moment M of the eigentriple rearrangement is $M = 322$ (but strong separability is valid everywhere).

The mosaic structure of the top part of the H-matrix also has an explanation: for $m \geq M$ the choice $r = 1$ corresponds to selection of only one of the two leading eigentriples describing the series $F^{(1)}_{m-B+1,m}$. We thus deal with the effect of the dimension increase discussed in Section 3.2.1 in a different context.

The following example is similar to Example 3.14, but it has some peculiarities.

Example 3.15 *Eigentriple rearrangement: conservation of dimension*
Consider the time series F_N with $N = 400$ and general term $f_n = f_n^{(1)} + f_n^{(2)}$, where

$$f_n^{(1)} = 10\, a^{-n} \cos(2\pi n/10), \quad f_n^{(2)} = a^n \cos(2\pi n/20),$$

$a = 1.005$. The series F_N is depicted in the bottom graph of Fig. 3.20.

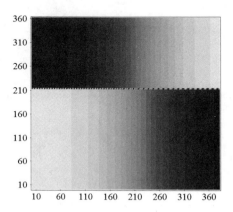

Figure 3.19 *Eigentriple rearrangement and conservation of dimension: H-matrix.*

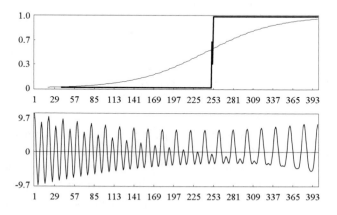

Figure 3.20 *Eigentriple rearrangement: detection functions and the time series.*

The choice $B = 39$ and $L = 20$ provides weak separability of the subseries $F_{i,i+B-1}^{(1)}$ and $F_{i,i+B-1}^{(2)}$ for all i; moreover, both subseries are described by two

HETEROGENEITY AND SEPARABILITY

eigentriples. As above, the series $F_{1,B}^{(1)}$ and $F_{1,B}^{(2)}$ are strongly separable. The series $F_{1,B}^{(1)}$ corresponds to two leading eigentriples (that is, $r = 2$ and $I = \{1, 2\}$).

The heterogeneity matrix is depicted in Fig. 3.19 (the parameter is $T = 20$); plots of the row (thin line) and the column (thick line) detection functions are given in the top graph of Fig. 3.20.

These figures show two features that are different from what we have seen in the previous example. First, the big jump in the values of the column detection function is not instantaneous; at the point $m = 250$ the function is almost zero, and at $m = 253$ it is almost one, but in the two points between, its values are far from 0 and 1. Second, the rows in the top part (and the bottom part as well) of the H-matrix are (theoretically) equal, but the values of the row heterogeneity functions increase at the bottom and decrease at the top.

The first feature can be explained by the fact that the rearrangement of the eigentriples is not instantaneous. Instead, there are two stages. Indeed, for $m < M = 251$ the series $F_{m-B+1,m}^{(1)}$ are described by the first two eigentriples; starting at $m = 253$ they are described by the third and the fourth. For the two intermediate points ($m = 251, 252$) the series correspond to the eigentriples 2 and 4. (Note that strong separability holds again.)

The second feature becomes clear as well: since we have $r = 2$ and $I = \{1, 2\}$, but after the time when the eigentriple rearrangement is completed, the series $F_{i,i+B-1}^{(2)}$ is described by the two leading eigentriples; this means that the series $F_{i,i+B-1}^{(1)}$ are $F_{i,i+B-1}^{(2)}$ just change places. The opposite behaviour of the row heterogeneity functions is caused by the difference in behaviour of the nonperiodic series $F_N^{(1)}$ and $F_N^{(2)}$: the first of them tends to zero, while the second has an increasing amplitude.

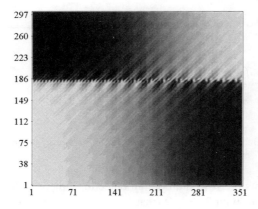

Figure 3.21 *General deviation from stable separability: H-matrix.*

Finally, let us briefly illustrate the summary effect of both deviations from stable separability.

Example 3.16 *General deviation from stable separability*
Consider the series F_N with $N = 400$, general term $f_n = f_n^{(1)} + f_n^{(2)}$ and

$$f_n^{(1)} = 10\,a^{-n}\cos(2\pi n/11), \quad f_n^{(2)} = a^n \cos(2\pi n/20),$$

$a = 1.005$. The series F_N is depicted in the bottom graph of Fig. 3.22. Thus, $r = 2$ and $I = \{1,2\}$. Let $B = 100$, $L = T = 50$. The resulting H-matrix is depicted in Fig. 3.21; it looks like a slightly perturbed matrix of Fig. 3.19. The detection functions are plotted in the top graph of Fig. 3.22 (the types of lines are the same as in Fig. 3.20).

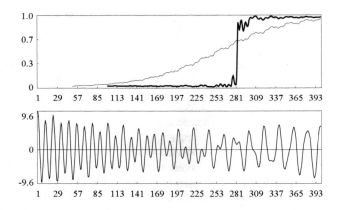

Figure 3.22 *General deviation: detection functions and the time series.*

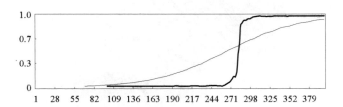

Figure 3.23 *General deviation: detection functions for large T.*

Fig. 3.23 presents the row and column detection functions for the same values of B and L, but with $T = 69$ (that is, the sums in the numerator and the denominator in the definition of g_{ij} use 20 L-lagged vectors rather than just one, when $T = L$). We can see that the increase of T does not remove the essential differences between the row and column detection functions, although it makes the functions smoother.

Remark 3.3 Suppose that time series $F_N^{(1)}$ and $F_N^{(2)}$ are approximately (B, L)-stably separable and the terms of the time series $F_N^{(2)}$ are essentially smaller than that of $F_N^{(1)}$. Then we have (i) all the elements of the heterogeneity matrix have small values; and (ii) if the subseries $F_{i,i+B-1}^{(1)}$ of the time series $F_N^{(1)}$ are described by a small number of the leading eigentriples in the SVD of the L-trajectory matrices of the series $F_{i,i+B-1}$, then the effect of the eigentriple rearrangement is small.

3.3.2 Heterogeneity in separable series

Until now, we have assumed that the series $F_N^{(1)}$ is homogeneous. Our next problem is to discuss the case when it has intervals of heterogeneity. The heterogeneity matrix **G** is, as always, our main point of interest. The series $F_N^{(1)}$ and $F_N^{(2)}$ themselves can be either stably separable on the homogeneity intervals or have discrepancies from this ideal situation.

(a) Stable separability

Suppose that the subseries $F_{1,Q-1}^{(1)}$ of the series $F_N^{(1)}$ is a homogeneous series governed by a minimal LRF of dimension $r < Q$. Then it can be continued with the help of this LRF up to the time N. Let us denote by $\widehat{F}_N^{(1)}$ the result of this continuation.

Let Q be the change-point for the series $F_N^{(1)}$. We consider the heterogeneities of the types described in Section 3.2.

As in the previous section, the series $F_N^{(1)}$ and $F_N^{(2)}$ are related to each other. More precisely, we assume that the series $\widehat{F}_N^{(1)}$ and $F_N^{(2)}$ are (B, L)-stably separable under the choice of proper parameters $L < B < Q$. In other words, the series $\widehat{F}_{i,i+B-1}^{(1)}$ and $F_{i,i+B-1}^{(2)}$ are stably L-separable for all $i = 1, \ldots, N - B + 1$, and for any i in the SVD of the L-trajectory matrix of the series

$$\widehat{F}_{i,i+B-1} \stackrel{\text{def}}{=} \widehat{F}_{i,i+B-1}^{(1)} + F_{i,i+B-1}^{(2)}$$

the subseries $\widehat{F}_{i,i+B-1}^{(1)}$ is described by r eigentriples with the indices in a fixed set $I = (i_1, \ldots, i_r)$.

This means, in particular, that $r < L$ and the series $F_{i,i+B-1}^{(1)}$ and $F_{i,i+B-1}^{(2)}$ are (B, L)-stably separable for all $i < Q - B$. Note that generally we do not assume any separability of the series $F_{i,i+B-1}^{(1)}$ and $F_{i,i+B-1}^{(2)}$ for $i \geq Q + S$. Formally, this situation differs from that of Section 3.2 in that here the series $F_N^{(2)}$ is not a zero series.

In accordance with our assumptions, for any T and a single heterogeneity of the series $F_N^{(1)}$, the H-matrix must generally have the same structure as in the case of the zero series $F_N^{(2)}$. This matrix consists of the heterogeneity cross corresponding to the transition period $[Q, Q + S]$, and four blocks (see Fig. 3.2): block $\mathcal{A}$ refers to the series $F_{1,Q-1}$, block $\mathcal{D}$ – to the series $F_{Q+S,N}$, and blocks $\mathcal{B}$ and $\mathcal{C}$ describe the relations between these series.

The main difference is that the block $\mathcal{A}$ is not a zero block in this case and also the other blocks may have a more complex structure than in the case of Section 3.2.

Let us describe some general features of the heterogeneity matrices corresponding to the series under consideration.

1. Since the block $\mathcal{A}$ corresponds to the series $F_{1,Q-1}$, which is a sum of stably separable homogeneous series, this block has the same structure as the H-matrices of stably separable series discussed in Section 3.3.1.

    More precisely, the elements g_{ij} of the heterogeneity matrix $\mathbf{G}$ have the form (3.15) for $i < Q - B$ and $j < Q - T$. In particular, each column of the block $\mathcal{A}$ consists of equal elements.

2. The row heterogeneity functions $H_{N-T+1}^{(r,i)}$ (in other words, the rows of the matrix $\mathbf{G}$) coincide for $i < Q - B$. This is caused by the coincidence of the base spaces $\mathfrak{L}_{I,B}^{(L,i)}$ for this range of i. In particular, all the columns of the block $\mathcal{C}$ consist of equal elements.

3. If both $F_{1,Q-1}^{(1)}$ and $F_{1,Q-1}^{(2)}$ are periodic series, then all the elements g_{ij} in the block $\mathcal{A}$ are equal. Otherwise the block $\mathcal{A}$ has an equal-row structure similar to that of Fig. 3.14.

4. If the heterogeneity is temporary, then the matrix is cross-structured; the heterogeneity cross is located either on a constant background (for the periodic series) or on an equal-row background.

    The elements g_{ij} of the matrix $\mathbf{G}$ have the form (3.15) in all four blocks.

5. In the case of permanent heterogeneity, there is no general rule for the elements in the blocks $\mathcal{B}$, $\mathcal{C}$ and $\mathcal{D}$. Moreover, these elements can be both larger or smaller than the elements of the block $\mathcal{A}$.

6. If the series $F_N^{(2)}$ is sufficiently small, then the entire classification of Section 3.2.1 is valid. The difference is that all zero blocks of Examples 3.1-3.7 now correspond to blocks with small g_{ij} (see Remark 3.3).

Summarizing these considerations, we can state that, in general, the problem of finding and identifying heterogeneities in the series $F_N^{(1)}$ under the influence of an addend $F_N^{(2)}$ (which is stably separable until a change-point) is more complex in view of the nonzero background in the heterogeneity matrix $\mathbf{G}$. However, in some cases (for example, in the case of temporary heterogeneity in periodic series and in the case of small $F_N^{(2)}$) this matrix gives a good description of the entire situation.

HETEROGENEITY AND SEPARABILITY

Remark 3.4 The case when the whole series $F_N^{(1)}$ is homogeneous and the heterogeneity occurs in the series $F_N^{(2)}$ does not need a special study. Indeed, this corresponds (up to a re-indexation of both series) to the case when we take I for the set of eigentriples describing $F_{i,i+B-1}^{(2)}$ rather than $F_{i,i+B-1}^{(1)}$. Then the elements g'_{ij} of the new heterogeneity matrix $\mathbf{G}'$ relate to the elements g_{ij} of $\mathbf{G}$ via the equality $g'_{ij} = 1 - g_{ij}$. Thus, the matrix $\mathbf{G}'$ is complementary to $\mathbf{G}$.

Now let us consider two examples, the first one relating to a harmonic series with permanent heterogeneity and the second one describing a nonperiodic situation with a temporary heterogeneity.

Example 3.17 *Permanent violation in a component of a periodic series*
Consider the series $F_N = F_N^{(1)} + F_N^{(2)}$ with

$$f_n^{(1)} = \begin{cases} \sin(2\pi n/3) & \text{for } n < Q - 1, \\ \sin(2\pi n/4) & \text{for } n \geq Q - 1, \end{cases}$$

$$f_n^{(2)} = 0.5 \sin(2\pi n/5),$$

where $N = 100$, $Q = 51$. Thus, we have a permanent violation in the harmonic series $F_N^{(1)}$; the series $F_N^{(2)}$ is also harmonic.

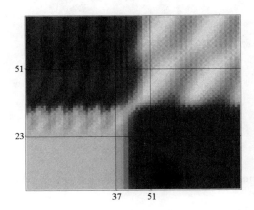

Figure 3.24 *Permanent violation in a component of a periodical series: H-matrix.*

Take $B = 29$, $T = L = 15$. Since L and $K = B - L + 1$ are proportional to 3 (period of the first series up to the change-point Q) and 5 (period of the second series), it follows that the SVD of any subseries $F_{i,i+B-1}$ of the series $F_{1,Q-1}$ provides weak separability of the series $F_{i,i+B-1}^{(1)}$ and $F_{i,i+B-1}^{(2)}$. Since the amplitudes of both series differ, the separation is a strong one. In addition, the first two eigentriples of the SVD of the L-trajectory matrix of the series $F_{i,i+B-1}$ correspond to the series $F_{i,i+B-1}^{(1)}$; the third and fourth eigentriples describe the series $F^{(2)}_{i,i+B-1}$. Thus, $r = 2$ and $I = \{1, 2\}$. Fig. 3.24 shows the corresponding H-matrix.

The matrix is similar to that of Example 3.1, where $F^{(2)}_N$ is the zero series. The range of elements of the H-matrix is approximately $[0.02, 0.27]$.

Example 3.18 *Temporary violation in a component of a nonperiodic series*
This example shows the form of a typical heterogeneity matrix in the case of nonperiodic series $F^{(1)}_N$ and $F^{(2)}_N$ and a temporary violation in $F^{(1)}_N$.

Consider the series $F^{(1)}_N$ with general term

$$f^{(1)}_n = \begin{cases} 10\,a^{-n} \cos(2\pi n/10) & \text{for } n < Q - 1, \\ 10\,a^{-n} \cos(2\pi n/10 + \phi) & \text{for } n \geq Q - 1, \end{cases}$$

and the series $F^{(2)}_N$ with $f^{(2)}_n = a^n$. The parameters are $N = 150$, $Q = 100$, $a = 1.005$ and $\phi = \pi/4$. The series is depicted in the bottom graph of Fig. 3.26.

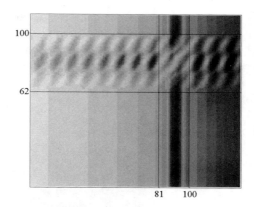

Figure 3.25 *Temporary violation in a component of a nonperiodic series: H-matrix.*

Thus, we deal with the change in the phase of the exponential-cosine series $F^{(1)}_N$. This is the case of a temporary violation.

If we take $B = 39$, $L = 20$, $r = 2$ and $I = \{1, 2\}$, then we can see that the series $F^{(1)}_{i,i+B-1}$ and $F^{(2)}_{i,i+B-1}$ are stably separable up to the change-point.

The heterogeneity matrix for the case $T = L$ is presented at Fig. 3.25. The 'cross structure' of the matrix is apparent, though less distinct in view of the nonconstant background, typical for the nonperiodic situation.

Fig. 3.26 depicts the column (thick line) and row (thin line) detection functions corresponding to the heterogeneity matrix of Fig. 3.25.

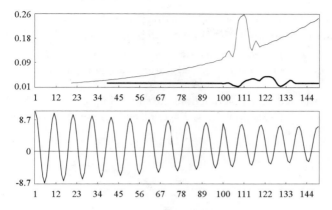

Figure 3.26 *Temporary violation: detection functions and the time series.*

(b) Deviations from stable separability

Let us now give examples of H-matrices corresponding both to heterogeneity and deviations from stable separability of the series $F^{(1)}_{1,Q-1}$, $F^{(2)}_{1,Q-1}$. The case of the eigentriple rearrangement has been illustrated in the previous section; therefore, we do not treat this deviation separately but restrict ourselves to approximate separability and the general case of this deviation.

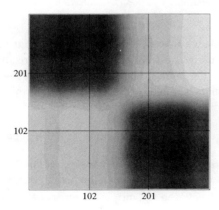

Figure 3.27 *Heterogeneous noisy periodicity: H-matrix.*

Example 3.19 *Permanent violation in a harmonic signal corrupted by noise*
Let us add Gaussian white noise with variance 0.04 to the series of Example 3.1 (the parameters stay the same).

Since the noise is relatively small, and the parameters B and L are sufficiently large, the series of Example 3.1 (in our terminology, the series $F_N^{(1)}$) is approximately weakly separated from the noise series (that is, from $F_N^{(2)}$) on the time interval $[1, Q-1]$, $Q = 201$. The eigentriples are not rearranged for the same reason.

The H-matrix and detection functions are shown in Figs. 3.27 and 3.28, respectively. The detection functions (row, diagonal, and symmetric) are depicted in the manner of Fig. 3.4.

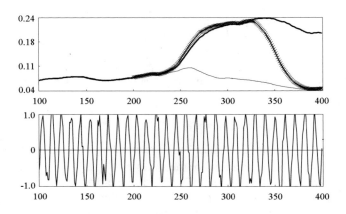

Figure 3.28 *Heterogeneous noisy periodicity: detection functions and the time series.*

Since in this example we have a permanent violation of the signal with conservation of dimension and the noise is small, it is not surprising that the matrix of Fig. 3.27 has generally the same structure as the matrix of Fig. 3.3.

Fig. 3.28 shows that the symmetric detection function (thin line) can easily fail to detect a structural change in the presence of noise.

Let us now consider the general deviation from stable separability.

Example 3.20 *General deviation from stable separability*
This example corresponds to the case when the heterogeneity occurs in the series $F_N^{(1)}$, but the set I of the chosen eigentriples (approximately) describes the initial part of the series $F_N^{(2)}$. In view of Remark 3.4, this is not an essential modification, but it makes the figures clearer for interpretation.

Thus, we take $F_N^{(2)} = F_N^{(2,1)} + F_N^{(2,2)}$, where $F_N^{(2,1)}$ has the general term

$$f_n^{(2,1)} = 10\, a^{-n} \cos(2\pi n/20)$$

and the series $F_N^{(2,2)}$ is the standard Gaussian white noise series.

The series $F_N^{(1)}$ has the form

$$f_n^{(1)} = \begin{cases} a^n \cos(2\pi n/10) & \text{for } n < Q - 1, \\ a^n \cos(2\pi n/7) & \text{for } n \geq Q - 1, \end{cases}$$

with $N = 400$, $Q = 301$ and $a = 1.006$. The series F_N is plotted in Fig. 3.30 (bottom graph).

The series $F_N^{(1)}$ has a permanent violation and the series $F_N^{(2)}$ has a noise component. On the other hand, if we consider the sum of the series $F_{1,Q-1}^{(1)} + F_{1,Q-1}^{(2,1)}$, then we come to the situation of eigentriple rearrangement, which is similar to that of Example 3.15.

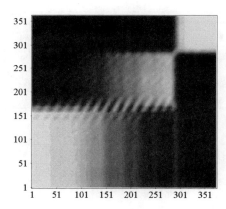

Figure 3.29 *General deviation: H-matrix.*

If we take $B = 39$, $L = 20$, then for $m < M = 210$ the series $F_{m-B+1,m}^{(1)}$ and $F_{m-B+1,m}^{(2,1)}$ are stably separable, and the second of them is described by the two leading eigentriples. The eigentriple rearrangement takes place during three successive moments of time, and then the series becomes stably separable again, with a permutation of their eigentriple numbers.

As a result, we have both types of deviations from stable separability and a permanent heterogeneity.

The heterogeneity matrix is depicted in Fig. 3.29. The parameters of the matrix are $B = T = 39$. The window length L is equal to 20. The choice $r = 2$ and $I = \{1, 2\}$ provides the (approximate) stable separability of $F_{i,i+B-1}^{(1)}$ and $F_{i,i+B-1}^{(2)}$ for small i. Let us discuss the structure of this heterogeneity matrix.

Three light rectangles can be seen close to the matrix main diagonal. The bottom left rectangle corresponds to the subseries preceding the interval of the eigentriple rearrangement. (Note that due to the noise influence, this interval is rather large.) The top right rectangle indicates the part of the series that follows the heterogeneity interval. The middle one describes the period between the end of the eigentriple rearrangement and the beginning of the heterogeneity interval.

Small values of the heterogeneity indices g_{ij} in these rectangles can be explained by (approximate) weak separability and by the fact that the subseries that are approximately described by the third and fourth eigentriples are relatively small at these 'nonpermuted' time intervals.

If we would consider the series only after the rearrangement period, then the heterogeneity matrix will be represented by the top right matrix rectangle of Fig. 3.29, including two light and two dark rectangles. This part of the matrix is similar to the matrix of Example 3.19 (see Fig. 3.27).

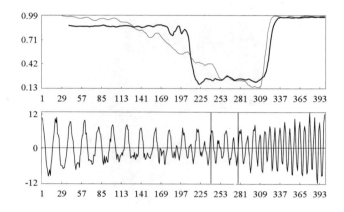

Figure 3.30 *General deviation: detection functions and the time series.*

The top graph of Fig. 3.30 contains two heterogeneity functions, corresponding to the 240th row and the 240th column of the matrix. Both heterogeneity functions are depicted in the manner of detection functions in order to align them with the moments of the rearrangement and the change-point (see Section 3.1.2). We can see that both the row (thin line) and column (thick line) heterogeneity functions perfectly indicate the change-point. As for the moment of rearrangement, the column function shows a rapid decrease there, while the row one continues to behave as a slowly decreasing function.

The subseries $F_{i,i+B-1}$, corresponding to the 240th row heterogeneity function is marked at the bottom graph of Fig. 3.30 by two vertical lines.

Note that if we were to use the standard row and column detection functions, corresponding to the first row and the first column of the heterogeneity matrix, then we should miss the change-point because of the nonconstant background heterogeneity indices g_{ij} appearing after the eigentriple rearrangement. As regards the rearrangement itself, the column detection function indicates it well, while the row one fails once again.

Note also that if we were to choose the four leading eigentriples instead of two, that is if we were to take $r = 4$ and $I = \{1, 2, 3, 4\}$, then the eigentriple rearrangement would have no influence on the heterogeneity matrix, and therefore the heterogeneity of the series will be seen much more clearly. The corresponding matrix is depicted in Fig. 3.31.

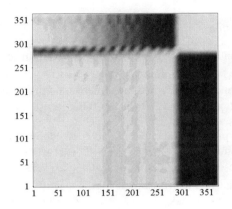

Figure 3.31 *The choice $r = 4$: eigentriple rearrangement is not indicated.*

3.4 Choice of detection parameters

As always, we consider the series $F_N = F_N^{(1)} + F_N^{(2)}$.

Let us assume that the series $F_N^{(1)}$ is homogeneous within the time intervals $[1, Q_1 - 1], [Q_1 + S_1, Q_2 - 1], \ldots, [Q_l + S_l, N]$, but we do not have information about the number l of change-points and the parameters Q_j, S_j. (In particular, it is possible that $Q_1 = N + 1$ and there is no heterogeneity at all.)

Then we have several problems to be solved. First, our aim is to determine whether a violation of homogeneity did actually occur; if it did, then we have to

- find the number of change-points;
- find their location (that is, to determine parameters Q_j and S_j);
- classify each heterogeneity in the manner of Section 3.2.1.

Here we make several remarks that may help in making the appropriate decisions. We mostly concentrate on the situation when the search for heterogeneities begins at the initial part of the series (forward search).

Let us describe the choice of parameters and interpretation of the detection results for different classes of time series. We begin with the simplest situation and then move to more and more complicated ones.

3.4.1 Single heterogeneity

Detection of a single heterogeneity is the basis for general detection. It helps to understand the peculiarities of multiple heterogeneity. However, even in the case of detection of a single heterogeneity, we have both simple and complicated situations. Let us start with the 'ideal' case.

(a) The 'ideal' detection

The clearest situation is when $F_N^{(2)}$ is the zero series. In this case we deal with a single series $F_N^{(1)} = F_N$ and assume that there exists $Q' < N$ such that the series $F_{1,Q'-1}$ is a homogeneous series, and the dimension d of its minimal LRF is less than $Q'/2$ (see Sections 3.1 and 3.2 for the full description). By definition, the maximal Q' coincides with Q.

Let us choose the proper parameters B and L. (According to our assumptions, any $B > 2d$ and L such that $d < \min(L, B - L + 1)$ will suffice.)

If we take any subseries $F_{i,i+B-1}$, such that $i \leq Q - B$, and consider the SVD of its L-trajectory matrix, then for some r all the eigenvalues $\lambda_s^{(i)}$ with $s > r$ are equal to zero. Therefore, we take $I = \{1, 2, \ldots, r\}$. Evidently, $r = d$.

In this 'ideal' situation, the nonzero elements of the H-matrix indicate the existence of some heterogeneity in the series. A classification of the types of heterogeneity is presented in Section 3.2.1. This classification allows us to characterize permanent and temporary heterogeneities. It also helps to understand differences in H-matrices depending on the dimensions of the LRFs before and after the change-point.

(b) Nonzero $F_N^{(2)}$: identification

If $F_N^{(2)}$ is not the zero series, then the problem of the choice of the detection parameters is more difficult. In particular, the representation

$$F_N = F_N^{(1)} + F_N^{(2)} \qquad (3.18)$$

is fundamental in theory in view of the different roles of both series in the detection problems. (For instance, $F_N^{(2)}$ is usually regarded as a noise series.)

In practice, however, we deal with the entire series F_N and the difference between $F_N^{(1)}$ and $F_N^{(2)}$ becomes relative. Thus, the practical problem of identifying the terms in the sum (3.18) arises; we assume that the representation (3.18) does hold, but we do not know the addends.

Let us assume that the parameters B and L are selected in such a way that the subseries $F_{i,i+B-1}^{(1)}$ and $F_{i,i+B-1}^{(2)}$ are (approximately) stably separable for some (maximal) range $1 \leq i \leq Q' - B$ of i. Then the difference between $F_N^{(1)}$ and $F_N^{(2)}$ can been explained as follows (see also Remark 3.4).

1. We assume that the series $F_{1,Q'-1}^{(1)}$ is homogeneous.

2. The chosen set I of eigentriples corresponds to the subseries $F_{i,i+B-1}^{(1)}$ of the series $F_{1,Q'-1}^{(1)}$ for all $i = 1, \ldots, Q' - B$; as a rule, it is a set of several leading eigentriples.

3. The heterogeneity under detection is assumed to happen at the series $F_N^{(1)}$.

Let us turn now to the identification problem. By our assumptions, for the selected B and L certain eigentriples (the same for all i) of the trajectory matrices of the subseries $F_{i,i+B-1}$ ($i \leq Q' - B$) are stably interpretable as (approximately) describing subseries of the same homogeneous series.

Then, until the moment of time Q' the series $F_N^{(1)}$ is (approximately) identified by the obtained set I of the eigentriples; we thus have the detection parameters r and I. Remark 3.4 yields that we do not need to know at which series the heterogeneity actually happens, whether it is $F_N^{(1)}$ or its residual.

Formally, we have the inequality $Q' \leq Q$. Since Q is the first change-point, the 'ideal' case is the equality $Q' = Q$. However, in practice the inequality $Q' < Q$ is more realistic. This is due to various deviations from stable separability which may happen before the moment Q. In the situation of the eigentriple rearrangement we may get $Q' = M$, where M is the time that the rearrangement occurs; deviations from weak separability may also lead to the inequality $Q' < Q$.

The identification procedure can easily be performed in some cases, and it is difficult in others. Moreover, for certain F_N no suitable parameters $Q' > 1$, B and L can be found.

However, there is a situation when the described procedure is valid with $Q' \approx Q$. This is the case of small noisy-like $F_N^{(2)}$.

(c) Small noisy-like $F_N^{(2)}$

In view of Remark 3.3 the case of small noisy-like $F_N^{(2)}$ is similar to the 'ideal' case of the zero series $F_N^{(2)}$. A natural example of this situation is the problem of the change-point detection in a homogeneous signal $F_N^{(1)}$ in the presence of a small noise $F_N^{(2)}$. Let us consider this example.

To obtain (approximate) separability of the series $F_{i,i+B-1}^{(1)}$ and $F_{i,i+B-1}^{(2)}$ until the change-point, we must take a relatively large B. (Though it must not to be larger than the expected value of Q.)

Concerning the choice of L, we can take it approximately equal to $B/2$ (see Section 1.6 for details of the parameter choice for separation).

Since the series $F_N^{(2)}$ is small enough and in view of the (approximate) separability obtained, several r leading singular values produced by the series $F_{i,i+B-1}$ must be large enough and describe the series $F_{i,i+B-1}^{(1)}$, while the other singular values are expected to be small. Thus, an abrupt decrease of the singular values, placed in order of decrease of their magnitudes, may help in finding the number r

and the set $I = \{1, 2, \ldots, r\}$. The other (and even more important) method is the eigenvector analysis, which was thoroughly described in Chapter 1.

If the parameters are chosen in a proper way, then the corresponding heterogeneity matrix has small elements in the left bottom block $\mathcal{A}$ (see Fig. 3.2).

For example, if $F^{(1)}_{1,Q-1}$ is a harmonic with amplitude C and $F^{(2)}_N$ is a white noise series with variance σ^2, then asymptotically in N and other parameters, the elements of the block $\mathcal{A}$ are close to $\sigma^2/(0.5C^2 + \sigma^2)$; see Example 3.13.

If $\sigma^2 \ll C^2$, then the block $\mathcal{A}$ is zero-like. The values of the heterogeneity indices for the other blocks depend on the type of heterogeneity.

Evidently, the case of noisy-like $F^{(2)}_N$ is not the only simple case of heterogeneity detection. Generally, $F^{(2)}_N$ should be a 'small' series and there ought to exist parameters B, L providing (approximate) stable separability of the subseries $F^{(1)}_{i,i+B-1}$ and $F^{(2)}_{i,i+B-1}$ up to a point $Q' \approx Q$. This is the case of Example 3.17. However, there is no general rule for the choice of B and L in all 'simple' situations.

As a result, if the detection is performed in a situation close to 'ideal', then large values of the heterogeneity index indicate a heterogeneity, and the general form of the H-matrix can help to identify both the change-point and the type of the heterogeneity.

(d) General $F^{(2)}_N$

By choosing the eigentriples for $F^{(1)}_N$ we are trying to obtain good separability features and to collect all stable eigentriples with relatively large singular values to describe $F^{(1)}_N$ until the moment of time Q'. Since Q' depends on B, L, r and I; in practice we try to obtain Q' as large as we can.

If the time series $F^{(2)}_N$ is noisy-like and small (large signal/noise ratio), these goals can often be realized at least for large N. In particular, we can obtain the equality $Q' \approx Q$. In the case of the general $F^{(2)}_N$, the goals may contradict. We, however, have to try our best to satisfy these goals, paying primary attention to the value of Q'.

Examples of this situation are described in Section 3.3.2. For instance, in Example 3.20 we have that for $r = 2$ and $I = \{1, 2\}$ the moment of time Q' corresponds to the eigentriple rearrangement moment $M = 210$, while the choice $r = 4$ and $I = \{1, 2, 3, 4\}$ gives $Q' \approx Q = 301$.

It follows from these examples that the detection problem is complicated in view of the possibility that the detection background (that is, block $\mathcal{A}$) may contain large elements in certain columns. If separability is approximate, then the equal-row background is perturbed, and it is difficult to recognize a possible heterogeneity on the perturbed nonconstant background.

Moreover, if the entire series is generally increasing or decreasing, then the detection background also varies in a monotone manner, and the heterogeneity recognition is even more complicated.

3.4.2 Detection functions

Let us make several remarks concerning the interpretation of the behaviour of the detection functions (see Section 3.1.2 for their formal definition), which is useful in the case of forward detection.

1. As a rule, the row detection function is best if we want to detect the first change-point of the series. In the ideal situation, the change-point coincides with the first point of sharp increase of this function.

2. In principle, the diagonal detection function may indicate change-points more clearly than the other ones if a sharp heterogeneity takes place on the background of a slowly varying structure of the time series. On the other hand, its use can be inconvenient since a single heterogeneity may give rise to several sharp peaks on the plot of the diagonal function.

3. The symmetric detection function has bad detection power. Even so, it is useful since it characterizes the local description of the series F_N by a fixed set of eigentriples.

4. The column detection function is, as a rule, a weaker detector of the heterogeneity than the row one; however, it is often informative when we try to distinguish the heterogeneity from the eigentriple rearrangement.

   The difference between the row and column detection functions is a good indicator of the true heterogeneity. If the column function has a sharp increase and the row function is slowly varying, then we have an argument that we are dealing with the eigentriple rearrangement.

3.4.3 Multiple heterogeneity

Until now we have discussed the case of a single heterogeneity and its forward detection. This means that we were looking for a single change-point (and its heterogeneity type), starting at the initial subseries $F_{1,B}$ of the time series F_N. Let us make several remarks concerning the specifics of general heterogeneity detection, when both suppositions are not assumed.

For simplicity we deal only with the case of zero series $F_N^{(2)}$. In particular, this means that the set I is always of the kind $\{1, \ldots, r\}$ for certain r.

The case of small noisy-like $F_N^{(2)}$ is analogous. As usual, other $F_N^{(2)}$ may significantly complicate the detection problem.

1. Since the selection of the eigentriples (that is, the choice of the set I) is typically performed by analyzing a single base subseries $F_{i,i+B-1}$ (we call it the *starting base subseries*) of the time series F_N, the corresponding H-matrix is adapted to detect heterogeneities relative to this base subseries. If we do not deal with the forward detection, then the starting base subseries can differ from $F_{1,B}$. For example, the backward H-matrix can essentially differ from the forward one due to the difference between their base eigentriple sets.

2. If all the homogeneity intervals of the series F_N have the same difference dimension r (equal to that of their minimal LRFs), then the H-matrix does not depend on the choice of the starting base subseries belonging to the intervals of homogeneity.

3. If the difference dimensions are different, then we generally cannot take a single starting base subseries for the entire series F_N and therefore cannot get a single H-matrix of a simple and understandable structure.

   If the sequential difference dimensions are decreasing (increasing), then the forward (backward) detection would give the entire picture. For instance, to understand the type of the heterogeneity of Example 3.4, it would be useful to apply the backward H-matrix with $r = 4$, which for appropriate values of parameters has a form similar to that of the matrix of Fig. 3.3.

4. In the general case, it is useful to search sequentially for the change-points and heterogeneities. This can be done as follows.

   Starting, for instance, with the forward detection, we choose the parameters B and L, take $F_{1,B}$ as a starting base subseries, select the base set of the eigentriples, and obtain (at least approximately) the interval $[Q_1, Q_1 + S_1]$ of the first heterogeneity with the help of the (forward) H-matrix. Then we take the second homogeneous interval of the series and produce the second H-matrix, and so on until the end of the series is reached. The collection of H-matrices obtained in this way would give the entire description of the situation.

5. The row and column detection functions can be used at each step of the sequential detection procedure in the usual way. The use of the diagonal detection functions should be done with care; these functions give adequate information only while searching for the first forward heterogeneity. At subsequent steps, the straightforward use of all the sharp peaks of the diagonal detection function may lead to misinterpretations.

3.4.4 Heterogeneities in trends and periodicities

Let us consider two separate problems related to heterogeneity detection in trends and periodicities.

(a) Trends

In accordance with our definition, the trend of the series is associated with its low-frequency (but nonperiodic) component. To separate it from the other components of all the series $F_{i,i+B-1}$ with the help of a stable set of eigentriples, we must take a relatively small B. The choice of the window length L depends on the concrete series and should be made so that the best separation is achieved.

Under these conditions, a sufficiently large trend is described by the single leading eigentriple of SVD of the trajectory matrix of the series $F_{i,i+B-1}$. (Otherwise, if the trend is not large enough, we can add a constant to the initial series.)

CHOICE OF DETECTION PARAMETERS

Therefore, the series $\widetilde{F}_{i,i+B-1}$, reconstructed from the leading eigentriple, must be similar to the exponential series with some rate α_i, that is to the series of the form $g_n = c_i e^{\alpha_i n}$ with some c_i and $n = i - 1, \ldots, i + B - 2$. If the trend is the exponential-like series on some time interval, the corresponding rates α_i (and c_i) are almost equal to each other. (Note that any reasonable monotone series can be locally approximated by some exponential series.)

On time intervals where the trend changes its behaviour, the rates also change, and we come to the case of (approximate) permanent violations in the piecewise exponential series. Therefore, sharp changes in the trend behaviour will be detected via the increase of the detection functions.

Let us now turn to the structure of the H-matrix detecting changes in trend.

As was mentioned above, a single permanent heterogeneity does not change the dimension of the series and is characterized by small values of the heterogeneity index in blocks $\mathcal{A}$ and $\mathcal{D}$ (see Examples 3.1, 3.17 and 3.19). Thus, we can expect that all the rectangles along the main diagonal of the H-matrix have small elements.

Besides, computational experiments show that, as a rule, other blocks of homogeneity (for the case of a single heterogeneity, these are the blocks $\mathcal{B}$ and $\mathcal{C}$) of the H-matrix also have small elements.

This effect can be explained by the specific nature of the H-matrices corresponding to relatively small variations in the rate of the exponential series. Indeed, even large changes in trend behaviour correspond to relatively small changes in the exponential rate (say, from 0.005 to 0.01). Therefore, despite the fact that the heterogeneity under consideration is of a permanent type, the heterogeneity matrix is going to be 'cross-structured'.

(b) Periodicities

Let us assume that we deal with the forward heterogeneity detection in a series $F_N = F_N^{(1)} + F_N^{(2)}$, and the component $F_N^{(1)}$ of the series is homogeneous until time Q. We also assume that $F_N = G_N^{(1)} + G_N^{(2)}$, where $G_N^{(1)}$ (the signal) is a periodic series until time Q and $G_N^{(2)}$ is a small noise (as earlier, this is assumed for simplicity).

Since $F_N^{(1)}$ is a homogeneous series, it is a component of $G_N^{(1)}$, and the (first) heterogeneity occurs in one or several harmonic components of the series $G_N^{(1)}$. Our aim is to discuss the choice of parameters in the manner of Section 3.4.1.

In view of the periodic feature of the signal, the detection parameters B, L and T ought to be proportional to the period of the series $G_N^{(1)}$. Thus, we can hope that the harmonic components of the signal are (approximately) separated from each other (and from the noise) in the SVD of the L-trajectory matrices of the subseries $F_{i,i+B-1}$ for all $i \leq Q' - B \approx Q - B$.

By collecting the leading harmonic components with stable eigentriple numbers we build $G_{1,Q'-1}^{(1)}$. If we wish, we can assume that $F_N^{(1)} = G_N^{(1)}$. Otherwise, we reduce the set of chosen stable eigentriples determining $F_N^{(1)}$. At any rate, at least the block $\mathcal{A}$ of the heterogeneity matrix will consist of approximately equal elements.

If, additionally, the series F_N has a trend component and we do not look for its heterogeneities, then it would be better to extract the trend (say, with the help of Basic SSA) and perform heterogeneity detection for the detrended series.

3.4.5 Role of parameter T

Generally, the detection parameter T is a smoothing parameter. Small values of T imply both a large contrast in the detection and a high sensitivity to small perturbations of the series. By enlarging T, we reduce the contrast between small and large values of the detection functions and make these functions smoother.

The minimal value of T is $T = L$. This corresponds to the case of a single addend in both the numerator and denominator of the formula (3.1), which defines the heterogeneity index. In the general case, the number of addends in (3.1) is $T - L + 1$.

If we deal with periodic series, then there are additional constraints for choosing proper values of parameter T. In this case, as was explained in the previous subsection, a properly chosen T must be proportional to the period of the series.

3.5 Additional detection characteristics

Various additional detection characteristics can help to identify and interpret heterogeneities in time series. Let us discuss three groups of these characteristics.

The first group of characteristics is based on the so-called renormalized heterogeneity matrices (and the corresponding detection functions). Renormalization can help, for example, when we deal with monotone series or its monotone components.

The second group (a collection of polynomial root functions) is related to the variations in linear spaces $\mathfrak{L}_{I,B}^{(L,i)}$. The third group (characteristics associated with the moving periodograms) describes changes in the spectral structure of the initial series in time.

3.5.1 Renormalized heterogeneity matrices

The normalization of the heterogeneity index, produced by the denominator in the formula (3.1), is natural; indeed, the range of the index is $[0, 1]$, and the extreme values 0 and 1 indicate the opposite situations: pure homogeneity and pure heterogeneity (see examples in Section 3.2.1).

However, in some cases this normalization can be a disadvantage. Let us discuss the problem by the example of the row detection functions.

Assume that the series is positive and monotone increasing. Then the denominator of the row detection function – see (3.4) and (3.1) – increases as well. There-

ADDITIONAL DETECTION CHARACTERISTICS

fore, the heterogeneity index of the last part of the series is generally smaller than the analogous index for the initial interval of the series only because of the increase of the entire series.

This effect has two unpleasant consequences:

1. The background for the detection is nonconstant (see Example 3.9). This produces difficulties in identification of possible heterogeneities.
2. Two 'equivalent' heterogeneities occurring at the beginning and at the end of the series will produce different (in absolute scale) increases of the heterogeneity characteristics. Thus, we shall not be able to compare the heterogeneities by their 'power'.

To avoid these difficulties, we can denormalize the heterogeneity index by omitting the denominator in the formula (3.1). However, it is more convenient to have a denominator (not depending on j) in the definition of this index. Thus, we define the *renormalized heterogeneity index* by

$$\widetilde{g}_{ij} \stackrel{\text{def}}{=} \frac{\frac{1}{(T-L+1)L} \sum_{l=1}^{T-L+1} \text{dist}^2\left(X_{l,j}, \mathfrak{L}_{I,B}^{(L,i)}\right)}{\frac{1}{N} \sum_{k=0}^{N-1} f_k^2} \qquad (3.19)$$

where B, L, T, r and I are fixed detection parameters and $X_{l,j}$ are the L-lagged vectors of the series $F_{j,j+T-1}$. In this definition, we use the squared sum of all the elements of the series F_N as the denominator and take averaging coefficients in agreement with the total number of the terms of the series in all the sums.

Let us explain the features of this renormalization via the following example.

Example 3.21 *Renormalized heterogeneity indices*
Consider a time series $F_N = F_N^{(1)} + F_N^{(2)}$ with two temporary heterogeneities of the 'outlier' type (see Example 3.7). To describe F_N, we start with the time series $F_N^* = F_N^{(*,1)} + F_N^{(*,2)}$, where

$$f_n^{(*,1)} = 2a^n, \quad a = 1.008,$$

and $f_n^{(*,2)}$ is Gaussian white noise with zero mean and unit variance.

Let temporary violations be caused by local shifts of the series F_N^*. That is, let f_n and f_n^* coincide for all n except the time intervals $100 \leq n \leq 104$ and $300 \leq n \leq 304$, where $f_n = f_n^* + 5$. The initial series is depicted in the bottom graph of Fig. 3.32.

As usual, we cannot formally determine which one of the two series, either the signal $F_N^{(*,1)}$ or noise $F_N^{(*,2)}$, is affected by this shift. For definiteness, we apply the shifts to $F_N^{(*,1)}$.

Let $B = 60, L = 30, r = 1, I = \{1\}$ and $T = 39$. The top graph of Fig. 3.32 shows the row (thin line) and column (thick line) detection functions corresponding to the standard heterogeneity matrix **G**.

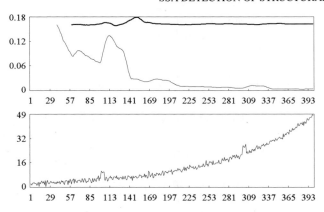

Figure 3.32 *Renormalization: standard detection functions and the time series.*

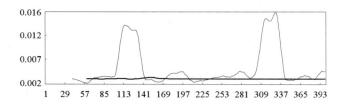

Figure 3.33 *Renormalization: renormalized detection functions.*

The row detection function clearly shows the first change-point, but since the signal increases, it poorly indicates the second one. It can be checked that the (absolute) perturbations of the row detection function caused by the second heterogeneity and the random fluctuation occurred around the time 170 are approximately equal.

Unlike the case depicted in Fig. 3.32, the row detection function corresponding to the renormalized heterogeneity matrix $\widetilde{\mathbf{G}}$ (thin line in Fig. 3.33) properly reflects the situation. Two peaks of the row function have approximately equal values; this is in agreement with the equality of the shifts for the unperturbed time series.

The behaviour of the renormalized column detection function (Fig. 3.33, thick line) is generally similar to that of the standard column function.

Remark 3.5 If the heterogeneity occurs in a stationary time series (especially if the series is a white noise) and means a change in its amplitude, then the standard heterogeneity index defined by (3.2) and (3.1) may have no sense.

Indeed, since the noise is generally 'structureless', there are no stable eigentriples determining the base spaces $\mathfrak{L}_{I,B}^{(L,i)}$. Therefore, the only reasonable version of $\mathfrak{L}_{I,B}^{(L,i)}$ is the zero space $\mathfrak{L}_0^{(L)}$ corresponding to $r = 0$ and $I = \varnothing$.

This choice leads to $g_{ij} \equiv 1$ implying that the standard heterogeneity index is meaningless. By contrast, the renormalized heterogeneity index defined in (3.19) can be very useful in detection of change-points in the variance of the noise.

3.5.2 Roots of characteristic polynomials

When considering the heterogeneity matrices, we relate the lagged vectors of the test subseries $F_{j,j+T-1}$ and the base linear spaces $\mathfrak{L}_{I,B}^{(L,i)}$ generated by the base subseries $F_{i,i+B-1}$ of the series F_N.

It may also be reasonable to study the sequence of spaces $\mathfrak{L}_{I,B}^{(L,i)}$ irrespectively of the other objects. Such a study may be useful for several purposes. For example,

1. The problem of the eigentriple rearrangement (see Section 3.3) is formulated in terms of the base subseries, rather than in terms of the test subseries. By definition, the spaces $\mathfrak{L}_{I,B}^{(L,i)}$ relate to the base subseries of the series F_N and reflect their important properties.

2. Since heterogeneities usually affect the base spaces $\mathfrak{L}_{I,B}^{(L,i)}$, these spaces can help in studying the nature of heterogeneities.

Note that the symmetric heterogeneity function describes the spaces $\mathfrak{L}_{I,B}^{(L,i)}$ only as the approximation spaces. Therefore, this function is not very informative in view of its low sensitivity to different heterogeneities. The root functions of the characteristic polynomial seem to be preferable for the purpose of monitoring the homogeneity of the series.

Let $\mathfrak{L}_{I,B}^{(L,i)}$ be a certain base space with fixed $i \leq Q - B$. In the case of stable separability, $\mathfrak{L}_{I,B}^{(L,i)}$ is chosen as the trajectory space of a homogeneous additive component $F_{i,i+B-1}^{(1)}$ of the series $F_{i,i+B-1}$.

Theorem 5.2 of Section 5.2 shows how to construct the LRF of dimension $L-1$, which governs the series $F_{i,i+B-1}^{(1)}$. This LRF is 'extracted' from the space $\mathfrak{L}_{I,B}^{(L,i)}$. The roots of the characteristic polynomial of this LRF give essential information concerning both the series $F_{i,i+B-1}^{(1)}$ and the space $\mathfrak{L}_{I,B}^{(L,i)}$.

Let d be the difference dimension of the series $F_{i,i+B-1}^{(1)}$ (evidently, $d = \operatorname{card} I$). Then, only d roots of the characteristic polynomial affect the behaviour of the series (as a rule, they have maximal moduli); the other roots are extraneous.

When i varies, the LRFs obtained give a dynamical description of the sequence of subseries $F_{i,i+B-1}^{(1)}$ and the associated linear spaces. All essential modifications of the spaces must be reflected in this description. These modifications can be caused by the eigentriple rearrangements and other structural changes in the series.

In the case of approximate separability these considerations are also valid. Then d, the number of selected main polynomial roots, has the sense of the difference dimension of a certain series approximating the series $F_{i,i+B-1}^{(1)}$.

Let us consider the characteristic polynomial of some LRF and assume for simplicity that it does not have multiple roots. In accordance with the agreement of Section 2.2.1, we describe a nonzero polynomial root with a modulus ρ in the following way:

1. Every positive root has the representation $(\rho, 0)$.
2. Every negative root is represented as $(\rho, 1/2)$.
3. Every pair of complex conjugate roots has a representation (ρ, ω), where $\pm 2\pi\omega$ are their polar angles ($0 < \omega < 1/2$).

For a fixed polynomial, the collection of pairs (ρ, ω) is called the *modulus-frequency representation* of the polynomial roots.

Note that every pair (ρ, ω), $\omega \in [0, 1/2]$, corresponds to a separate real (exponential-cosine) additive component of the series governed by the LRF. (This component is equal to $A\rho^n \cos(2\pi\omega n + \phi)$ for some A and ϕ.) Therefore, we do not distinguish the conjugate roots from each other and treat them as a single object with the representation (ρ, ω). Evidently, the number of modulus-frequency root representations does not exceed the degree of the polynomial.

Having a sequence of characteristic polynomials which correspond to the base linear spaces $\mathcal{L}_{I,B}^{L,i}$, we choose the number m of the module-root representations to be investigated.

The selection procedure for m is as follows. Since the number r of eigentriples is already chosen, we take the number d of the main polynomial roots equal to r. Then, analyzing the base subseries $F_{1,B}$ (we are describing the forward detection) and perhaps some other base subseries $F_{i,i+B-1}$, we understand whether these d roots are real or complex and thus obtain the number m.

Having obtained a collection of polynomial root representations, we arrange them in order of decreasing moduli. Thus, we come to m two-dimensional piecewise linear curves with nodes (ρ_i, ω_i) on the modulus-frequency plane ($i = 1, \ldots, N - B + 1$). Evidently, the curves may intersect and produce an unclear picture.

To make the situation clearer, we reorder the root representations in the following manner. For fixed i we arrange the selected m leading root representations in order of decreasing frequencies. Then we have m *root-frequency functions* with nonintersecting plots and m *root-modulus functions*, which are generally not ordered with respect to their values.

Abrupt changes in the behaviour of the root-frequency and the root-modulus functions can give a lot of information concerning the behaviour of the series F_N. Let us consider an example.

Example 3.22 *Root functions*
Consider the time series described in Example 3.20. The choice $r = 2$ and $I = \{1, 2\}$ gives us two violations: the eigentriple rearrangement and a permanent

ADDITIONAL DETECTION CHARACTERISTICS

heterogeneity. They divide the entire time interval into three parts (homogeneity intervals).

Since for any i the two leading polynomial roots are complex conjugate and correspond to a single harmonic approximating the series $F^{(1)}_{i,i+B-1}$, they give a single modulus-frequency representation (ρ_i, ω_i) and therefore produce a single root-modulus function and a single root-frequency function.

Fig. 3.34 depicts the root-modulus function (top graph) and the root-frequency function (bottom graph). Note that in what follows, the terms of the root functions are indexed in the same manner as the detection functions to simplify the comparisons.

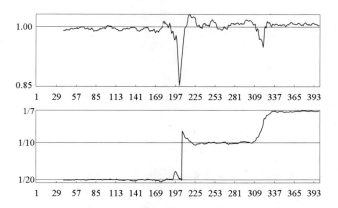

Figure 3.34 *Root functions: two leading eigentriples.*

The root functions give additional information to Figs. 3.29 and 3.30. The root-modulus and root-frequency functions correspond to the series component $F^{(2,1)}_N$ on the first homogeneity interval. Then the root-modulus function oscillates around the constant $1/a \approx 0.994$, and the values of the root-frequency function are approximately equal to 1/20.

On the second and third homogeneity intervals these functions correspond to the series component $F^{(1)}_N$. Therefore, the root-modulus function oscillates around the constant $a = 1.006$, and the values of the root-frequency function are close to 1/10 in front of the second heterogeneity interval and are close to 1/7 behind it.

Fig. 3.34 (as well as the moving periodograms of Example 3.23) demonstrates the difference between the eigentriple rearrangement (sharp alternation of the frequency) and the true change-point (smooth transition from $\omega = 1/10$ to $\omega = 1/7$).

A different situation will hold if we choose $r = 4$ and $m = 2$ (see Fig 3.31). In this case, we have a single perturbation corresponding to the true change-point.

Fig. 3.35 demonstrates two pairs of the root-modulus and root-frequency functions corresponding to the two harmonics approximating the series $F^{(1)}_{i,i+B-1}$ (and corresponding to two pairs of the complex conjugate polynomial roots). Note that the representations are arranged in order of decreasing frequencies.

The root-frequency functions are presented in the bottom graph of Fig.3.34, and the root-modulus functions are shown in the top graph. The leading root-frequency function (thick line) demonstrates a smooth transition from $\omega = 1/10$ to $\omega = 1/7$ on the heterogeneity interval. The corresponding root-modulus function (also marked by the thick line) is stable.

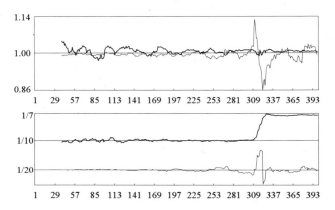

Figure 3.35 *Root functions: four leading eigentriples.*

The second root-frequency function (thin line) has small deviations from $\omega = 1/20$ everywhere except for the heterogeneity interval. The corresponding root-modulus function has a similar behaviour.

Thus, the heterogeneity is detected by both modulus-frequency representations. Naturally, the eigentriple rearrangement has no influence on the root functions for $r = 4$.

3.5.3 Moving periodograms

Spectral properties of 'moving' subseries $F_{i,i+B-1}$ can be characterized by a sequence of the corresponding periodograms $\Pi_f^{(i,i+\widetilde{B}-1)}$ (see Section 1.4.1 for the definition and a discussion).

Formally, to relate the periodograms and the detection/root functions, we must take $\widetilde{B} = B$, but, in view of the discreteness of the periodogram grid, it can be more convenient to take $\widetilde{B}$ slightly different from B. Define

$$\pi_{jk} \stackrel{\text{def}}{=} \Pi_f^{(k,k+\widetilde{B}-1)}(j/\widetilde{B}).$$

Then, the *periodogram matrix* is defined as

$$\Pi = \{\pi_{jk}\}, \quad k = 1, \ldots, N - \widetilde{B} + 1, \quad j = 0, \ldots, [\widetilde{B}/2],$$

ADDITIONAL DETECTION CHARACTERISTICS

where $[x]$ denotes the integer part of x.

If we depict the periodogram matrix in the same manner as the heterogeneity matrices, then we will obtain a figure of the local spectral behaviour of the series F_N. In such a figure we should be able to see a relation between the subseries $F_{i,i+\widetilde{B}-1}$ and the powers of their main periodogram frequencies (calculated up to the errors caused by the discreteness of the grid $\{j/\widetilde{B}\}$). This can help in explaining the nature of the heterogeneity.

For a fixed frequency $\omega = j/\widetilde{B}$, the *frequency-power function* (*f-power function*) is defined as

$$p_n^{(\omega)} \stackrel{\text{def}}{=} \pi_{j,n+1}, \quad n = 0, \ldots, N - \widetilde{B}.$$

This function provides us with useful information concerning the dynamics of a single frequency component of the series.

Example 3.23 *Moving periodograms*

Consider the time series described in Examples 3.20 and 3.22. In these examples, the length B of the base intervals was taken as 39 to achieve a better separability. Since the periodograms relate to the entire series and are independent of separability of its components, we take the length $\widetilde{B}$ of the moving interval as 42 to get a better correspondence with the main frequencies (1/20, 1/10 and 1/7) of the series.

The part of the periodogram matrix corresponding to the low frequencies is depicted in Fig. 3.36.

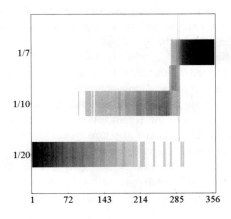

Figure 3.36 *Moving periodograms: periodogram matrix.*

Three frequency rows can be easily seen in the matrix plot of Fig. 3.36. They relate to the frequencies 2/42, 4/42 and 6/42 (from bottom to top), which are approximately equal to 1/20, 1/10 and 1/7. The changes in the frequency structure of the series become apparent. To understand the nature of these modifications we use the f-power functions corresponding to these frequencies. Note that in what follows, the terms of the f-power functions are indexed in the same way as the detection and root functions to simplify the comparisons.

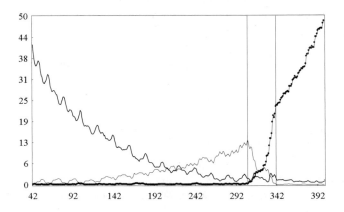

Figure 3.37 *Moving periodograms: f-power functions.*

Three f-power functions corresponding to $\omega = 2/42 \approx 1/20$ (thick line), $\omega = 4/42 \approx 1/10$ (thin line) and $\omega = 6/42 = 1/7$ (thin line marked by dots) are shown in Fig. 3.37. They demonstrate that there is no heterogeneity until time 300; all the f-power functions demonstrate smooth behaviour. (Small fluctuations in the f-power functions are related to the approximation quality of the true frequencies by the nodes of the grid $\{j/42\}$.)

The interval of heterogeneity, marked by the two vertical lines, is clearly seen as well; it is characterized by the abrupt changes in f-power functions. At the same time, the intersection of the f-power functions, corresponding to $\omega \approx 1/20$ and $\omega \approx 1/10$, indicates the region of the eigentriple rearrangement.

In summary, the intervals of the eigentriple rearrangement and the heterogeneity can be (approximately) calculated by three methods: with the help of the detection functions, root functions and by means of the f-power functions. All three methods give similar results, though the intervals obtained are slightly different in width and have small shifts relative to each other as a result of the noise.

3.6 Examples

3.6.1 'Coal sales': detection of outliers

We use the example 'Coal sales' (U.S. monthly sales of coal, from January 1971 to December 1991) discussed in Makridakis, Wheelwright and Hyndman (1998), to demonstrate the capabilities of SSA detection of outliers in slowly varying noisy series having periodic components.

EXAMPLES

A plot of 'Coal' time series F_N is given in Fig. 3.38 (bottom graph). Several abrupt local changes of its behaviour can be easily seen there. Basic SSA with double centring, applied to the entire 'Coal' series, clearly shows the general structure of the series: it has a linear-like trend, several components describing the annual periodicity, and the residuals.

Let us select parameters to perform the SSA heterogeneity detection. Since we are interested in 'outliers', we must not take B larger than the base time interval between the arrivals of the 'outliers'. Also, small values of B provide a good description of the trend by a single leading eigentriple.

On the other hand, B and L ought to be large enough to ensure separation of smooth trends of the subseries $F_{i,i+B-1}$ from their annual periodicities. The resulting decision is $B = 23$ and $L = 12$.

Since B is rather small, for any i the trend of the series $F_{i,i+B-1}$ can be perfectly approximated by an exponential series and therefore described by just one leading eigentriple of the SVD of the corresponding trajectory matrix. Thus, we select $r = 1$ and $I = \{1\}$.

The heterogeneity matrix with $T = 16$ is shown in Fig. 3.39, the corresponding row detection function is presented at the top graph of Fig. 3.38.

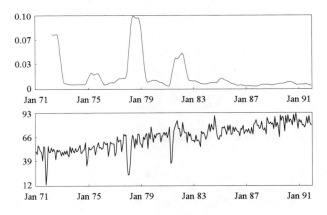

Figure 3.38 *Coal sales: row detection function and the time series.*

Fig. 3.39 demonstrates an approximate equal-row structure of the H-matrix. Therefore, approximate separability takes place all the time. We are thus dealing with several temporary violations of the kind of Example 3.7, but the 'cross-structure' of the H-matrix is hardly recognizable due to the stability of the linear spaces $\mathfrak{L}_{I,B}^{(L,i)}$ (see formulae (3.2) and (3.1)) with respect to the influence of the outliers.

Sharp peaks of the row detection function correspond to outliers in the 'Coal' series. The correspondence can be easily seen by comparing the top and bottom graphs of Fig. 3.38. Note that the first 'outlier' (October 1971) is detected from

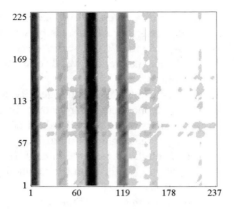

Figure 3.39 *Coal sales: H-matrix.*

the sharp decrease of the row function; the time of its occurrence belongs to the first series interval $F_{1,B}$ selected as the base one.

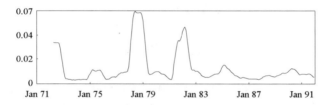

Figure 3.40 *Coal sales: renormalized row detection function.*

Since the 'Coal' series is generally increasing and the leading eigentriple describes just the increasing part of the series, the detection above is made on the decreasing background. Thus (see Section 3.5.1), it might be worthwhile to use the renormalized heterogeneity measures, i.e., to apply the alternative normalization in the definition of the heterogeneity indices.

The resulting row detection function is presented in Fig. 3.40. Comparing this plot with the analogous plot of Fig. 3.38, we can see that the two last peaks of the detection function (August 1984 and July 1989) became more distinct and therefore compatible, say, with an earlier peak in November 1974, which corresponds to the 'outlier' of similar discrepancy relative to the general tendency of the series.

3.6.2 *'Petroleum sales': detection of trend changes*

The series 'Petroleum sales' (U.S. monthly sales of petroleum, from January 1971 to December 1991, Makridakis, Wheelwright and Hyndman, 1998) has a trend of

EXAMPLES

complex form and a single outlier-like sharp peak corresponding to autumn 1990. The series can be found in the bottom graph of Fig. 3.42.

If we take relatively small B and L, choose $r = 1$ and $I = \{1\}$ and use the H-matrix and the detection functions of Section 3.4.4, then in addition to detecting the 'outlier' in the manner of the example of Section 3.6.1, we shall be able to indicate the changes in the series trend.

Here we deal with multiple permanent heterogeneities in the exponential series in the presence of a small noisy-like additive component and a single outlier region.

The H-matrix of the 'Petroleum' series computed for $B = T = 23$ and $L = 12$, is depicted in Fig. 3.41. The row detection function can be found at the top graph of Fig. 3.42 (thick line). The sharp peaks of the row detection function and the corresponding dark vertical and horizontal lines in Fig. 3.41 indicate the main changes in the trend behaviour.

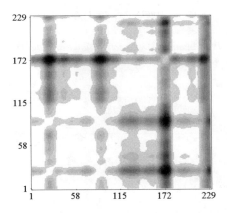

Figure 3.41 *Petroleum sales: H-matrix.*

The first large peak of the row detection function describes two changes that occurred during the period from autumn 1973 to summer 1974. The chosen B is too large to distinguish them. The second peak corresponds to a sharp increase of the series around October 1978. The end of this increase (approximately February 1981) is indicated by the next smoother peak of the row detection function. Note that the renormalization of the heterogeneity indices (see Section 3.5.1) would make this peak even more distinct.

The beginning (November 1985) and the end (October 1986) of a sharp decrease of the 'Petroleum' series are indicated by the (biggest) single peak of the row detection function (for large values of B these peaks are not distinguishable). Lastly, the outlier region is also apparent.

A plot of the leading root-modulus function of the characteristic polynomials corresponding to the linear spaces $\mathfrak{L}_{I,B}^{(L,i)}$ (see Fig. 3.43) confirms our conclusions concerning the approximation of the trend by a piecewise exponential series and the interpretation of the behaviour of the row detection function.

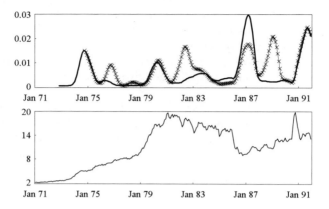

Figure 3.42 *Petroleum sales: detection functions and the time series.*

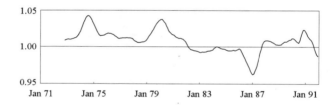

Figure 3.43 *Petroleum sales: leading root-modulus function.*

Indeed, all the roots are real and therefore they correspond to (locally) exponential series. Thus, the values of the roots determine the local rates of exponentials and their increasing/decreasing behaviour, which describe the behaviour of the series trend. It is the perfect correspondence between the peaks of the row detection function and the local extremes of the root-modulus function that confirms our considerations.

The thin line marked with crosses at the top graph of Fig. 3.42 depicts the diagonal detection function. As was mentioned in Remark 3.1 (see also Example 3.6), the diagonal detection function may have several peaks related to the same change-point. We can see that in this case, the diagonal detection function has four 'true' peaks, agreeing with the peaks of the row detection function as well as three spurious peaks, which mirror the true ones.

The structure of the H-matrix (Fig. 3.41) corresponds to the general description of Section 3.4.4, relating to trends. Thus, we have an explanation of both the light

EXAMPLES

rectangles located along the main diagonal of the matrix of Fig. 3.41 (the case of a permanent heterogeneity with conservation of dimension) and the other light rectangles (the specifics of a piecewise exponential series).

Though the H-matrix of Fig. 3.41 has a 'multiple cross' structure, it depicts permanent violations rather than temporary ones (the 'outlier' region is an exception).

3.6.3 'Sunspots': detection of changes in amplitudes and frequencies

The 'Sunspots' series (annual sunspot numbers, from 1700 to 1988, Hipel and McLeod, 1994) is a standard test-bed for numerous methods of time series analysis. The series (thin line in Fig. 3.44) is an amplitude-modulated oscillatory sequence of a rather complex form.

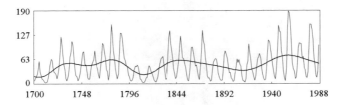

Figure 3.44 *Sunspots: time series and its low-frequency component.*

Its behaviour has obviously changed in the interval somewhere between the end of the eighteenth century to the thirties of the nineteenth century. (We do not discuss here the reasons for this well-known heterogeneity.) The thick line in Fig. 3.44 depicts the low-frequency component of the 'Sunspots' series.

The component is reconstructed with the help of Toeplitz SSA with window length 132 by the average triple and the eigentriples 5,6 and 9-11. It includes the slowly varying trend and two periodicities of approximately 100 and 55 years.

We can see that this component generally corresponds to the modulation of the amplitude of the 'Sunspots' series and reflects some specifics of the heterogeneity interval of interest. Despite this we shall investigate the 'Sunspots' series as a whole, trying to detect (and describe) the amplitude and frequency heterogeneities altogether.

Since the main frequency of the 'Sunspots' series is approximately $1/11$, we take $B = 43$ and $L = 22$ to separate the corresponding component from the other components and choose $r = 3$ and $I = \{1, 2, 3\}$ to join the trend and the main-frequency periodicity.

The resulting row detection function and the H-matrix for $T = 43$ are depicted in Fig. 3.45 and Fig. 3.46, respectively. The column detection function is similar to the row one.

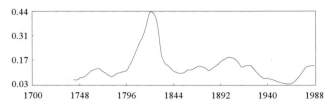

Figure 3.45 *Sunspots: row detection function.*

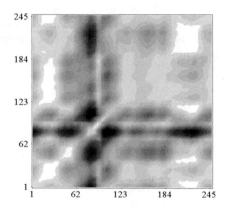

Figure 3.46 *Sunspots: H-matrix.*

A sharp peak of the row detection function and the cross-structure of the H-matrix precisely correspond to the heterogeneity region of the 'Sunspots' series.

The roots of the characteristic polynomial help us to understand the nature of the heterogeneity. Since the base spaces $\mathfrak{L}_{I,B}^{(L,i)}$ correspond to approximation of the time series $F_{i,i+B-1}$ by the sum of one exponential (the first eigentriple) and one harmonic (the second and third eigentriples), let us consider the two leading modulus-frequency root representations. The top graph of Fig. 3.47 presents the two leading root-modulus functions. The thick line corresponds to the real root (exponential component) and the thin line relates to the amplitude of the harmonic component of the series. The corresponding frequencies can be found in the bottom graph of Fig. 3.47.

We can see that both root-modulus functions are in perfect correspondence with each other. Since one function relates to the slowly varying part of the series and the other presents its main harmonic component, the correspondence of the root-modulus functions means that the amplitude modulation of the main harmonic component of the series is in agreement with its low-frequency component (see Fig. 3.44).

EXAMPLES

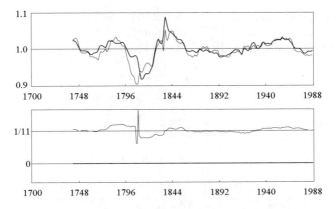

Figure 3.47 *Sunspots: two pairs of the leading root functions.*

Both series in the top graph of Fig. 3.47 have low values precisely at the heterogeneity region of the 'Sunspots' series, which is characterized by small amplitudes. The same region can be indicated by the abrupt changes of the leading root-frequency function (the thin line in the top graph of Fig. 3.47; the thick line is the zero one since it corresponds to the real root).

On the whole, the frequency values are about the main 'Sunspots' frequency (approximately, $1/11$), but the heterogeneity region is characterized by large discrepancies from this frequency. This means that the row detection function (see Fig. 3.45) detects both the heterogeneity in the amplitude modulation and the heterogeneity in frequency.

3.6.4 'Demands': rearrangement or heterogeneity?

Sometimes the distinction between the eigentriple rearrangement and heterogeneity of the series is difficult to pick up by a straightforward heterogeneity analysis. But it is important to discriminate between these effects since the rearrangement is a 'false' heterogeneity. The following example shows how additional detection techniques help in these situations.

The 'Demands' series (demand for a double-knit polyester fabric, sequential observations, Montgomery and Jonson, 1976) is an example of this kind. The series (after standardization) is shown at the bottom graph of Fig. 3.48. Its periodogram (see Fig. 3.49) shows a very regular frequency-structure of the series: the peaks of the periodogram are located on the grid $\{j/30\}$, though the series length is $N = 240$. Note that the true main frequencies of the series are slightly greater than $1/30$, $1/15$, etc.

A more detailed investigation shows that the most powerful frequency of the beginning of the series is $\omega = 1/2$. Therefore, we take $B = 59, L = 30$ (to achieve the best separation) and $r = 1, I = \{1\}$ (to start with the main frequency).

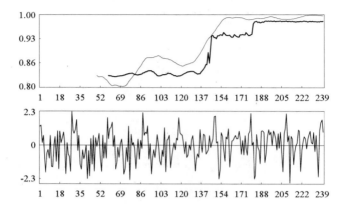

Figure 3.48 *Demands: detection functions and the time series.*

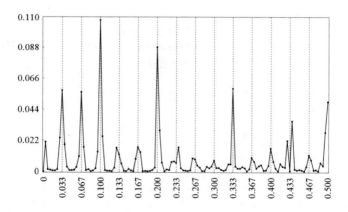

Figure 3.49 *Demands: periodogram of the series.*

The corresponding H-matrix (Fig. 3.50) and the row/column detection functions (top graph of Fig. 3.48, the column function is represented by a thick line) show two explicit rearrangement-like intervals of the series.

The root functions (Fig. 3.51) clarify the situation: while the root-modulus function (top graph) is generally stable (apart from the interval of the first rearrangement), the root-frequency function (bottom graph) indicates transitions from the frequency $\omega = 1/2$ to $\omega \approx 1/10$ and then to $\omega \approx 1/5$.

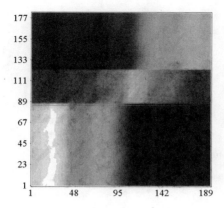

Figure 3.50 *Demands: H-matrix.*

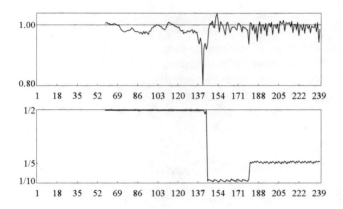

Figure 3.51 *Demands: root functions.*

The case can be clarified in an even better way. The periodogram matrix (see Section 3.5.3), calculated for $\widetilde{B} = 60$ (note that $B = 59$) shows that there is a time interval when all the powerful frequencies (except for the stable frequency $\omega \approx 1/10$) either lose their power or start gaining it. In particular, the frequency $\omega = 1/2$ vanishes almost at the same time as the power of the frequency $\omega \approx 1/5$ starts to increase.

The f-power functions (see Fig. 3.53) corresponding to the frequencies $\omega = 1/2$ (thin line), $\omega \approx 1/10$ (thick line) and $\omega \approx 1/5$ (thin line marked with dots) indicate both the two rearrangement intervals and a hidden heterogeneity between them.

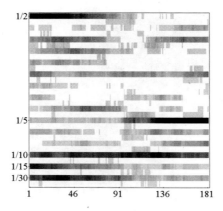

Figure 3.52 *Demands: periodogram matrix.*

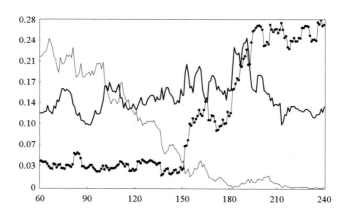

Figure 3.53 *Demands: f-power functions.*

The powers of the frequency $\omega = 1/2$ are slowly decreasing in time, while the powers of $\omega \approx 1/10$ are relatively stable. The interval where their values are approximately equal indicates the first eigentriple rearrangement, though not in terms of eigenvalues but rather in terms of the amplitudes of the Fourier decomposition of the series. After this interval $\omega \approx 1/10$ becomes the main frequency (see Fig. 3.51).

As it decreases, the power of the frequency $\omega = 1/2$ becomes as small as the frequency $\omega \approx 1/5$, which was almost constant until this time. Then the power of the frequency $\omega \approx 1/5$ has an abrupt increase, which indicates heterogeneity. When the values of the powers corresponding to $\omega \approx 1/5$ and $\omega \approx 1/10$ become close, the second rearrangement is indicated.

EXAMPLES

Lastly, let us consider the components of the 'Demands' series, corresponding to the three frequencies of interest. For this purpose, we apply the Toeplitz SSA with $L = 60$ and take eigentriples 1, 2-3 and 4-5 for the reconstruction. The result can be seen in Fig. 3.54: the component of $\omega = 1/2$ has a slowly decreasing amplitude (top graph), and the component of $\omega \approx 1/10$ is almost harmonic (middle graph), while the third component, which corresponds to $\omega \approx 1/5$, has a constant amplitude in the beginning of the series and then becomes modulated by the amplitude. Despite the fact that the abrupt change of the amplitude is smoothed, as is evident from Fig. 3.53, the whole picture corresponds to the indicated heterogeneity.

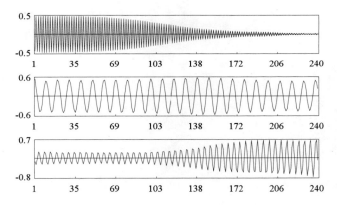

Figure 3.54 *Demands: three main components of the series.*

PART II
SSA: Theory

CHAPTER 4

Singular value decomposition

This chapter is devoted to a description of the singular value decomposition (SVD) of real matrices, which is the main mathematical tool in the Basic SSA method. Most features of SVD discussed below are used in different parts of the book and clarify either theoretical constructions or interpretation of results in various examples.

4.1 Existence and uniqueness

Let $\mathbf{X}$ be a nonzero matrix with $L > 1$ rows and $K > 1$ columns. Then $\mathbf{S} \stackrel{\text{def}}{=} \mathbf{X}\mathbf{X}^\mathrm{T}$ is a symmetric $L \times L$ matrix. Like every symmetric matrix, the matrix $\mathbf{S}$ has L linearly independent *eigenvectors*, that is, linearly independent vectors $U_1, \ldots, U_L$ such that

$$\mathbf{S} U_i = \lambda_i U_i,$$

where the λ_i are real numbers called the *eigenvalues* of the matrix $\mathbf{S}$. The linear space spanned by the collection of eigenvectors U_i is called the *eigenspace*.

We can choose the eigenvectors U_i to be *orthonormal*, that is, $(U_i, U_j) = 0$ for $i \neq j$ (the orthogonality property) and $\|U_i\| = 1$ (the unit norm property), where (X, Y) is the inner product of vectors X and Y, and $\|X\| = \sqrt{(X, X)}$ is the *norm* of the vector X.

Furthermore, the matrix $\mathbf{S}$ is *positive semidefinite*, i.e., $\lambda_i \geq 0$ for all $i = 1, \ldots, L$. We assume that the eigenvalues λ_i are placed in decreasing order: $\lambda_1 \geq \lambda_2 \geq \ldots \geq \lambda_L \geq 0$.

We denote by d the number of nonzero eigenvalues of the matrix $\mathbf{S}$. If $d < L$, $\lambda_d > 0$ and $\lambda_{d+1} = 0$, then all the other eigenvalues with indices larger than d are also zero. If $\lambda_L > 0$, then $d = L$. Since d is equal to the rank of the matrix $\mathbf{X}$, we obtain $d \leq \min(L, K)$.

For $1 \leq i \leq d$ we set

$$V_i = \frac{1}{\sqrt{\lambda_i}} \mathbf{X}^\mathrm{T} U_i. \tag{4.1}$$

Proposition 4.1 *Vectors U_i and V_i have the following properties.*
1. *Let $1 \leq i, j \leq d$. Then $(V_i, V_j) = 0$ for $i \neq j$ and $\|V_i\| = 1$. If $i > d$, then $\mathbf{X}^\mathrm{T} U_i = \mathbf{0}_\mathrm{K} \in \mathbf{R}^K$, where $\mathbf{0}_\mathrm{K}$ is the zero vector.*
2. *V_i is an eigenvector of the matrix $\mathbf{X}^\mathrm{T}\mathbf{X}$ corresponding to the eigenvalue λ_i.*

3. *If $1 \leq i \leq d$, then*

$$U_i = \frac{1}{\sqrt{\lambda_i}} \mathbf{X} V_i .$$

4. *If $K > d$ then all the other $K - d$ eigenvectors of the matrix $\mathbf{X}^T\mathbf{X}$ correspond to the zero eigenvalue.*

5. *The following equality is valid:*

$$\mathbf{X} = \sum_{i=1}^{d} \sqrt{\lambda_i} U_i V_i^T . \qquad (4.2)$$

Proof.

1. The first statement follows from the equality

$$(\mathbf{X}^T U_i, \mathbf{X}^T U_j) = (U_i, \mathbf{X}\mathbf{X}^T U_j) = \lambda_j (U_i, U_j)$$

and the orthonormality of the vectors U_i. In particular, if $i = j > d$, then $\lambda_j = 0$ and $\|\mathbf{X}^T U_j\| = 0$.

2. Consider the vector $\mathbf{X}^T\mathbf{X} V_i$:

$$\mathbf{X}^T\mathbf{X} V_i = \frac{1}{\sqrt{\lambda_i}} \mathbf{X}^T\mathbf{X}(\mathbf{X}^T U_i)$$

$$= \frac{1}{\sqrt{\lambda_i}} \mathbf{X}^T(\mathbf{X}\mathbf{X}^T) U_i = \sqrt{\lambda_i} \mathbf{X}^T U_i = \lambda_i V_i.$$

3. The proof is straightforward:

$$\mathbf{X} V_i = \frac{1}{\sqrt{\lambda_i}} \mathbf{X}\mathbf{X}^T U_i = \sqrt{\lambda_i} U_i.$$

4. Let V be an eigenvector of the matrix $\mathbf{X}^T\mathbf{X}$ orthogonal to V_i, $1 \leq i \leq d$. Then

$$0 = \sqrt{\lambda_i}(V_i, V) = (\mathbf{X}^T U_i, V) = (U_i, \mathbf{X} V)$$

for any $1 \leq i \leq d$. We set $U = \mathbf{X} V$. Then $\mathbf{X}\mathbf{X}^T U = \mathbf{0}_L$, $\mathbf{X}^T U = \mathbf{0}_K$ and

$$\mathbf{X}^T\mathbf{X} V = \mathbf{X}^T U = \mathbf{0}_K.$$

5. Since the eigenvectors $U_1, \ldots, U_L$ of the matrix $\mathbf{S} = \mathbf{X}\mathbf{X}^T$ form an orthonormal basis of $\mathbf{R}^L$, it follows that

$$\mathbf{E}_L = \sum_{i=1}^{L} U_i U_i^T, \qquad (4.3)$$

where $\mathbf{E}_L$ is the identity $L \times L$ matrix. Since $\mathbf{E}_L \mathbf{X} = \mathbf{X}$, we obtain

$$\mathbf{X} = \sum_{i=1}^{L} U_i U_i^T \mathbf{X} = \sum_{i=1}^{L} U_i (\mathbf{X}^T U_i)^T$$

$$= \sum_{i=1}^{d} \sqrt{\lambda_i} U_i V_i^T + \sum_{i=d+1}^{L} U_i (\mathbf{X}^T U_i)^T.$$

EXISTENCE AND UNIQUENESS   221

Since $\mathbf{X}^\mathrm{T} U_i = \mathbf{0}_\mathrm{K}$ for $i > d$, the proof is complete. □

The equality (4.2) is called the *singular value decomposition (SVD)* of the matrix $\mathbf{X}$. Standard terminology calls numbers $\sqrt{\lambda_i}$ *singular values* of the matrix $\mathbf{X}$, while the vectors U_i and V_i are called the *left* and *right singular vectors* of the matrix $\mathbf{X}$. The collection $(\sqrt{\lambda_i}, U_i, V_i)$ is called ith *eigentriple* of the matrix $\mathbf{X}$.

Let us discuss the uniqueness properties of SVD (4.2). Of course this uniqueness could not be treated *ad litteram*. First, since $-U_i$ is also the eigenvector of $\mathbf{S}$ corresponding to the eigenvalue λ_i, in (4.2) we can replace one or more pairs (U_i, V_i) by $(-U_i, -V_i)$ with each term on the right side of (4.2) remaining the same.

Moreover, if, for example, $\lambda \stackrel{\text{def}}{=} \lambda_1 = \lambda_2 > \lambda_3$, then the choice of the basis in the two-dimensional eigenspace corresponding to the eigenvalue λ is not well-defined and the vectors U_1, U_2 (as well as the vectors V_1, V_2) are not determined uniquely. This means that if $\lambda_1 = \lambda_2 > \lambda_3$, then both matrices $\sqrt{\lambda_1} U_1 V_1^\mathrm{T}$ and $\sqrt{\lambda_2} U_2 V_2^\mathrm{T}$ have a sense only through their sum (which does not depend on the choice of U_1 and U_2), but not individually.

Bearing these considerations in mind, the uniqueness property of SVD can be formulated as follows.

Let $P_1, \ldots, P_L$ and $Q_1, \ldots, Q_L$ be some orthonormal systems in $\mathbf{R}^L$ and $\mathbf{R}^K$, respectively. Assume that there exist nonnegative constants $c_1 \geq \ldots \geq c_L \geq 0$ such that

$$\mathbf{X} = \sum_{i=1}^L c_i P_i Q_i^\mathrm{T}. \tag{4.4}$$

Consider an SVD (4.2) of the matrix $\mathbf{X}$.

Proposition 4.2
1. $c_d > 0$ and $c_{d+1} = \ldots = c_L = 0$.
2. $c_i^2 = \lambda_i$ for $1 \leq i \leq d$.
3. *For each* $i = 1, \ldots, d$ *the vector* P_i *is an eigenvector of the matrix* $\mathbf{X}\mathbf{X}^\mathrm{T}$ *corresponding to the eigenvalue* λ_i.
4. $Q_i = \mathbf{X}^\mathrm{T} P_i / \sqrt{\lambda_i}$ $(i = 1, \ldots, d)$.
5. *If all the numbers c_i are different, then* (4.4) *coincides with* (4.2) *up to the signs of U_i and V_i.*

Proof.
Since $\mathbf{X}$ admits the decomposition (4.4) and the vectors $Q_1, \ldots, Q_L$ are orthonormal,

$$\mathbf{S} = \mathbf{X}\mathbf{X}^\mathrm{T} = \sum_{i,j=1}^L c_i c_j P_i Q_i^\mathrm{T} Q_j P_j^\mathrm{T} = \sum_{i=1}^L c_i^2 P_i P_i^\mathrm{T}.$$

Multiplying the last equality by P_i on the right, we obtain that $\mathbf{S} P_i = c_i^2 P_i$, and the first three statements of the proposition are proved. The fourth one is obtained by transposing the equality (4.4), which gives

$$\mathbf{X}^T = \sum_{j=1}^{L} c_j Q_j P_j^T,$$

and then multiplying it by P_i on the right. The last statement is an evident consequence of the properties of eigenvectors. □

Corollary 4.1 *Let* $I \subset \{1,\ldots,d\}$, $J = \{1,\ldots,d\} \setminus I$. *Set*

$$\mathbf{X}_I = \sum_{i \in I} \sqrt{\lambda_i} U_i V_i^T$$

and $\mathbf{X}_J \stackrel{\text{def}}{=} \mathbf{X} - \mathbf{X}_I$. *Then the decomposition*

$$\mathbf{X}_J = \sum_{i \in J} \sqrt{\lambda_i} U_i V_i^T$$

is the SVD of the matrix $\mathbf{X}_J$.

4.2 SVD matrices

Singular value decomposition (4.2) can be rewritten in matrix form as follows. Set $\mathbf{U}_d = [U_1 : \ldots : U_d]$, $\mathbf{V}_d = [V_1 : \ldots : V_d]$, and let Λ_d be the diagonal $d \times d$ matrix with the eigenvalue λ_i as the ith diagonal element. Then (4.2) takes the form

$$\mathbf{X} = \mathbf{U}_d \Lambda_d^{1/2} \mathbf{V}_d^T, \qquad (4.5)$$

which is the standard matrix form of the singular value decomposition.

The equality (4.5) can be rewritten in the form known as the quasi-diagonal representation of the matrix $\mathbf{X}$. It is well known that for a suitable choice of an orthonormal basis in $\mathbf{R}^L$, any symmetric $L \times L$ matrix has a diagonal representation. It follows from (4.5) that we can select proper bases both in $\mathbf{R}^L$ and $\mathbf{R}^K$ and obtain an analogous representation for any rectangular matrix $\mathbf{X}$.

Let $\mathbf{U} = [U_1 : \ldots : U_L]$ and $\mathbf{V} = [V_1 : \ldots : V_K]$. (Note that in the case $d < K$ we take $V_{d+1},\ldots,V_K$ as an orthonormal system of eigenvectors corresponding to the zero eigenvalue of the matrix $\mathbf{X}^T \mathbf{X}$.)

The matrices $\mathbf{U}$ and $\mathbf{V}$ are $L \times L$ and $K \times K$ *unitary* (or *rotation*) matrices. For the matrix $\mathbf{U}$, this means that for all vectors $X, Y \in \mathbf{R}^L$ the equality $(\mathbf{U}X, \mathbf{U}Y) = (X, Y)$ is valid and therefore the matrix $\mathbf{U}$, treated as a linear mapping $\mathbf{R}^L \mapsto \mathbf{R}^L$, conserves both the vector norms and the angles between vectors. Another characterization of the rotation property is the identity $\mathbf{U}^{-1} = \mathbf{U}^T$.

Denote by Λ the matrix of the size of the initial matrix $\mathbf{X}$ with the diagonal elements $\lambda_{ii} = \lambda_i$ for $1 \leq i \leq d$ and all the other elements equal to zero. Then (4.5) can be rewritten as

$$\mathbf{X} = \mathbf{U} \Lambda^{1/2} \mathbf{V}^T \quad \text{or} \quad \Lambda^{1/2} = \mathbf{U}^T \mathbf{X} \mathbf{V}. \qquad (4.6)$$

The equalities (4.6) have the meaning of the *quasi-diagonal representation* of the matrix $\mathbf{X}$. For suitable bases $U_1, \ldots, U_L$ in $\mathbf{R}^L$ and $V_1, \ldots, V_K$ in $\mathbf{R}^K$ (in other words, under two proper rotations), any rectangular $L \times K$ matrix has a quasi-diagonal representation $\Lambda^{1/2}$. The term 'rotation' is used here since a transition from one orthonormal basis in a linear space to another one is performed with the help of a rotation matrix similar to $\mathbf{U}$ and $\mathbf{V}$.

4.2.1 Matrix orthogonal decompositions and SVD

Consider the linear space $\mathcal{M}_{L,K}$ of real $L \times K$ matrices equipped with the standard operations of matrix addition and multiplication by constants. Evidently, the dimension of this space is LK. Define the *inner product of matrices* as follows. Let $\mathbf{X} = (x_{ij})_{i,j=1}^{L,K}$ and $\mathbf{Y} = (y_{ij})_{i,j=1}^{L,K}$ be matrices in $\mathcal{M}_{L,K}$. Then

$$\langle \mathbf{X}, \mathbf{Y} \rangle_{\mathcal{M}} = \sum_{i=1}^{L} \sum_{j=1}^{K} x_{ij} y_{ij}. \tag{4.7}$$

In the standard manner the equality

$$\|\mathbf{X}\|_{\mathcal{M}}^2 = \langle \mathbf{X}, \mathbf{X} \rangle_{\mathcal{M}} = \sum_{i=1}^{L} \sum_{j=1}^{K} x_{ij}^2 \tag{4.8}$$

defines the square of the matrix norm (usually called the *Frobenius* matrix norm), $\text{dist}_{\mathcal{M}}(\mathbf{X}, \mathbf{Y}) = \|\mathbf{X} - \mathbf{Y}\|_{\mathcal{M}}$ has the sense of the distance between matrices $\mathbf{X}$ and $\mathbf{Y}$, and so on.

The inner product (4.7) is the usual inner product of vectors in $\mathbf{R}^{LK}$ (with elements x_{ij} and y_{ij}) and does not depend on the rectangular structure of the matrices. In particular, (4.7) does not depend on mutual permutation of matrix elements and therefore does not take into consideration many important matrix characteristics such as their rank.

On the other hand, the definition (4.7) tells us that the inner product of matrices $\mathbf{X}$ and $\mathbf{Y}$ is equal to the inner product of $\mathbf{X}^T$ and $\mathbf{Y}^T$, which seems to be natural. (Of course these inner products act in different spaces: $\mathbf{X}, \mathbf{Y} \in \mathcal{M}_{L,K}$, while $\mathbf{X}^T, \mathbf{Y}^T \in \mathcal{M}_{K,L}$.) The other useful property of the matrix inner product is that proximity of two matrices can be considered as proximity of their columns; if $X_1, \ldots, X_K \in \mathbf{R}^L$ and $Y_1, \ldots, Y_K \in \mathbf{R}^L$ are columns of matrices $\mathbf{X}$ and $\mathbf{Y}$, respectively, then

$$\|\mathbf{X} - \mathbf{Y}\|_{\mathcal{M}}^2 = \sum_{i=1}^{K} \|X_i - Y_i\|^2.$$

The analogous equality can obviously be written in terms of the matrix rows rather than columns.

Matrices $\mathbf{X}$ and $\mathbf{Y}$ are *orthogonal* if $\langle \mathbf{X}, \mathbf{Y} \rangle_{\mathcal{M}} = 0$. Though the concept of orthogonality is too general to be useful in all cases, there exist sufficient conditions that relate orthogonality of matrices to other properties of these matrices, such as the span of their columns or rows. Indeed, in terms of the matrix columns,

$$\langle \mathbf{X}, \mathbf{Y} \rangle_{\mathcal{M}} = \sum_{i=1}^{K}(X_i, Y_i)$$

and, therefore, if $\operatorname{span}(\mathbf{X}) \stackrel{\text{def}}{=} \operatorname{span}(X_1, \ldots, X_K)$ is orthogonal to $\operatorname{span}(\mathbf{Y}) = \operatorname{span}(Y_1, \ldots, Y_K)$, then $\mathbf{X}$ and $\mathbf{Y}$ are orthogonal. An analogous sufficient condition can be formulated in terms of the matrix rows; i.e., linear spaces $\operatorname{span}(\mathbf{X}^T)$ and $\operatorname{span}(\mathbf{Y}^T)$. Note that the orthogonality $\operatorname{span}(\mathbf{X}) \perp \operatorname{span}(\mathbf{Y})$ can be expressed as $\mathbf{X}^T \mathbf{Y} = \mathbf{0}_{KK}$, while the condition $\operatorname{span}(\mathbf{X}^T) \perp \operatorname{span}(\mathbf{Y}^T)$ is equivalent to $\mathbf{X}\mathbf{Y}^T = \mathbf{0}_{LL}$ (here $\mathbf{0}_{NN}$ stands for the $N \times N$ zero matrix).

Orthogonality of the span spaces in any combination does not give necessary conditions for matrix orthogonality. However, there exists a class of matrices where matrix orthogonality is expressed in terms of orthogonality of the span spaces. Such matrices are *unit-rank* or *elementary* matrices.

Every elementary matrix has proportional (nonzero) columns and proportional (nonzero) rows. This means that every elementary $L \times K$ matrix $\mathbf{X}$ has a representation

$$\mathbf{X} = cPQ^T \qquad (4.9)$$

where $P \in \mathbf{R}^L, Q \in \mathbf{R}^K, ||P|| = ||Q|| = 1$ and $c > 0$. The vector P constitutes the basis of $\operatorname{span}(\mathbf{X})$; Q plays the same role for $\operatorname{span}(\mathbf{X}^T)$. The representation (4.9) is unique up to signs of P and Q which can be modified simultaneously.

A useful feature of the elementary matrices is that for any matrix $\mathbf{Y} = PQ^T$ and any $\mathbf{X} \in \mathcal{M}_{L,K}$

$$\langle \mathbf{X}, \mathbf{Y} \rangle_{\mathcal{M}} = (\mathbf{X}^T P, Q) = (\mathbf{X}Q, P). \qquad (4.10)$$

Indeed, if $Q = (q_1, \ldots, q_K)^T$, then $PQ^T = [q_1 P : \ldots : q_K P]$ and therefore

$$\langle \mathbf{X}, \mathbf{Y} \rangle_{\mathcal{M}} = \sum_{i=1}^{K} q_i(X_i, P) = \sum_{i=1}^{K} q_i X_i^T P = (\mathbf{X}^T P, Q)$$

where $\mathbf{X} = [X_1 : \ldots : X_K]$. Thus, if $\mathbf{X} = c_1 P_1 Q_1^T$ and $\mathbf{Y} = c_2 P_2 Q_2^T$, then

$$\langle \mathbf{X}, \mathbf{Y} \rangle_{\mathcal{M}} = c_1 c_2 (P_1, P_2)(Q_1, Q_2),$$

and, therefore, $\mathbf{X}$ and $\mathbf{Y}$ are orthogonal if and only if $P_1 \perp P_2$ or $Q_1 \perp Q_2$.

When dealing with linear spaces equipped with an inner product, the usual decomposition of an element of such a space is its orthogonal decomposition into a sum of 'simple' elements. If we express the 'simplicity' of a matrix in terms of the value of its rank, then we shall consider the decomposition of a matrix into a sum of orthogonal elementary matrices:

$$\mathbf{X} = \sum_{i=1}^{m} \mathbf{X}_i = \sum_{i=1}^{m} c_i P_i Q_i^T \qquad (4.11)$$

with $\mathbf{X}_i = c_i P_i Q_i^T$, $c_i > 0$, $P_i \in \mathbf{R}^L$, $Q_i \in \mathbf{R}^K$, $||P_i|| = ||Q_i|| = 1$ and $(P_i, P_j)(Q_i, Q_j) = 0$ for $i \neq j$.

It can be easily proved that the decomposition (4.11) has the following properties:

1. $||\mathbf{X}_i||_\mathcal{M} = c_i$;
2. $||\mathbf{X}||_\mathcal{M}^2 = \sum_{i=1}^m c_i^2 = \sum_{i=1}^m ||\mathbf{X}_i||_\mathcal{M}^2$;
3. $\mathbf{X} Q_i = c_i P_i + \sum_{j:(Q_i,Q_j) \neq 0} c_j(Q_i, Q_j) P_j$;
4. $\mathbf{X}^T P_i = c_i Q_i + \sum_{j:(P_i,P_j) \neq 0} c_j(P_i, P_j) Q_j$;
5. $(\mathbf{X} Q_i, P_i) = c_i$.

Any matrix $\mathbf{X}$ of rank d can be decomposed in many ways into the orthogonal sum of elementary matrices.

Since the dimension of the linear space $\mathcal{M}_{L,K}$ is LK, there exist LK pairwise orthogonal $L \times K$ matrices. For example, suppose that one of the standard orthonormal bases of $\mathcal{M}_{L,K}$ consists of the matrices with a single element equal to 1 and all the other elements zero. These matrices have rank 1, and therefore they are elementary matrices. If we denote by $E_i^{(k)} \in \mathbf{R}^k$ the vector with all zeros apart from the ith component, which is equal to 1, then any matrix $\mathbf{X} \in \mathcal{M}_{L,K}$ with elements x_{ij} has the orthogonal decomposition

$$\mathbf{X} = \sum_{i,j=1}^{L,K} x_{ij} E_i^{(L)} \left(E_j^{(K)} \right)^T.$$

This decomposition is universal, but it can have a lot of terms, even when $\mathbf{X}$ is itself an elementary matrix.

If we consider an orthonormal basis $P_1, \ldots, P_L$ in $\mathbf{R}^L$, then

$$\mathbf{X} = \sum_{i=1}^L P_i P_i^T \mathbf{X} = \sum_{i=1}^L P_i S_i^T \quad (4.12)$$

with $S_i = \mathbf{X}^T P_i$. Thus, taking all the nonzero vectors S_i and setting $c_i = ||S_i||$ and $Q_i = S_i/c_i$ we obtain (4.11). A similar decomposition holds if we take an orthonormal basis $Q_1, \ldots, Q_K$ in $\mathbf{R}^K$ and multiply $\mathbf{X}$ by

$$\mathbf{E}_K = \sum_{i=1}^K Q_i Q_i^T$$

on the right.

Decomposition (4.11) shows that each column of the matrix $\mathbf{X}$ treated as a vector is a linear combination of the vectors $P_1, \ldots, P_m$. Therefore, $m \geq d = \text{rank } \mathbf{X}$. It is easy to construct the decomposition (4.12) with d nonzero terms.

For example, if $P_1, \ldots, P_d$ form an orthonormal basis of span $(\mathbf{X})$, then $\mathbf{X}^T P_i$ is a zero vector for $i \geq d$ and (4.12) turns into

$$\mathbf{X} = \sum_{i=1}^{d} P_i S_i^T$$

with linearly independent S_i.

Let us characterize all the decompositions of the kind

$$\mathbf{X} = \sum_{i=1}^{m} P_i Q_i^T \qquad (4.13)$$

with $m = \operatorname{rank} \mathbf{X}$, without the orthogonality restrictions on P_i and Q_i.

Proposition 4.3
1. *The equality $m = \operatorname{rank} \mathbf{X}$ holds if and only if the vectors $P_1, \ldots, P_m$ are linearly independent and belong to* span $(\mathbf{X})$.
2. *The equality $m = \operatorname{rank} \mathbf{X}$ holds if and only if both vector systems $P_1, \ldots, P_m$ and $Q_1, \ldots, Q_m$ are linearly independent.*

Proof.
1. Since span $(\mathbf{X}) \subset$ span $(P_1, \ldots, P_m)$,

$$\dim (\operatorname{span}(\mathbf{X})) \leq \dim (\operatorname{span}(P_1, \ldots, P_m)) \leq m \qquad (4.14)$$

and the equality $m = \operatorname{rank} \mathbf{X}$ holds if and only if both inequalities in (4.14) become equalities. This means that a) the vectors $P_1, \ldots, P_m$ are linearly independent and b) span $(\mathbf{X}) =$ span $(P_1, \ldots, P_m)$.
2. As has already been mentioned, $m \geq d = \operatorname{rank} \mathbf{X}$. Moreover if $P_1, \ldots, P_m$ are linearly dependent, then we can express each P_i as a linear combination of the basis vectors of span $(P_1, \ldots, P_m)$. Recalculation of Q_i then leads to a decomposition similar to (4.13), but with a smaller number of terms on its right side.

Now let us assume that both systems $P_1, \ldots, P_m$ and $Q_1, \ldots, Q_m$ are linearly independent and $m > d$. Then span $(\mathbf{X})$ is a subspace of span $(P_1, \ldots, P_m)$, and these linear spaces do not coincide. Denote by $Y_1, \ldots, Y_m$ the orthonormal basis of span $(P_1, \ldots, P_m)$ such that $Y_1, \ldots, Y_d$ is a basis of span $(\mathbf{X})$. Then $\mathbf{X}^T Y_k = \mathbf{0}_K$ for $k > d$.

Since

$$P_i = \sum_{j=1}^{m} c_{ij} Y_j$$

for some c_{ij},

$$\mathbf{X} = \sum_{i=1}^{m} \left(\sum_{j=1}^{m} c_{ij} Y_j \right) Q_i^T = \sum_{j=1}^{m} Y_j \left(\sum_{i=1}^{m} c_{ij} Q_i^T \right).$$

Thus, for $k > d$

$$\mathbf{0}_K = \mathbf{X}^T Y_k = \sum_{i=1}^{m} c_{ik} Q_i$$

and the Q_i are linearly dependent. □

A decomposition (4.13) with $m = \operatorname{rank} \mathbf{X}$ is called *minimal*.

Corollary 4.2 *The decomposition* (4.13) *is a minimal orthogonal decomposition if and only if both vector systems* $P_1, \ldots, P_m$ *and* $Q_1, \ldots, Q_m$ *are linearly independent and* $(P_i, P_j)(Q_i, Q_j) = 0$ *for all* $i \neq j$.

As a result, we have obtained a class of matrix decompositions into a minimal number of orthogonal elementary matrices. Of course, the singular value decompositions belong to this class.

We set $\mathbf{X}_i = \sqrt{\lambda_i} U_i V_i^T$ in SVD (4.2). Then (4.2) can be rewritten in the form

$$\mathbf{X} = \mathbf{X}_1 + \ldots + \mathbf{X}_d. \tag{4.15}$$

The matrices $\mathbf{X}_i$ have unit ranks and are orthogonal to each other. Moreover, the matrices $\mathbf{X}_i$ are *biorthogonal* in the sense that $\mathbf{X}_i \mathbf{X}_j^T = \mathbf{0}_{LL}$ and $\mathbf{X}_i^T \mathbf{X}_j = \mathbf{0}_{KK}$ for $i \neq j$. This means that the SVD is not only a decomposition of a matrix into the minimal system of orthogonal elementary matrices, but also this decomposition is biorthogonal. In view of Proposition 4.2, it is a unique (up to multiplicity of the eigenvalues λ_i) biorthogonal elementary decomposition. Evidently,

$$\|\mathbf{X}\|_{\mathcal{M}}^2 = \lambda_1 + \ldots + \lambda_d. \tag{4.16}$$

4.3 Optimality of SVDs

The optimal features of SVD are based on two extreme properties of the eigenvalues/eigenvectors of symmetric matrices. These properties are well known; their proofs can be found in Gantmacher (1998).

Theorem 4.1 *Let* $\mathbf{C}$ *be a symmetric* $L \times L$ *matrix. Denote by* $\lambda_1 \geq \ldots \geq \lambda_L$ *the eigenvalues of the matrix* $\mathbf{C}$ *and by* $U_1, \ldots, U_L$ *the corresponding orthonormal system of its eigenvectors. Then*

1. *a)*

$$\lambda_1 = \max_{P} (\mathbf{C}P, P) = (\mathbf{C}U_1, U_1),$$

where the maximum is taken over all $P \in \mathbf{R}^L$ *with* $\|P\| = 1$;

b) for $2 \leq k \leq L$

$$\lambda_k = \max_{P}{}^{(k)} (\mathbf{C}P, P) = (\mathbf{C}U_k, U_k),$$

where the maximum is taken over all $P \in \mathbf{R}^L$ *with* $\|P\| = 1$ *and* $(P, U_i) = 0$ $(1 \leq i < k)$;

2. *for any* $1 \leq k \leq L$

$$\max_{P_1,\ldots,P_k} \sum_{i=1}^{k}(\mathbf{C}P_i, P_i) = \sum_{i=1}^{k}(\mathbf{C}U_i, U_i) = \sum_{i=1}^{k}\lambda_i,$$

where the maximum is taken over all orthonormal systems $P_1, \ldots, P_k \in \mathbf{R}^L$.

We now turn to SVDs starting with an auxiliary proposition. Assume that a matrix $\mathbf{X}$ has a rank $d > 0$. Fix k such that $1 \leq k \leq d$. We consider the problem of approximation of the matrix $\mathbf{X}$ with respect to the matrix norm by matrices $\mathbf{Y}$ of the form

$$\mathbf{Y} = \sum_{i=1}^{k} P_i Q_i^{\mathrm{T}}, \qquad (4.17)$$

where $P_i \in \mathbf{R}^L$ and $Q_i \in \mathbf{R}^K$. Of course, we can assume that the vectors P_i are orthonormal. We, however, do not assume the orthonormality of the Q_i.

Let us fix the orthonormal vectors $P_1, \ldots, P_k$ and denote by $\mathcal{M}_{k,P}$ the collection of matrices (4.17). Our problem is to find the matrix $\mathbf{Y}_0 \in \mathcal{M}_{k,P}$ such that

$$\min_{\mathbf{Y} \in \mathcal{M}_{k,P}} \|\mathbf{X} - \mathbf{Y}\|_{\mathcal{M}} = \|\mathbf{X} - \mathbf{Y}_0\|_{\mathcal{M}}.$$

To find the optimal matrix $\mathbf{Y}_0$ it is sufficient to find the corresponding matrices $Q_1, \ldots, Q_k$ in (4.17).

Proposition 4.4 *Optimal* Q_i *have the form* $Q_i = \mathbf{X}^{\mathrm{T}} P_i$.

Proof.
Since for all vectors $Q_1, \ldots, Q_k$ the matrices $\mathbf{Y}_i = P_i Q_i^{\mathrm{T}}$ are orthogonal,

$$\left\langle \mathbf{X} - \sum_{i=1}^{k} P_i \mathbf{X}^{\mathrm{T}} P_i, \sum_{j=1}^{k} P_j Q_j^{\mathrm{T}} \right\rangle_{\mathcal{M}} = 0. \qquad (4.18)$$

Indeed, by (4.10) we have

$$\left\langle \mathbf{X}, \sum_{j=1}^{k} P_j Q_j^{\mathrm{T}} \right\rangle_{\mathcal{M}} = \sum_{j=1}^{k} (\mathbf{X}^{\mathrm{T}} P_j, Q_j),$$

and in view of the orthonormality of the vectors $P_1, \ldots, P_k$ the same result is valid for the inner product

$$\left\langle \sum_{i=1}^{k} P_i \mathbf{X}^{\mathrm{T}} P_i, \sum_{j=1}^{k} P_j Q_j^{\mathrm{T}} \right\rangle_{\mathcal{M}}.$$

The equality (4.18) shows that the matrix $\sum_{i=1}^{k} P_i \mathbf{X}^{\mathrm{T}} P_i$ is the orthogonal projection of the matrix $\mathbf{X}$ on $\mathcal{M}_{k,P}$; this completes the proof. □

An analogous statement holds if we fix the orthonormal system $Q_1, \ldots, Q_k$. Then the optimal P_i (which are not necessarily orthonormal in this case) are equal to $\mathbf{X} Q_i$.

Now let $\mathcal{M}_k$ be the set of matrices of the form

$$\mathbf{Y} = \sum_{i=1}^{k} P_i Q_i^{\mathrm{T}} \tag{4.19}$$

with $k < d$. Take $\mathbf{X}$, a matrix of a rank d, and consider the problem of optimal approximation with respect to the matrix norm of this matrix by matrices in $\mathcal{M}_k$. In this case, both P_i and Q_i are arbitrary, while Proposition 4.4 deals with a fixed collection of orthonormal vectors P_i.

Let $(\sqrt{\lambda_i}, U_i, V_i)$ be eigentriples of the SVD (4.2). The optimal features of the SVD can be formulated as follows.

Proposition 4.5

1. $\min\limits_{\mathbf{Y} \in \mathcal{M}_k} \|\mathbf{X} - \mathbf{Y}\|_{\mathcal{M}}^2 = \sum\limits_{i=k+1}^{d} \lambda_i$.
2. *If we take*

$$\mathbf{Y}_0 = \sum_{i=1}^{k} \sqrt{\lambda_i} U_i V_i^{\mathrm{T}} \in \mathcal{M}_k, \tag{4.20}$$

then

$$\|\mathbf{X} - \mathbf{Y}_0\|_{\mathcal{M}}^2 = \min_{\mathbf{Y} \in \mathcal{M}_k} \|\mathbf{X} - \mathbf{Y}\|_{\mathcal{M}}^2. \tag{4.21}$$

Proof.
We can assume that $P_1, \ldots, P_k$ are linearly independent and form an orthonormal system. Indeed, if the P_i are linearly dependent, then we decompose several P_i into linear combinations of the others and recalculate the Q_i. The matrix (4.19) will remain the same, while k will be reduced. In the same manner, if $P_1, \ldots, P_k$ are linearly independent but not pairwise orthogonal, then we can find an orthonormal basis of $\mathrm{span}(P_1, \ldots, P_k)$, decompose each P_i in terms of this basis, and recalculate the Q_i.

By Proposition 4.4 we know that for fixed orthonormal P_i ($1 \le i \le k$), the optimal Q_i have the form $Q_i = \mathbf{X}^{\mathrm{T}} P_i$. Therefore, the problem is to find the optimal P_i. Since

$$\left\|\mathbf{X} - \sum_{i=1}^{k} P_i \mathbf{X}^{\mathrm{T}} P_i\right\|_{\mathcal{M}}^2 = \|\mathbf{X}\|_{\mathcal{M}}^2 - \sum_{i=1}^{k} \left(\mathbf{X}\mathbf{X}^{\mathrm{T}} P_i, P_i\right), \tag{4.22}$$

we have to find orthonormal $P_1, \ldots, P_k$ such that the right side of (4.22) is minimal. But the answer to this problem is well-known (see Theorem 4.1); these vectors can be selected as the leading k eigenvectors of the matrix $\mathbf{S} = \mathbf{X}\mathbf{X}^{\mathrm{T}}$. This means that $P_i = U_i$. In view of the equality (4.16), both statements are proved. □

The optimal features of SVD can be re-expressed differently for different purposes. For example, the set $\mathcal{M}_k$ can be looked at from another viewpoint. Since any matrix $\mathbf{Y} \in \mathcal{M}_k$ has rank not exceeding k, the problem (4.21) is the problem of approximation of the matrix $\mathbf{X}$ by a matrix of smaller rank. Therefore, Proposition 4.5 tells us that *the sum of the first k SVD terms makes the best approximating matrix $\mathbf{Y}_0$ of rank not larger than k.*

On the other hand, it follows from (4.19) that any column Y_i of the matrix $\mathbf{Y} \in \mathcal{M}_k$ belongs to the linear space span $(P_1, \ldots, P_k)$ of dimension not larger than $k < d$, whereas the columns $X_1, \ldots, X_K$ of the matrix $\mathbf{X}$ span the linear space span $(\mathbf{X})$ of dimension d. Since

$$\|\mathbf{X} - \mathbf{Y}\|_\mathcal{M}^2 = \sum_{i=1}^{K} \|X_i - Y_i\|^2,$$

the optimization problem of Proposition 4.5 can be regarded as the problem of simultaneous approximation of the vectors $X_1, \ldots, X_K$, spanning the d-dimensional vector space span $(\mathbf{X})$ by some vectors $Y_1, \ldots, Y_K$ spanning a linear space $\mathcal{L}$ of dimension not exceeding $k < d$. Of course, Proposition 4.5 gives the solution: the *optimal linear space $\mathcal{L}$ is equal to* span $(U_1, \ldots, U_k)$, *while the columns of the matrix* (4.20) *are equal to the optimal* Y_i.

A natural characteristic of these (equivalent) optimal approximations is defined by

$$\frac{\|\mathbf{X} - \mathbf{Y}_0\|_\mathcal{M}^2}{\|\mathbf{X}\|_\mathcal{M}^2} = \frac{\lambda_{k+1} + \ldots + \lambda_d}{\lambda_1 + \ldots + \lambda_d}.$$

If we do not deal with the optimal approximation and set

$$\mathbf{X}_I = \sum_{i \in I} \sqrt{\lambda_i} U_i V_i^\mathrm{T}$$

with $I = \{j_1, \ldots, j_k\} \subset \{1, \ldots, d\}$, $j_1 > \ldots > j_k$, and $k < d$, then

$$1 - \frac{\|\mathbf{X} - \mathbf{X}_I\|_\mathcal{M}^2}{\|\mathbf{X}\|_\mathcal{M}^2} = \frac{\lambda_{j_1} + \ldots + \lambda_{j_k}}{\lambda_1 + \ldots + \lambda_d}. \tag{4.23}$$

The characteristic (4.23) can be called the *eigenvalue share of the eigentriples with numbers $j_1, \ldots, j_k$*.

Another description of the optimal features of SVD is related to the so-called principal vectors of the collection $X_1, \ldots, X_K \in \mathbf{R}^L$. Let $X, P \in \mathbf{R}^L$, $X \neq \mathbf{0}_L$, $\|P\| = 1$. Then $(X, P)P$ is the projection of X onto the one-dimensional linear space $\mathcal{L}_P = \text{span}(P)$, and $c = |(X, P)|$ is the norm of this projection. The value $c = c(P)$ can be regarded as a measure of the quality of the approximation of the vector X by $\mathcal{L}_P$; the larger $c = c(P)$ is, the better X is approximated by span (P).

If we want to find P such that $\mathcal{L}_P$ approximates the collection of vectors $X_1, \ldots, X_K$ in the best way, then we arrive at the following optimization problem: find the vector P_0 such that $\|P_0\| = 1$ and

$$\nu_1 \stackrel{\text{def}}{=} \sum_{i=1}^{K} (X_i, P_0)^2 = \max_{P} \sum_{i=1}^{K} (X_i, P)^2 \tag{4.24}$$

OPTIMALITY OF SVDS

where the maximum on the right side of (4.24) is taken over all $P \in \mathbf{R}^L$ with $||P|| = 1$.

The solution to this problem can also be described in terms of SVD.

Proposition 4.6 *Consider the matrix* $\mathbf{X} = [X_1 : \ldots : X_K]$ *and its SVD* (4.2). *Then*
1. *The vector* $P_0 = U_1$ *is the solution of the problem* (4.24) *with* $\nu_1 = \lambda_1$.
2. *Let* P_0 *be the solution of the following optimization problem*

$$\nu_k \stackrel{\text{def}}{=} \sum_{i=1}^{K}(X_i, P_0)^2 = \max_{P}{}^{(k)} \sum_{i=1}^{K}(X_i, P)^2, \quad (4.25)$$

where the maximum on the right side of (4.25) *is taken over all* $P \in \mathbf{R}^L$ *such that* $||P|| = 1$ *and* $(P, U_i) = 0$ *for* $1 \le i < k$. *If* $k \le d$, *then the vector* $P_0 = U_k$ *is the solution of the problem* (4.25), *and* $\nu_k = \lambda_k$. *If* $k > d$, *then* $\nu_k = 0$.

Proof.
Since

$$\sum_{i=1}^{K}(X_i, P)^2 = ||\mathbf{X}^T P||^2 = (\mathbf{X}\mathbf{X}^T P, P),$$

both statements follow from Theorem 4.1. $\square$

Proposition 4.6 enables us to call the vector U_i the *ith principal vector* of the collection $X_1, \ldots, X_K$. We set $c_j(U_i) = (X_j, U_i)$. Since

$$X_j = \sum_{i=1}^{d} c_j(U_i) U_i,$$

the coefficient $c_j(U_i)$ is called the *ith principal component* of the vector X_j and the vector

$$Z_i = (c_1(U_i), \ldots, c_K(U_i))^T = \mathbf{X}^T U_i$$

is the *vector of ith principal components*. Note that, in view of (4.1), $Z_i = \sqrt{\lambda_i} V_i$ and SVD (4.2) can be treated as a simultaneous decomposition of the columns of the matrix $\mathbf{X}$ with respect to the basis of their principal vectors. Such an interpretation is standard in principal component analysis where the columns of the matrix $\mathbf{X}$ form an L-dimensional sample of size K.

In the same way, the V_i are the principal vectors for the rows of the matrix $\mathbf{X}$, the vectors $\sqrt{\lambda_i} U_i$ are the vectors of their principal components and the decomposition (4.2) produces two systems of principal vectors and two related decompositions with respect to these systems.

Now let us consider one more optimization problem related to SVD. Let us fix $1 \le k < d = \text{rank}\,\mathbf{X}$ and an orthonormal system $W_1, \ldots, W_k \in \mathbf{R}^L$ and consider the matrix

$$\mathbf{Y} = \sum_{i=1}^{k} W_i Q_i^{\mathrm{T}} + \sum_{i=k+1}^{d} P_i Q_i^{\mathrm{T}} \qquad (4.26)$$

under the assumption that the collection of vectors $W_1, \ldots, W_k, P_{k+1}, \ldots, P_d$ forms an orthonormal system. (Here the P_i are arbitrary up to this restriction, and the Q_i are arbitrary vectors in $\mathbf{R}^L$.) Denote by $\mathcal{M}_{k,W}$ the set of such matrices and consider the problem of finding the matrix $\mathbf{Y}_0 \in \mathcal{M}_{k,W}$ such that

$$\|\mathbf{X} - \mathbf{Y}_0\|_{\mathcal{M}} = \min_{\mathbf{Y} \in \mathcal{M}_{d,W}} \|\mathbf{X} - \mathbf{Y}\|_{\mathcal{M}}. \qquad (4.27)$$

Proposition 4.7 *The solution $\mathbf{Y}_0$ to the problem (4.27) has the following structure: $Q_i = \mathbf{X}^{\mathrm{T}} W_i$ for $1 \leq i \leq k$; if $d \geq i > k$, then the vectors P_i coincide with the first $d - k$ orthonormal eigenvectors of the matrix*

$$\mathbf{X}_W \stackrel{\text{def}}{=} \mathbf{X} - \sum_{i=1}^{k} W_i (\mathbf{X}^{\mathrm{T}} W_i)^{\mathrm{T}}$$

and $Q_i = \mathbf{X}^{\mathrm{T}} P_i$.

Proof.
The expression for Q_i immediately follows from Proposition 4.4. Therefore, setting

$$\mathbf{Z}_W = \sum_{i=1}^{k} W_i Q_i^{\mathrm{T}},$$

we can take

$$\mathbf{Y} = \mathbf{Z}_W + \sum_{i=k+1}^{d} P_i Q_i^{\mathrm{T}} = \mathbf{Z}_W + \mathbf{Y}_0^*$$

with $\mathbf{Y}_0^* \in \mathcal{M}_{d-k}$. Since $\|\mathbf{X} - \mathbf{Y}\|_{\mathcal{M}} = \|\mathbf{X}_W - \mathbf{Y}_0^*\|_{\mathcal{M}}$, the proof is complete. □

4.4 Centring in SVD

Centring is not a standard procedure in SVD. Two versions of centring will be discussed; single centring is usual in principal component analysis, while double centring is a specific version of SSA aimed at extracting linear-like signals.

4.4.1 Single centring

For the initial matrix $\mathbf{X}$ with K columns and L rows, we set

$$\mathcal{A}_1(\mathbf{X}) = \frac{1}{K} \mathbf{X} \mathbf{1}_K \mathbf{1}_K^{\mathrm{T}}, \qquad (4.28)$$

where $\mathbf{1}_K = (1, \ldots, 1)^T \in \mathbf{R}^K$. If we consider the vector

$$\mathcal{E}_1(\mathbf{X}) = \frac{1}{K}\mathbf{X}\mathbf{1}_K \in \mathbf{R}^L, \tag{4.29}$$

then $\mathcal{E}_1(\mathbf{X})$ is the result of averaging the elements of $\mathbf{X}$ over its rows. In other words, if $\mathbf{X} = [X_1 : \ldots : X_K]$, then $\mathcal{E}_1(\mathbf{X}) = (X_1 + \ldots + X_K)/K$.

Since $\mathcal{A}_1(\mathbf{X}) = \mathcal{E}_1(\mathbf{X})\mathbf{1}_K^T$, the matrix $\mathcal{A}_1(\mathbf{X})$ has K identical columns which are equal to $\mathcal{E}_1(\mathbf{X})$. The transition $\mathbf{X} \mapsto \mathbf{X}' \stackrel{\text{def}}{=} \mathbf{X} - \mathcal{A}_1(\mathbf{X})$ has the meaning of row centring.

Let us consider the SVD of the matrix $\mathbf{X}'$:

$$\mathbf{X}' = \sum_{i=1}^{d} \sqrt{\lambda_i} U_i V_i^T. \tag{4.30}$$

If $\mathcal{E}_1(\mathbf{X}) = \mathbf{0}_L$, then the equality (4.30) is the SVD of the initial matrix $\mathbf{X}$ and therefore coincides with (4.2). Otherwise (4.30) can be rewritten as

$$\mathbf{X} = \sqrt{\lambda_{0(1)}} U_{0(1)} V_{0(1)}^T + \sum_{i=1}^{d} \sqrt{\lambda_i} U_i V_i^T = \mathcal{A}_1(\mathbf{X}) + \sum_{i=1}^{d} \mathbf{X}_i \tag{4.31}$$

with

$$U_{0(1)} = \mathcal{E}_1(\mathbf{X})/\|\mathcal{E}_1(\mathbf{X})\|, \quad V_{0(1)} = \mathbf{1}_K/\sqrt{K}$$

and $\sqrt{\lambda_{0(1)}} = \|\mathcal{E}_1(\mathbf{X})\|\sqrt{K}$. Therefore, (4.31) is a decomposition of the matrix $\mathbf{X}$ into a sum of elementary matrices. Note that here d is the order of the SVD of the matrix $\mathbf{X}'$ and therefore not of $\mathbf{X}$ in general.

We call decomposition (4.31) the *single centring SVD* of the matrix $\mathbf{X}$. The triple $(\sqrt{\lambda_{0(1)}}, U_{0(1)}, V_{0(1)})$ is called the *first average triple*.

Let us discuss the properties of this version of SVD.

Proposition 4.8
1. *The decomposition* (4.31) *is a decomposition of the matrix* $\mathbf{X}$ *into a sum of orthogonal elementary matrices.*
2. *The decomposition* (4.31) *is an SVD of the matrix* $\mathbf{X}$ *if and only if* $\mathbf{1}_K$ *is an eigenvector of the matrix* $\mathbf{X}^T\mathbf{X}$.
3. *The decomposition* (4.31) *is the minimal decomposition if and only if either* $\mathbf{1}_K$ *belongs to the space* $\mathrm{span}\,(\mathbf{X}^T)$ *or* $\mathbf{1}_K$ *is orthogonal to this space.*

Proof.
If $\mathcal{E}_1(\mathbf{X}) = \mathbf{0}_L$, i.e., if $\mathbf{1}_K$ is orthogonal to $\mathrm{span}\,(\mathbf{X}^T)$, then all the statements obviously hold. Assume that $\mathcal{E}_1(\mathbf{X}) \neq \mathbf{0}_L$.
1. Note that

$$\mathbf{X}'\mathbf{1}_K = \mathbf{X}\mathbf{1}_K - \mathcal{E}_1(\mathbf{X})\mathbf{1}_K^T\mathbf{1}_K = \mathbf{0}_L$$

and therefore $V_{0(1)}$ is an eigenvector of the matrix $(\mathbf{X}')^T\mathbf{X}'$ corresponding to the zero eigenvalue. This means that all right singular vectors V_i are orthogonal to $\mathbf{1}_K$. (In other words, the sums of their coordinates are zeros.) Thus, $V_{0(1)}, V_1, \ldots, V_d$

form an orthonormal system, and the corresponding elementary matrices are orthogonal.

2. Suppose that $\mathbf{1}_K$ is an eigenvector of the matrix $\mathbf{X}\mathbf{X}^T$. Since

$$\mathcal{A}_1(\mathbf{X}) = \mathbf{X}^T V_{0(1)} V_{0(1)}^T,$$

$\mathcal{A}_1(\mathbf{X})$ is one of the matrix components of the SVD of the matrix $\mathbf{X}$. This is a direct consequence of Corollary 4.1.

3. All we need (see Proposition 4.3) is to demonstrate that the vector V_i belongs to the span $(\mathbf{X}^T)$ if and only if $\mathbf{1}_K \in \mathrm{span}\,(\mathbf{X}^T)$.

If $\mathbf{1}_K \in \mathrm{span}\,(\mathbf{X}^T)$, then all the rows of the matrix (4.30) are linear combinations of the rows of the matrix $\mathbf{X}$ and vice versa. Therefore, $\mathbf{1}_K \in \mathrm{span}\,(\mathbf{X}^T)$ if and only if $V_i \in \mathrm{span}\,(\mathbf{X}^T)$ for $1 \leq i \leq d$. □

Corollary 4.3

1. Since the decomposition (4.31) is orthogonal, the equality

$$||\mathbf{X}||_{\mathcal{M}}^2 = \lambda_{0(1)} + \sum_{i=1}^{d} \lambda_i \qquad (4.32)$$

is valid.

2. The decomposition (4.31) is optimal in the sense of Proposition 4.7 applied to the matrix $\mathbf{X}^T$ with $d_0 = 1$ and a fixed vector $W_1 = V_{0(1)}$.

Remark 4.1 The equality (4.32) means that the number $\lambda_{0(1)}/||\mathbf{X}||_{\mathcal{M}}^2$ can be regarded as the *share of the first average triple* in the sense of (4.23).

Analogous to the row centring of the matrix $\mathbf{X}$, the column centring SVD can be performed as well. Then, instead of $\mathcal{E}_1(\mathbf{X})$ defined by (4.29), and $\mathcal{A}_1(\mathbf{X})$ defined by (4.28), we set

$$\mathcal{E}_2(\mathbf{X}) = \frac{1}{L} \mathbf{X}^T \mathbf{1}_L, \qquad (4.33)$$

$$\mathcal{A}_2(\mathbf{X}) = \mathbf{1}_L (\mathcal{E}_2(\mathbf{X}))^T = \frac{1}{L} \mathbf{1}_L \mathbf{1}_L^T \mathbf{X} \qquad (4.34)$$

and apply the SVD to the matrix $\mathbf{X} - \mathcal{A}_2(\mathbf{X})$ rather than to the matrix $\mathbf{X}'$. Then the decomposition (4.31) becomes

$$\mathbf{X} = \sqrt{\lambda_{0(2)}} U_{0(2)} V_{0(2)}^T + \sum_{i=1}^{d} \sqrt{\lambda_i} U_i V_i^T$$

with $U_{0(2)} = \mathbf{1}_L/\sqrt{L}$, $V_{0(2)} = \mathcal{E}_2(\mathbf{X})/||\mathcal{E}_2(\mathbf{X})||$, $d = \mathrm{rank}\,(\mathbf{X} - \mathcal{A}_2(\mathbf{X}))$ and the corresponding $\lambda_{0(2)}$. Of course, the analogous statements to Proposition 4.8 and Corollary 4.3 are valid in this case as well.

4.4.2 Double centring

Let us consider the double centring SVD. Let the matrix $\mathbf{X}''$ be defined as

$$\mathbf{X}'' \stackrel{\text{def}}{=} \mathbf{X}' - \mathcal{A}_2(\mathbf{X}'),$$

which is the same as

$$\mathbf{X}'' = \mathbf{X} - \mathcal{A}^{(12)}(\mathbf{X})$$

with

$$\mathcal{A}^{(12)}(\mathbf{X}) = \mathcal{A}_1(\mathbf{X}) + \mathcal{A}_2(\mathbf{X}) - \mathcal{A}_1(\mathcal{A}_2(\mathbf{X})).$$

The matrix $\mathcal{A}_{12}(\mathbf{X}) \stackrel{\text{def}}{=} \mathcal{A}_1(\mathcal{A}_2(\mathbf{X})) = \mathcal{A}_2(\mathcal{A}_1(\mathbf{X}))$ is the matrix with each element equal to the average value of all the elements of the matrix $\mathbf{X}$. Formally this matrix average is equal to

$$\mathbf{a_x} = \frac{1}{KL}\mathbf{1}_L^T \mathbf{X} \mathbf{1}_K$$

and the matrix $\mathcal{A}_{12}(\mathbf{X})$ itself is

$$\mathcal{A}_{12}(\mathbf{X}) = \mathbf{a_x} \mathbf{1}_L \mathbf{1}_K^T.$$

In the double centring version we deal with the SVD of the matrix

$$\mathbf{X}'' = \sum_{i=1}^{d} \sqrt{\lambda_i} U_i V_i^T$$

and thus obtain the *double centring SVD* of the matrix $\mathbf{X}$:

$$\mathbf{X} = \mathcal{A}^{(12)}(\mathbf{X}) + \sum_{i=1}^{d} \sqrt{\lambda_i} U_i V_i^T = \mathcal{A}_1(\mathbf{X}) + \mathcal{A}_2(\mathbf{X}') + \sum_{i=1}^{d} \mathbf{X}_i. \quad (4.35)$$

Since $\mathbf{X}'' \mathbf{1}_K = \mathbf{0}_L$ and $(\mathbf{X}'')^T \mathbf{1}_L = \mathbf{0}_K$, both U_i and V_i are centred in the sense that $(U_i, \mathbf{1}_L) = 0$ and $(V_i, \mathbf{1}_K) = 0$.

Note that

$$\mathcal{A}_1(\mathbf{X}) = \sqrt{\lambda_{0(1)}}\, U_{0(1)} V_{0(1)}^T$$

with $V_{0(1)} = \mathbf{1}_K/\sqrt{K}$, $U_{0(1)} = \mathcal{E}_1(\mathbf{X})/\|\mathcal{E}_1(\mathbf{X})\|$ and $\lambda_{0(1)}^2 = \|\mathcal{E}_1(\mathbf{X})\|^2 K$. In the same way,

$$\mathcal{A}_2(\mathbf{X}') = \sqrt{\lambda_{0(2)}^{(1)}}\, U_{0(2)}^{(1)} \left(V_{0(2)}^{(1)}\right)^T$$

with $U_{0(2)}^{(1)} = \mathbf{1}_L/\sqrt{L}$, $V_{0(2)}^{(1)} = \mathcal{E}_2(\mathbf{X}')/\|\mathcal{E}_2(\mathbf{X}')\|$ and $\lambda_{0(2)}^{(1)} = \|\mathcal{E}_2(\mathbf{X}')\|^2 L$.

This means that the squared matrix norm of $\mathbf{X}$ can also be obtained as the sum of the squared matrix norms of the matrices on the right side of (4.35)

$$\|\mathbf{X}\|_{\mathcal{M}}^2 = \|\mathcal{A}_1(\mathbf{X})\|_{\mathcal{M}}^2 + \|\mathcal{A}_2(\mathbf{X}')\|_{\mathcal{M}}^2 + \sum_{i=1}^{d} \|\mathbf{X}_i\|_{\mathcal{M}}^2$$

or, in terms of λ_i, $\lambda_{0(1)}$ and $\lambda_{0(2)}^{(1)}$,

$$||\mathbf{X}||_{\mathcal{M}}^2 = \lambda_{0(1)} + \lambda_{0(2)}^{(1)} + \sum_{i=1}^{d} \lambda_i.$$

The triple $\left(\sqrt{\lambda_{0(2)}^{(1)}}, U_{0(2)}^{(1)}, V_{0(2)}^{(1)}\right)$ is called the *second average triple* of the decomposition (4.35). The numbers $\lambda_{0(1)}/||\mathbf{X}||_{\mathcal{M}}^2$ and $\lambda_{0(2)}^{(1)}/||\mathbf{X}||_{\mathcal{M}}^2$ can be regarded as the *shares of the first and second average triples* in the double centring SVD.

CHAPTER 5

Time series of finite rank

In practice, a time series F_N is usually a result of some measurements, and therefore the order d of the SVD decomposition of the trajectory matrix of F_N typically equals $\min(L, K)$, $K = N - L + 1$. However, there are series such that d does not depend on L or K, when L and K are large enough.

As was mentioned in Section 1.6.1, this property is important for the eigentriple grouping in Basic SSA. Moreover, such time series (and series sufficiently close to them) admit relatively simple and efficient continuation procedures discussed in Chapter 2. The same class of series is the basis for the change detection problems of Chapter 3.

The present chapter is devoted to a formal mathematical description of this class of time series and the linear recurrent formulae that govern these series. Several results that are useful for the construction of the continuation procedures are established.

5.1 General properties

Consider a real-valued sequence (series) $F_N = (f_0, \ldots, f_{N-1})$ with $N \geq 3$ and fix the window length L $(1 < L < N)$.

The result of the embedding procedure (see Section 1.2.1) is a sequence of the L-lagged vectors of the series F_N:

$$X_i^{(L)} = X_i = (f_{i-1}, \ldots, f_{i+L-2})^\mathrm{T}, \qquad i = 1, \ldots, K.$$

Denote by $\mathfrak{L}^{(L)}(F_N) \stackrel{\text{def}}{=} \mathrm{span}(X_1, \ldots, X_K)$ the trajectory space of the series F_N (this space will be denoted by $\mathfrak{L}^{(L)}$ for short).

Let $0 \leq d \leq L$. If $\dim \mathfrak{L}^{(L)} = d$, then we shall say that *the series F_N has L-rank d* and write this as $\mathrm{rank}_L(F_N) = d$. We shall assume that $d \neq 0$, which means that not all the f_n are zero.

It is clear that the equality $\mathrm{rank}_L(F_N) = d$ can hold only if

$$d \leq \min(L, K). \tag{5.1}$$

When dealing with equalities of the kind $\mathrm{rank}_L(F_N) = d$ we shall always assume that the condition (5.1) is met. If the equality $\mathrm{rank}_L(F_N) = d$ holds for all the appropriate L, then we say that *the series F_N has rank d* $(\mathrm{rank}(F_N) = d)$.

Let $\mathbf{X} = [X_1 : \ldots : X_K]$ be the L-trajectory matrix of the series F_N. Obviously

$$\mathrm{rank}_L(F_N) = \mathrm{rank}\,\mathbf{X} = \mathrm{rank}\,\mathbf{X}\mathbf{X}^\mathrm{T} = \mathrm{rank}\,\mathbf{X}^\mathrm{T}\mathbf{X}.$$

Moreover, the orthonormal system of eigenvectors $U_1, \ldots, U_d$ corresponding to the positive eigenvalues $\lambda_1 \geq \ldots \geq \lambda_d$ of the matrix $\mathbf{X}\mathbf{X}^\mathrm{T}$ constitute a basis of the space $\mathfrak{L}^{(L)}$. Therefore, $\mathrm{rank}_L(F_N)$ is the order of the SVD decomposition of the trajectory matrix $\mathbf{X}$.

Let us give examples of series of a finite rank and infinite series F such that $\mathrm{rank}_L(F) = L$ for any L.

Example 5.1 *Series of finite rank*
1. *Exponential-cosine sequences*
These series have the form

$$f_n = Ae^{\alpha n}\cos(2\pi\omega n + \phi) \qquad (5.2)$$

with $A \neq 0$, $\omega \in [0, 1/2]$ and $\phi \in [0, 2\pi)$. The exponential-cosine sequences can have rank either 1 or 2:

a) If $\omega = 0$ and $\cos(\phi) \neq 0$, then f_n is proportional to $e^{\alpha n}$ and we have the exponential series, which has rank 1. For all $L \geq 1$ and $N \geq L$, the space $\mathfrak{L}^{(L)}$ is spanned by the vector $(1, a, \ldots, a^{L-1})^\mathrm{T}$, $a = e^\alpha$. The same result takes place for the exponentially modulated saw-tooth sequence with $\omega = 1/2$, $\cos(\phi) \neq 0$ and $a = -e^\alpha$.

b) If $\omega \in (0, 1/2)$, then the sequence (5.2) has rank 2. For all L, $2 \leq L \leq N-1$, the space $\mathfrak{L}^{(L)}$ is spanned by the vectors Y_1 and Y_2 with components $y_k^{(1)} = e^{\alpha(k-1)}\cos(2\pi\omega(k-1))$ and $y_k^{(2)} = e^{\alpha(k-1)}\sin(2\pi\omega(k-1))$, respectively ($1 \leq k \leq L$).

2. *Polynomial sequences*

a) The linear series $f_n = an + b$, $a \neq 0$, is a series of rank 2. For $L \geq 2$ and $N \geq L + 1$, the space $\mathfrak{L}^{(L)}$ is spanned by the vectors

$$Y_1 = (1, \ldots, 1)^\mathrm{T}, \quad Y_2 = (0, 1, \ldots, L-1)^\mathrm{T}. \qquad (5.3)$$

b) The quadratic sequence $f_n = n^2$ has rank 3. For $3 \leq L \leq N-2$, the linear space $\mathfrak{L}^{(L)}$ is spanned by the vectors Y_1 and Y_2 defined in (5.3) and by $Y_3 = (0, 1^2, 2^2, \ldots, (L-1)^2)^\mathrm{T}$.

The case of a general polynomial is considered later (see Example 5.3).

3. *A series that does not have a finite rank*
Let us first take $N = 2L - 1$ and consider the series F_N with $f_0 = f_L = 1$ and the other f_n being zero. It is easy to see that all L coordinate vectors are present among the lagged vectors $X_i^{(L)}$ of the series F_N. Hence $\mathrm{rank}_L(F_N) = L$. The required series F can now be constructed as follows: the first three terms of the series F are the terms of the series F_3 (taken in the same order), while the next five are the terms of F_5, and so on.

GENERAL PROPERTIES

Remark 5.1 Note that the SVD eigenvectors of the trajectory matrix of any series F_N are linear combinations of the basis vectors of $\mathfrak{L}^{(L)}(F_N)$; the same holds for the factor vectors if we deal with the space $\mathfrak{L}^{(K)}(F_N)$ rather than with $\mathfrak{L}^{(L)}(F_N)$. Therefore, the singular vectors corresponding to the series of Example 5.1 (items 1 and 2) have the form stated in Section 1.6.1.

Let us introduce one more concept related to series of finite rank. We shall say that a series F_N admits an *L-decomposition of order not larger than d* (we write this as $\mathrm{ord}_L(F_N) \leq d$), if there exist two systems of functions

$$\varphi_k : \{0, \ldots, L-1\} \mapsto \mathbf{R}, \quad \psi_k : \{0, \ldots, K-1\} \mapsto \mathbf{R}$$

$(k = 1, \ldots, d)$ such that

$$f_{i+j} = \sum_{k=1}^{d} \varphi_k(i)\,\psi_k(j), \qquad 0 \leq i \leq L-1,\ 0 \leq j \leq K-1. \tag{5.4}$$

If $\mathrm{ord}_L(F_N) \leq d$ and $\mathrm{ord}_L(F_N) \not\leq d-1$, then the series F_N admits an *L-decomposition of the order d* ($\mathrm{ord}_L(F_N) = d$).

Let us give an explicit form of the decomposition (5.4) for several series of Example 5.1.

Example 5.2 *Series of finite order*

a) The power sequence $f_n = a^n$, $a \neq 0$:

$$f_{i+j} = a^{i+j} = a^i \cdot a^j = \varphi_1(i)\,\psi_1(j);$$

b) the exponential-cosine sequence (5.2) with $\omega \in (0, 1/2)$:

$$f_{i+j} = Ae^{\alpha i}\cos(2\pi\omega i) \cdot e^{\alpha j}\cos(2\pi\omega j + \phi)$$
$$- Ae^{\alpha i}\sin(2\pi\omega i) \cdot e^{\alpha j}\sin(2\pi\omega j + \phi) = \varphi_1(i)\,\psi_1(j) + \varphi_2(i)\,\psi_2(j);$$

c) linear sequence $f_n = an + b$ with $a \neq 0$:

$$f_{i+j} = 1 \cdot (aj + b) + i \cdot a = \varphi_1(i)\,\psi_1(j) + \varphi_2(i)\,\psi_2(j);$$

d) quadratic sequence $f_n = n^2$:

$$f_{i+j} = i^2 \cdot 1 + 1 \cdot j^2 + (2i) \cdot j$$
$$= \varphi_1(i)\,\psi_1(j) + \varphi_2(i)\,\psi_2(j) + \varphi_3(i)\,\psi_3(j).$$

The following statement demonstrates that there is a close link between the notions of finite rank and finite order.

Proposition 5.1
1. *The conditions* $\mathrm{rank}_L(F_N) = d$ *and* $\mathrm{ord}_L(F_N) = d$ *are equivalent.*
2. *The equality* (5.4) *determines an L-decomposition of order d of the series F_N if and only if both systems of functions,* $(\varphi_1, \ldots, \varphi_d)$ *and* $(\psi_1, \ldots, \psi_d)$, *are linearly independent.*

Proof.
1. Evidently, it is sufficient to show that the inequalities $\mathrm{rank}_L(F_N) \leq d$ and $\mathrm{ord}_L(F_N) \leq d$ are equivalent. Let $\mathrm{rank}_L(F_N) \leq d$ and denote by

$$\Phi_k = (\varphi_k(0), \ldots, \varphi_k(L-1))^{\mathrm{T}}, \quad 1 \leq k \leq d, \tag{5.5}$$

some vectors that span the space $\mathfrak{L}^{(L)}$. Let $\psi_k(j)$ be the coefficients of the expansion of the vector $X_{j+1}^{(L)}$ with respect to these vectors. We then have the equalities

$$X_{j+1}^{(L)} = \sum_{k=1}^{d} \psi_k(j) \Phi_k, \quad 0 \leq j < K, \tag{5.6}$$

which in the component-wise notation have the form (5.4).

On the other hand, we can see that in the notation (5.5) the equations (5.4) become (5.6). This means that all the vectors $X_j^{(L)}$ belong to some linear space $\mathfrak{L}$ spanned by the vectors Φ_k ($k=1,\ldots,d$). Obviously, the dimension of this space does not exceed d.

2. Let the equalities (5.4) define $\mathrm{ord}_L(F_N)$. If the functions φ_k (or ψ_k) are not linearly independent, then replacing some of them by linear combinations of others we come to equalities of the same form (5.4), but with a smaller number of terms. This leads to a contradiction. To prove the converse, assume that both function systems are linearly independent. We introduce the vectors

$$\Psi_k = (\psi_k(0), \ldots, \psi_k(K-1))^{\mathrm{T}}, \quad 1 \leq k \leq d.$$

Then (5.4) can be rewritten in the form

$$\mathbf{X} = \sum_{k=1}^{d} \Phi_k \Psi_k^{\mathrm{T}}.$$

In view of the result of Proposition 4.3 of Section 4.2.1, $\mathrm{rank}\,\mathbf{X} = d$. □

Example 5.3 *Polynomial series*
Consider a general polynomial series $f_n = P_m(n)$, where $P_m(t)$, $t \in \mathbf{R}$, is a polynomial function of order m. Since

$$P_m(s+t) = \sum_{k=0}^{m} \frac{d^k P_m(s)}{ds^k} \frac{t^k}{k!},$$

it follows that $\mathrm{rank}_L(F_N) = m+1$ for $L > m$ and all sufficiently large N. Moreover, the singular vectors of the SVD for the corresponding trajectory matrices have a polynomial structure.

Remark 5.2 Consider the class of infinite series that admit decompositions of finite order. Then (5.4) implies that this class is a linear space and is closed with respect to term-by-term multiplication of series. Therefore, Examples 5.2 and 5.3 imply that any time series whose general term can be represented as a sum of products of an exponential ($e^{\alpha n}$), harmonic ($\cos(2\pi\omega n + \phi)$) and polynomial ($P_m(n)$)

GENERAL PROPERTIES

terms does admit a decomposition of finite order. In view of Proposition 5.1, its L-rank does not depend on L for large L and K.

Now let us turn to the SVD of the trajectory matrices of time series with finite rank. Consider the series F_N and assume that $\mathrm{rank}_L(F_N) = d$ for some L satisfying (5.1). Then for the L-trajectory matrix $\mathbf{X}$ of the series F_N we obtain that rank $\mathbf{X} = d$ and therefore the SVD of this matrix has d terms:

$$\mathbf{X} = \sqrt{\lambda_1} U_1 V_1^{\mathrm{T}} + \ldots + \sqrt{\lambda_d} U_d V_d^{\mathrm{T}}. \tag{5.7}$$

Proposition 5.1 (and Proposition 4.3 of Section 4.2.1) tells us that there are many representations

$$\mathbf{X} = \sum_{k=1}^{d} P_k Q_k^{\mathrm{T}} \tag{5.8}$$

with some $P_k \in \mathbf{R}^L$ and $Q_k \in \mathbf{R}^K$, where the vectors $P_1, \ldots, P_d$ (and the vectors $Q_1, \ldots, Q_d$ also) are linearly independent.

Suppose that we have obtained a decomposition (5.8) with $d = \mathrm{rank}\,\mathbf{X}$. The following proposition shows how to compute the SVD (5.7) in terms of the vectors P_k and Q_k. Evidently it is enough to compute only λ_k and U_k; see (4.1). We set

$$c_{ij}^{(p)} = (P_i, P_j), \quad c_{ij}^{(q)} = (Q_i, Q_j)$$

and consider $d \times d$ matrices $\mathbf{A}_p$, $\mathbf{A}_q$ with the elements $c_{ij}^{(p)}$ and $c_{ij}^{(q)}$, respectively.

Proposition 5.2 *The eigenvalues of the matrix $\mathbf{X}\mathbf{X}^{\mathrm{T}}$ coincide with the eigenvalues of the matrix $\mathbf{A}_q \mathbf{A}_p$. If $A = (a_1, \ldots, a_d)^{\mathrm{T}}$ is an eigenvector of the matrix $\mathbf{A}_q \mathbf{A}_p$, then $U = a_1 P_1 + \ldots + a_d P_d$ is an eigenvector of the matrix $\mathbf{X}\mathbf{X}^{\mathrm{T}}$ and vice versa.*

Proof.
Let $Z = a_1 P_1 + \ldots + a_d P_d$ with some $a_1, \ldots, a_d$. Since

$$\mathbf{X}^{\mathrm{T}} P_m = \sum_{j=1}^{d} c_{mj}^{(p)} Q_j, \quad \mathbf{X} Q_k = \sum_{j=1}^{d} c_{kj}^{(q)} P_j,$$

then

$$\mathbf{X}\mathbf{X}^{\mathrm{T}} Z = \sum_{k=1}^{d} \left(\sum_{m=1}^{d} a_m \left(\sum_{j=1}^{d} c_{kj}^{(q)} c_{jm}^{(p)} \right) \right) P_k.$$

As the P_k are linearly independent, the rest of the proof is obvious. □

Let us consider in detail the example of the exponential-cosine series F_N with $f_n = e^{\alpha n} \cos(2\pi \omega n + \phi)$, $\omega \in (0, 1/2)$. The exponential-cosine series has L-rank 2 for any $N \geq 3$ and $1 < L < N$, see Example 5.2. Here we are interested in the situation when both eigenvalues of the matrix $\mathbf{X}\mathbf{X}^{\mathrm{T}}$ are equal or approximately equal.

We set $\psi = \phi/2$. Since

$$f_{k+m} = e^{\alpha(k+m)}\cos(2\pi\omega k + \psi)\cos(2\pi\omega m + \psi)$$
$$-e^{\alpha(k+m)}\sin(2\pi\omega k + \psi)\sin(2\pi\omega m + \psi),$$

the equality (5.8) is valid with $d = 2$ and the vectors P_1, P_2, Q_1, Q_2 whose components are as follows:

$$p_{1k} = e^{\alpha k}\cos(2\pi\omega k + \psi), \quad p_{2k} = -e^{\alpha k}\sin(2\pi\omega k + \psi),$$

$$q_{1m} = e^{\alpha m}\cos(2\pi\omega m + \psi), \quad q_{2m} = e^{\alpha m}\sin(2\pi\omega m + \psi)$$

for $1 \leq k \leq L, 1 \leq m \leq K$. Let $\mathbf{p} = \|P_1\|/\|P_2\|$, $\mathbf{q} = \|Q_1\|/\|Q_2\|$,

$$\mathbf{c}_p = \frac{(P_1, P_2)}{\|P_1\|\|P_2\|}, \quad \mathbf{c}_q = \frac{(Q_1, Q_2)}{\|Q_1\|\|Q_2\|}$$

and $S = \|P_1\|\|P_2\|\|Q_1\|\|Q_2\|$.

Proposition 5.3 *Let $T = 1/\omega$.*

1. *If $L = K$, then $\lambda_1 = \lambda_2$.*
2. *Let $\alpha = 0$. If L/T and K/T are integers, then $\lambda_1 = \lambda_2$. If L/T is an integer and $K \to \infty$, then $\lambda_1/\lambda_2 \to 1$.*
3. *Let $\alpha \leq 0$. If $\min(L, K) \to \infty$, then $\lambda_1/\lambda_2 \to 1$.*

Proof.
1. Proposition 5.2 implies that if $d = 2$ and λ_1, λ_2 are the eigenvalues of the matrix $\mathbf{X}\mathbf{X}^T$, then $\widetilde{\lambda}_1 = \lambda_1/S$ and $\widetilde{\lambda}_2 = \lambda_2/S$ are the eigenvalues of the matrix

$$\mathbf{B} = \begin{pmatrix} \mathbf{pq} + \mathbf{c}_p\mathbf{c}_q & \mathbf{qc}_p + \mathbf{p}^{-1}\mathbf{c}_q \\ \mathbf{pc}_q + \mathbf{q}^{-1}\mathbf{c}_q & \mathbf{p}^{-1}\mathbf{q}^{-1} + \mathbf{c}_p\mathbf{c}_q \end{pmatrix}. \quad (5.9)$$

Let us show that both eigenvalues of the matrix $\mathbf{B}$ coincide if and only if $\mathbf{c}_p = -\mathbf{c}_q$ and $\|P_1\|\|Q_1\| = \|P_2\|\|Q_2\|$.

Since $\widetilde{\lambda}_{1,2}$ are the roots of the quadratic equation

$$\widetilde{\lambda}^2 - \left(\mathbf{pq} + (\mathbf{pq})^{-1} + 2\mathbf{c}_p\mathbf{c}_q\right)\widetilde{\lambda} + 1 + \mathbf{c}_p^2\mathbf{c}_q^2 - \mathbf{c}_p^2 - \mathbf{c}_q^2 = 0,$$

the eigenvalues coincide if and only if

$$D \stackrel{\text{def}}{=} \left(\mathbf{pq} + (\mathbf{pq})^{-1} + 2\mathbf{c}_p\mathbf{c}_q\right)^2 - 4(1 + \mathbf{c}_p^2\mathbf{c}_q^2 - \mathbf{c}_p^2 - \mathbf{c}_q^2) = 0. \quad (5.10)$$

Since $\mathbf{c}_p^2$ and $\mathbf{c}_q^2$ do not exceed 1 and $C \stackrel{\text{def}}{=} \mathbf{pq} + (\mathbf{pq})^{-1} \geq 2$, the minimal value of D for fixed $\mathbf{c}_p$ and $\mathbf{c}_q$ is equal to $4(\mathbf{c}_p + \mathbf{c}_q)^2$ and is achieved for $C = 2$. Therefore, the first statement is proved.

2. If $\alpha = 0$ and both L/T and K/T are integers, then $\|P_1\|^2 = \|P_2\|^2 = L/2$, $\|Q_1\|^2 = \|Q_2\|^2 = K/2$ and $\mathbf{c}_p = \mathbf{c}_q = 0$. Thus, $D = 0$ and the eigenvalues coincide.

Consider now the asymptotic case. Note that

$$\frac{\lambda_1}{\lambda_2} = \frac{\mathbf{pq} + (\mathbf{pq})^{-1} + 2\mathbf{c}_p\mathbf{c}_q + \sqrt{D}}{\mathbf{pq} + (\mathbf{pq})^{-1} + 2\mathbf{c}_p\mathbf{c}_q - \sqrt{D}},$$

where D is defined in (5.10). If $\alpha = 0$ and L/T is an integer, then $\mathbf{p} = 1$ and $\mathbf{c}_p = 0$. Since $K \to \infty$, it follows that both $||Q_1||^2$ and $||Q_2||^2$ are equivalent to $K/2$ and therefore $\mathbf{q} \to 1$. In view of the boundedness of (Q_1, Q_2), $\mathbf{c}_q \to 0$. Thus, $D \to 0$ and $\lambda_1/\lambda_2 \to 1$.

3. In the case $\alpha \leq 0$ and $\min(L, K) \to \infty$ we obtain

$$\lim ||P_1|| = \lim ||Q_1|| < \infty, \quad \lim ||P_2|| = \lim ||Q_2|| < \infty.$$

Moreover, there exist the finite limits $\lim (P_1, P_2) = -\lim (Q_1, Q_2)$. Thus, asymptotically the eigenvalues become close. This completes the proof. □

5.2 Series of finite rank and recurrent formulae

Let us give one more definition. We shall say that the series F_N has *difference dimension not larger than* d (fdim(F_N) $\leq d$), if $1 \leq d < N - 1$ and there are numbers $a_1, \ldots, a_d$ such that

$$f_{i+d} = \sum_{k=1}^{d} a_k f_{i+d-k}, \qquad 0 \leq i \leq N - d - 1, \quad a_d \neq 0. \qquad (5.11)$$

It is easy to see that for $d < N - 2$ the inequality fdim(F_N) $\leq d$ implies fdim(F_N) $\leq d + 1$.

The number $d = \min\{k : \text{fdim}(F_N) \leq k\}$ is called the *finite-difference dimension* of the series F_N (fdim(F_N) = d). For the zero series F_N with $f_n \equiv 0$, we set fdim(F_N) = 0. We shall typically assume that fdim(F_N) > 0.

The formula (5.11) will be called the *linear recurrent formula* (LRF). The LRF (5.11) with $d = $ fdim(F_N) is the *minimal* LRF. For infinite series the upper bounds on i and d disappear in the definition of the difference dimension.

If (5.11) is valid, then we shall say that the series F_N is *governed by* the LRF (5.11). If (5.11) holds without any restrictions on a_d, then F_N *satisfies* the LRF (5.11).

Example 5.4 *Series of finite difference dimension*
Let us specify the formula (5.11) for the series of Examples 5.1 and 5.2.

a) the power sequence $f_n = a^n$: $f_{i+1} = af_i$, $d = 1$;

b) the exponential-cosine sequence (5.2) with $\omega \in (0, 1/2)$:

$$f_{i+2} = 2e^\alpha \cos(2\pi\omega) f_{i+1} - e^{2\alpha} f_i, \quad d = 2;$$

c) the linear sequence $f_n = an + b$ with $a \neq 0$: $f_{i+2} = 2f_{i+1} - f_i$, $d = 2$;

d) the quadratic sequence $f_n = n^2$: $f_{i+3} = 3f_{i+2} - 3f_{i+1} + f_i$, $d = 3$.

These representations can be validated by direct insertion of the formulae for f_n into the corresponding LRFs.

The following proposition shows the relationship between the difference dimension and the L-rank of time series F_N.

Proposition 5.4 *Let $1 < L < N$, $K = N - L + 1$ and $d \leq \min(L, K)$.*
1. If $\mathrm{fdim}(F_N) \leq d$, then $\mathrm{rank}_L(F_N) \leq d$. Additionally, if the L-lagged vectors $X_1^{(L)}, \ldots, X_d^{(L)}$ are linearly independent, then $\mathrm{rank}_L(F_N) = d$.
2. If $\mathrm{fdim}(F_N) \leq d$ and the vectors $X_1^{(d)}, \ldots, X_d^{(d)}$ are linearly independent, then $\mathrm{fdim}(F_N) = d$. In this case, the LRF (5.11) is the unique minimal LRF governing the series F_N.

Proof.
1. The equalities (5.11) can be rewritten in the form

$$X_j^{(L)} = \sum_{k=1}^{d} a_k X_{j-k}^{(L)}, \qquad d+1 \leq j \leq K.$$

Thus, all the vectors $X_j^{(L)}$ are linear combinations of the vectors $X_1^{(L)}, \ldots, X_d^{(L)}$, which yields the inequality $\mathrm{rank}_L(F_N) \leq d$. In the case when these vectors are linearly independent, the last inequality becomes an equality.

2. Assume that $\mathrm{fdim}(F_N) = d_0 < d$. Then the first statement implies the inequality $\mathrm{rank}_d(F_N) \leq d_0$. On the other hand, $\mathrm{rank}_d(F_N) = d$ in view of the linear independence of the vectors $X_1^{(d)}, \ldots, X_d^{(d)}$. The uniqueness of the representation (5.11) is a consequence of the uniqueness of the representation of the vector $X_{d+1}^{(d)}$ as a linear combination of the $X_1^{(d)}, \ldots, X_d^{(d)}$. □

Remark 5.3 Proposition 5.4 implies that the LRFs of Example 5.4 are minimal.

Proposition 5.4 shows that any sequence governed by the LRF (5.11) is also a series of finite rank. It is easy to construct an example showing that the converse statement is false.

Example 5.5 *Counterexample*
Let $f_n = 1$ for $0 \leq n \leq N - 2$ and $f_{N-1} = 2$. It is clear that $\mathrm{rank}_L(F_N) = 2$ for $2 \leq L \leq N - 1$. At the same time, for $d < N - 1$ the equalities (5.11) cannot be satisfied for any set of coefficients a_k.

The following theorem (see Buchstaber, 1994) makes the situation clearer and also plays an important role in the problem of series continuation. This theorem implies that any (finite) series of L-rank $d < L$ is governed by an LRF of dimension not larger than d with, perhaps, an exception made for the first few and last few terms of the series.

Consider the series F_N and denote by $F_{i,j}$ ($1 \leq i \leq j \leq N$) the series consisting of $f_{i-1}, \ldots, f_{j-1}$. In the case $i = 1$ we shall write F_j instead of $F_{1,j}$. As before we set $X_i^{(L)} = (f_{i-1}, \ldots, f_{i+L-2})^{\mathrm{T}}$.

We set $e_i = (0, \ldots, 1, \ldots, 0)^{\mathrm{T}} \in \mathbf{R}^L$ for the vector all of whose components are zero except for the ith one which is equal to 1.

SERIES OF FINITE RANK AND RECURRENT FORMULAE

Theorem 5.1 *Let $1 \leq \operatorname{rank}_L(F_N) = d < L$. Then there are integers d_0 and M such that $0 \leq d_0 \leq d$, $0 \leq M \leq d - d_0$ and $\operatorname{fdim}(F_{M+1, M+K+d_0}) = d_0$.*

Proof.
Let $\mathcal{L}_d$ be the L-trajectory space of the series F_N. Denote by $\mathcal{L}_d^\perp$ the orthogonal complement of this space, and by $\mathcal{M}_i$ ($0 \leq i \leq L$) the linear vector space spanned by the vectors $e_1, \ldots, e_i$. Set $\mathcal{N}_i = \mathcal{M}_i \cap \mathcal{L}_d^\perp$. Obviously, $s \stackrel{\text{def}}{=} \dim \mathcal{L}_d^\perp = L - d$ and $\mathcal{N}_L = \mathcal{L}_d^\perp$. For $\Delta_i \stackrel{\text{def}}{=} \dim \mathcal{N}_i$ we have

$$0 = \Delta_0 \leq \Delta_1 \leq \ldots \leq \Delta_L = s$$

with $\Delta_{i+1} - \Delta_i \in \{0, 1\}$.

We define $n_i = \min\{k : \Delta_k = i\}$ for $1 \leq i \leq L - d$, set $r = n_1$ and observe that $r \leq d + 1$. (This follows from the fact that the maximal value of r corresponds to the equalities $\Delta_L = s$, $\Delta_{L-1} = s - 1, \ldots, \Delta_{L-s} = 0$.)

Consider first the case $r = 1$, that is $e_1 \in \mathcal{L}_d^\perp$. Then $(X_m^{(L)}, e_1) = 0$ for $1 \leq m \leq K$. This corresponds to the equalities $d = 1$ and $f_0 = \ldots = f_{K-1} = 0$. Hence $\operatorname{fdim}(F_{1,K}) = 0$, $M = 0$, and $d_0 = 0$.

Now let $r > 1$. We set $\rho = r - 1$. Choose the basis $Y_1, \ldots, Y_s$ in the space $\mathcal{L}_d^\perp$ so that $Y_i \in \mathcal{N}_{n_i}$. We can assume that Y_1 has the form

$$Y_1 = (a_\rho, a_{\rho-1}, \ldots, a_1, -1, 0, \ldots, 0)^\mathrm{T}, \tag{5.12}$$

and this representation of the vector Y_1 is unique. Since the vector Y_1 is orthogonal to all the vectors $X_m^{(L)}$, it follows that

$$f_{i+\rho} = \sum_{k=1}^{\rho} a_k f_{i+\rho-k} \tag{5.13}$$

for $0 \leq i \leq K - 1$. We now set $\rho_1 = \min\{k \geq 0 : a_j = 0, j > k\}$. Obviously, $\rho_1 \leq \rho$. If $\rho_1 = 0$, then $f_\rho = \ldots = f_{K+\rho-1} = 0$, $\operatorname{fdim}(F_{\rho+1, \rho+K}) = 0$, $M = \rho$ and $d_0 = 0$.

If $\rho_1 > 0$, then the formula (5.13) has the form

$$f_{i+\rho} = \sum_{k=1}^{\rho_1} a_k f_{i+\rho-k}, \qquad a_{\rho_1} \neq 0, \tag{5.14}$$

$0 \leq i \leq K - 1$. This is the LRF of order ρ_1 (i.e., $d_0 = \rho_1$) which governs the series $F_{M+1, M+K+d_0}$ with $M = \rho - d_0$. $\square$

Corollary 5.1
1. *If an infinite (in both directions) series*

$$F = (\ldots, f_{-n}, \ldots, f_{-1}, f_0, f_1, \ldots, f_n, \ldots)$$

satisfies the condition $\operatorname{rank}_L(F) = d > 0$ for some $L \geq d$, then $\operatorname{fdim}(F) = d$ and $\operatorname{rank}_{L_0}(F) = d$ for any $L_0 \geq d$.

2. *If an infinite series*

$$F = F_\infty = (f_0, \ldots, f_n, \ldots)$$

satisfies the condition $\mathrm{rank}_L(F) = d > 0$ for some $L \geq d$, then there exist integers d_0, M such that $0 \leq d_0 \leq d$, $0 \leq M \leq d - d_0$ and

$$\mathrm{fdim}(F_{M+1,\infty}) = \mathrm{rank}_{L_0}(F_{M+1,\infty}) = d_0$$

for any $L_0 \geq d$.

Proof.
1. Let us modify the proof of Theorem 5.1 for the series F. Since the series F is nonzero, $\rho \neq 0$ and $\rho_1 \neq 0$. Therefore, the equality (5.14) holds for all $i = 0, \pm 1, \ldots$ Hence $\mathrm{fdim}(F) \leq \rho \leq d$ and $\mathrm{fdim}(F) = \rho_1 \leq d$. According to Proposition 5.4 we have $\mathrm{rank}_L(F) \leq \rho_1$. This implies that $\rho_1 = d$ and $\mathrm{fdim}(F) = d$.

Let us turn to the second equality. First we note that the equality $\mathrm{fdim}(F) = d$ and Proposition 5.4 imply that $\mathrm{rank}_{L_0}(F) \leq d$ for any $L_0 \geq d$. Additionally, the inequality $\mathrm{rank}_{L_0}(F) < d$ cannot hold for L_0 since otherwise the inequality $\mathrm{fdim}(F) < d$ would hold.

2. The second statement of the corollary can be proved analogously. Since the series F is infinite, the restrictions on the end of the series cannot appear. On the other hand, the restriction on the beginning of the series F can occur. □

Remark 5.4 The second statement of Corollary 5.1 corresponds to Gantmacher (1998, Chapter XVI, Section 10, Theorem 7). The theorem asserts that an infinite (in both directions) trajectory (Hankel) matrix of an infinite series F has rank d if and only if (a) the series satisfies the LRF (5.11) for all i (without restrictions on a_d), and (b) it does not satisfy an analogous LRF with smaller d.

For instance, the infinite series with $f_0 = 0$ and $f_n = 1$, $n \geq 1$, has rank 2 for any $L \geq 2$. This series is governed by the LRF $f_{i+1} = f_i$ for $i \geq 1$ (see Corollary 5.1, item 2) and satisfies the LRF $f_{i+2} = f_{i+1} + 0 \cdot f_i$ for $i \geq 0$ (in correspondence with Gantmacher, 1998).

Remark 5.5 The basis $Y_1, \ldots, Y_s$ described in the proof of Theorem 5.1 is called a *Schubert basis*. We can always assume that each vector Y_i has a form similar to (5.12); i.e., the last $L - n_i$ components of the vector Y_i are zeros and the n_ith component is -1.

This means that each vector Y_i also defines an LRF describing some subseries of the series F_N. These LRFs have, however, larger order than the LRF (5.14). Moreover, the vectors $Y_2, \ldots, Y_s$, unlike Y_1, are not uniquely defined (to get the uniqueness we can consider, for instance, the orthogonal Schubert basis).

Remark 5.6 If we make an additional assumption that $e_L \notin \mathfrak{L}_d$, then among the basis vectors $Y_1, \ldots, Y_s$ of the space $\mathfrak{L}_d^\perp$ there should exist at least one whose last component is not zero. According to the construction of the Schubert basis, this can only be Y_s so that $Y_s = (a_{L-1}, a_{L-2}, \ldots, a_1, -1)^\mathrm{T}$. Since the vector Y_s is orthogonal to all the vectors $X_m^{(L)}$ ($1 \leq m \leq K$), we have

$$f_{i+L-1} = \sum_{k=1}^{L-1} a_k f_{i+L-1-k}, \quad 0 \leq i \leq K - 1. \quad (5.15)$$

SERIES OF FINITE RANK AND RECURRENT FORMULAE

We have thus arrived at an LRF of dimension smaller than or equal to $L - 1$. This LRF governs the series F_N with the exception, possibly, of its beginning. If $a_{L-1} \neq 0$, then the whole series F_N is governed by the LRF (5.15).

As was already mentioned, the LRF (5.15) is not uniquely defined since the Schubert basis is not unique. Let us give one of the variants of this LRF derived from totally different considerations (see Danilov, 1997a, 1997b).

For any vector $X \in \mathbf{R}^L$ we shall denote by $X^\nabla \in \mathbf{R}^{L-1}$ the vector consisting of the first $L - 1$ components of the vector X. Denote by $P_1, \ldots, P_d$ some orthonormal basis of the trajectory space $\mathcal{L}_d$ and consider the linear vector space $\mathcal{L}_d^\nabla$ spanned by $P_1^\nabla, \ldots, P_d^\nabla$. Denote by π_i the last component of the vector P_i.

Assuming that $e_L \notin \mathcal{L}_d$, we obtain the inequality $\nu^2 \stackrel{\text{def}}{=} \pi_1^2 + \ldots + \pi_d^2 < 1$. It is natural to call the quantity ν^2 the *verticality coefficient* of the space $\mathcal{L}_d$. Note that the verticality coefficient is equal to the squared cosine of the angle between e_L and the space $\mathcal{L}_d$, and therefore this coefficient does not depend on the choice of the basis in this space. Set

$$\mathcal{R} = \frac{1}{1-\nu^2} \sum_{i=1}^{d} \pi_i P_i^\nabla, \quad \mathcal{R} = (a_{L-1}, \ldots, a_1)^\mathrm{T}. \tag{5.16}$$

Theorem 5.2 *Let $1 \leq \mathrm{rank}_L(F_N) = d < L$ and $e_L \notin \mathcal{L}_d$. Then (5.15) holds, where the coefficients a_k are defined in (5.16).*

Proof.
The proof will be given in three steps.
1. Let us demonstrate that for any vector $V \in \mathcal{L}_d^\nabla$ there exists a vector $Y \in \mathcal{L}_d$ such that $Y^\nabla = V$, and, moreover, the condition $e_L \notin \mathcal{L}_d$ implies the uniqueness of this vector.

Indeed, since the vectors $P_1^\nabla, \ldots, P_d^\nabla$ span the space $\mathcal{L}_d^\nabla$, we have

$$V = h_1 P_1^\nabla + \ldots + h_d P_d^\nabla.$$

Also, the vectors $P_1, \ldots, P_d$ form a basis of the space $\mathcal{L}_d$, and therefore the vector $Y = h_1 P_1 + \ldots + h_d P_d$ belongs to $\mathcal{L}_d$. Since $Y^\nabla = V$, the existence is proved. Let $Y_1, Y_2 \in \mathcal{L}_d$ and $Y_1^\nabla = Y_2^\nabla = V$. Then the difference $Y_1 - Y_2$ obviously belongs to $\mathcal{L}_d$ and is proportional to e_L. Therefore, if $Y_1 \neq Y_2$, then $e_L \in \mathcal{L}_d$.

2. Let us now prove that the last component $y = y_L$ of the vector Y has the form $\mathcal{R}^\mathrm{T} V$. Let $V = (x_1, \ldots, x_{L-1})^\mathrm{T}$. Then there exist numbers h_k such that

$$(x_1, x_2, \ldots, x_{L-1}, 0)^\mathrm{T} + y\, e_L = \sum_{k=1}^{d} h_k P_k. \tag{5.17}$$

Taking the inner product of both sides of (5.17) and P_k, we obtain

$$h_k = \left(P_k^\nabla, V\right) + y\pi_k, \quad k = 1, \ldots, d. \tag{5.18}$$

On the other hand, taking the inner product of (5.17) and e_L and using (5.18), we have

$$y = \sum_{k=1}^{d} h_k \pi_k = \sum_{k=1}^{d} \left(P_k^\nabla, V\right) \pi_k + y \nu^2,$$

where $\nu^2 = \pi_1^2 + \ldots + \pi_d^2 < 1$. Since

$$\sum_{k=1}^{d} \left(P_k^\nabla, V\right) \pi_k = \left(\sum_{k=1}^{d} P_k^\nabla \pi_k, V\right) = (1 - \nu^2)\,(\mathcal{R}, V),$$

we have $y = y_L = \mathcal{R}^\mathrm{T} V$.

3. Consider now the L-lagged vector

$$X_{i+1} = X_{i+1}^{(L)} = (f_i, f_{i+1}, \ldots, f_{i+L-1})^\mathrm{T} \in \mathfrak{L}_d.$$

Clearly $V \stackrel{\mathrm{def}}{=} X_{i+1}^\nabla = (f_i, f_{i+1}, \ldots, f_{i+L-2})^\mathrm{T} \in \mathfrak{L}_d^\nabla$. Therefore, we obtain the equality $f_{i+L-1} = \mathcal{R}^\mathrm{T} X_{i+1}^\nabla$. This defines the required LRF. $\square$

Remark 5.7 The proof of Theorem 5.2 implies that the vector $Y \in \mathbf{R}^L$ belongs to a linear vector space $\mathfrak{L}$ with dim $\mathfrak{L} < L$ and $e_L \notin \mathfrak{L}$ if and only if $Y^\nabla \in \mathfrak{L}^\nabla$ and the last component y of the vector Y has the form $y = \mathcal{R}^\mathrm{T} Y^\nabla$.

Consider now the vector $\mathcal{R}$ from a different viewpoint. We set

$$\mathcal{Q} = (1 - \nu^2) \begin{pmatrix} -\mathcal{R} \\ 1 \end{pmatrix}. \tag{5.19}$$

Proposition 5.5 *The vector $\mathcal{Q}$ is the orthogonal projection of the vector e_L onto $\mathfrak{L}_d^\perp$. Moreover, $||\mathcal{Q}||^2 = 1 - \nu^2$ and $||\mathcal{R}||^2 = \nu^2/(1 - \nu^2)$.*

Proof.
Since $\nu^2 = \pi_1^2 + \ldots + \pi_d^2$, we have

$$\mathcal{Q} = e_L - \begin{pmatrix} \pi_1 P_1^\nabla + \ldots + \pi_d P_d^\nabla \\ \pi_1^2 + \ldots + \pi_d^2 \end{pmatrix} = e_L - \sum_{i=1}^{d} \pi_i P_i.$$

Using the equality $\pi_i = (e_L, P_i)$ we obtain

$$\mathcal{Q} = \sum_{i > d} \pi_i P_i \tag{5.20}$$

and thus the first statement of the proposition is proved. The required expression for $||\mathcal{Q}||^2$ is an immediate consequence of (5.20); the expression for $||\mathcal{R}||^2$ follows from (5.19). $\square$

Note that if $\nu^2 = 1$, then $\mathcal{Q} = \mathbf{0}_L$ (the vector $\mathcal{R}$ is not defined in this case). At the other extreme, when $\nu^2 = 0$, we have $\mathcal{Q} = e_L$ and $\mathcal{R} = \mathbf{0}_{L-1}$.

Let us give some examples of series of finite rank along with the corresponding LRFs. These examples serve to elucidate the role of the restrictions that cut off the beginning and the end of the series from the corresponding LRF of Theorem 5.1. The examples also illustrate Theorem 5.2.

Example 5.6 *LRF of Theorem* 5.1
Consider the series F_N of Example 5.5 with $N = 7$, and assume that $L = 3$. Then the L-trajectory space $\mathfrak{L}_2$ has a basis consisting of the vectors $e_1 + e_2$ and e_3 (this implies that Theorem 5.2 is not applicable to this series). The orthogonal complement $\mathfrak{L}_2^\perp$ is one-dimensional and has the basis $e_2 - e_1$. Furthermore, $\dim \mathcal{N}_0 = \dim \mathcal{N}_1 = 0$, $\mathcal{N}_2 = \mathcal{N}_3 = \mathfrak{L}_2^\perp$. Thus, $r = 2$, $\rho = d_0 = 1 < d = 2$ and $M = 0$. Since $Y_1 = e_1 - e_2 \in \mathcal{N}_2$, the formula (5.14) has the form $f_{i+1} = f_i$ and holds for all i except for the last one.

Example 5.7 *LRFs of Theorems* 5.1 *and* 5.2
Let $F_7 = (2, 1, 1, 1, 1, 1, 1)$ and $L = 4$. Obviously, $\text{rank}_4(F_7) = 2$. The basis of the space $\mathfrak{L}_2$ consists of the vectors e_1 and $e_2 + e_3 + e_4$. The space $\mathfrak{L}_2^\perp$ has the basis $e_2 - e_3, e_3 - e_4$, which is a Schubert basis. Consider the series F_7 from the viewpoint of Theorem 5.1. In this case, $\dim \mathcal{N}_0 = \dim \mathcal{N}_1 = \dim \mathcal{N}_2 = 0$, $\dim \mathcal{N}_3 = 1$, $r = 3$ and $\rho = 2$.

Since $Y_1 = e_2 - e_3$, we have $d_0 = 1$ and the LRF has the form $f_{i+2} = f_{i+1}$. This LRF holds for $i = 0, \ldots, 5$. Therefore, we have in this case $\rho = d = 2$, but the dimension of the LRF is one less (since $a_\rho = 0$); hence, $M = 1$. The LRF itself is valid for all the terms of the series apart from the first one.

The vector $Y_2 = e_3 - e_4$ generates the LRF $f_{i+3} = f_{i+2}$, which governs all the terms of the series except for the first two.

Let us check now what we can get from Theorem 5.2. In this case we can take $P_1 = e_1$ and $P_2 = (e_2 + e_3 + e_4)/\sqrt{3}$. This yields $\pi_1 = 0$, $\pi_2 = 1/\sqrt{3}$, $P_1^\nabla = (1, 0, 0)^T$ and $P_2^\nabla = (0, 1, 1)^T/\sqrt{3}$. Since $\nu^2 = 1/3$, we immediately obtain $\mathcal{R} = (0, 1, 1)^T/2$, which corresponds to the LRF $f_{i+3} = (f_{i+2} + f_{i+1})/2$ ($0 \leq i \leq 5$).

This formula holds for all the terms of the series apart from the first one. Note that this LRF corresponds to the second vector in the Schubert basis: $Y_2' = 0.5 e_1 + 0.5 e_2 - e_4 = Y_2 + 0.5 Y_1$.

If in this example we choose $L = 3$, then the space $\mathfrak{L}_2^\perp$ will become one-dimensional and the LRFs of both theorems will coincide.

Example 5.8 *LRFs of Theorems* 5.1
Consider the series $F_9 = (2, 1, 1, 1, 1, 1, 1, 1, 2)$ and the window length $L = 5$. Then $\text{rank}_4(F_9) = 3$ and the basis of the space $\mathfrak{L}_3 = \mathfrak{L}_3^{(5)}$ is $e_1, e_2 + e_3 + e_4$, e_5. Theorem 5.2 is again inapplicable. Correspondingly, $\dim \mathfrak{L}_3^\perp = 2$ and the Schubert basis of $\mathfrak{L}_3^\perp$ can be chosen as $e_2 - e_3, e_3 - e_4$. It is easy to see that $\dim \mathcal{N}_1 = \dim \mathcal{N}_2 = 0$, $\dim \mathcal{N}_3 = 1$, and $\dim \mathcal{N}_4 = 2$. Consequently, $r = n_1 = 3$, $n_2 = 4$, and $\rho = 2$.

We have $Y_1 = e_2 - e_3$ and $Y_2 = e_3 - e_4$. Therefore, either of the two equalities $(X_i, Y_1) = 0$ and $(X_i, Y_2) = 0$, which hold for $0 \leq i \leq 4$, generates the LRFs $f_{i+2} = f_{i+1}$ and $f_{i+3} = f_{i+2}$ of dimension 1. We thus again have $d_0 = 1$. The first of these LRFs describes the terms f_i with $1 \leq i \leq 7$ ($M = 1$); the second one describes the terms with $2 \leq i \leq 8$. Together these two LRFs describe the whole series, with the exception of the first and the last terms.

Example 5.9 *LRFs of Theorems* 5.1 *and* 5.2

Consider the series $F_7 = (1, 1, 1, 2, 1, 1, 1)$ and let the window length be $L = 5$. It is easy to see that in this case $d = \text{rank}_L(F_7) = 3$, which coincides with $K = N - L + 1$. The space $\mathfrak{L}_3^\perp$ has the basis $e_1 - e_5$, $-4e_1 + e_2 + e_3 + e_4$, and thus $Y_1 = 4e_1 - e_2 - e_3 - e_4$. Therefore, the LRF (5.14) of dimension $d_0 = 3$ has the form $f_{i+3} = -f_{i+2} - f_{i+1} + 4f_i$. It holds for $0 \leq i \leq 2$ and does not hold for $i = 3$.

It is easy to construct the LRF (5.15), (5.16) using Proposition 5.5. Simple calculations give $\mathcal{R} = (3, 4, 4, 4)/19$. The resulting formula of dimension 4 holds for $0 \leq i \leq 2$; i.e., it describes the whole series F_7.

Theorem 5.1 together with Corollary 5.1 (as well as Theorem 5.2) show that, despite the fact that the notions of a series of finite rank and of finite dimension are not generally equivalent, there is a close link between them. In this respect, the series governed by the LRFs are of particular importance since they possess analytic descriptions. Let us formulate the corresponding statement (the proof can be found, for instance, in Gelfond, 1967, Chapter V, §4).

Let the series $F = (f_0, \ldots, f_n, \ldots)$ be governed by the LRF (5.11) with $a_d \neq 0$ and $i \geq 0$. Consider the *characteristic polynomial* of the LRF (5.11):

$$P_d(\lambda) = \lambda^d - \sum_{k=1}^{d} a_k \lambda^{d-k}$$

and denote by k_m ($1 \leq m \leq p$) the multiplicities of its different roots $\lambda_1, \ldots, \lambda_p$ ($k_1 + \ldots + k_p = d$). Set

$$f_n(m, j) = n^j \lambda_m^n, \quad 1 \leq m \leq p, \quad 0 \leq j \leq k_m - 1.$$

Note that since $a_d \neq 0$, all the roots λ_m are different from zero.

Theorem 5.3 *A real-valued series F satisfies the LRF* (5.11) *if and only if*

$$f_n = \sum_{m=1}^{p} \sum_{j=0}^{k_m - 1} c_{mj} f_n(m, j) \tag{5.21}$$

with (complex) coefficients c_{mj} determined by the first d terms of the series, $f_0, \ldots, f_{d-1}$.

Remark 5.8 If the root λ_m is real, then its input into the sum (5.21) has the form of a polynomial in n, multiplied by λ_m^n. Moreover, the corresponding coefficients c_{mj} are real.

If the root λ_m is complex, then the polynomial has also a complex conjugate root λ_l of the same multiplicity. (Note that the coefficients a_k are real.) Rewriting one of these roots as $e^\alpha e^{i2\pi\omega}$ with $\omega \in (0, 1/2)$, we get that the joint input of both roots into the sum (5.21) has the form of a polynomial in n multiplied by the term $e^{\alpha n} \cos(2\pi \omega n + \phi)$. Moreover, in this case the corresponding coefficients c_{mj} and c_{lj} are conjugate complex numbers.

In both cases, if the multiplicity of the root λ_m is 1, then the polynomial in n becomes a constant. Therefore, any infinite series satisfying LRF (5.11) for any i is necessarily a sum of products of polynomial, exponential and harmonic series. This is in total agreement with Remark 5.2.

Example 5.10 *LRFs of dimension 1 and 2*

Let us consider the case $d = 1$ and $d = 2$ in detail. If $d = 1$, then the linear recurrent formula is $f_n = af_{n-1}$, $a \neq 0$, and the unique (up to a nonzero constant A) time series F satisfying this equation is $f_n = Aa^n$. This corresponds to the constant ($a = 1$), saw-tooth ($a = -1$), exponential ($a > 0$, $a \neq 1$) and exponential–saw-tooth ($a < 0$, $a \neq -1$) series.

For $d = 2$ we have to consider the polynomial $P_d(\lambda) = \lambda^2 - a_1 \lambda - a_2$ and its roots λ_1, λ_2 under the assumption $a_2 \neq 0$.

If these roots coincide, then they are real, and the series has the form

$$f_n = A_1 \lambda^n + A_2 n \lambda^n, \quad \lambda = \lambda_1 = \lambda_2.$$

Note that if F is a linear series, then $\lambda_1 = \lambda_2 = 1$. Two different real roots λ_1 and λ_2 generate a sum of two different power series:

$$f_n = A_1 \lambda_1^n + A_2 \lambda_2^n.$$

At the same time, two different (and conjugate) complex roots $\lambda_1 = e^{\alpha + i2\pi\omega}$, $\lambda_2 = e^{\alpha - i2\pi\omega}$ with $\alpha \in \mathbf{R}$ and $0 < \omega < 1/2$, generate a series of the form

$$f_n = Ae^{\alpha n} \cos(2\pi\omega n + \phi),$$

where A and ϕ are arbitrary constants, providing that the series is not a zero one. Note that a pure harmonic solution appears if $|\lambda_1| = |\lambda_2| = 1$.

Assume now that $\mathcal{L}_d \subset \mathbf{R}^L$ is some linear space of dimension $d < L$ and $e_L \notin \mathcal{L}_d$. Let us pose the question: 'Does there exist a nonzero series such that its L-lagged vectors belong to $\mathcal{L}_d$?' The following statement shows that for $d < L-1$ the answer to this question is typically negative.

Let us define the space $\mathcal{L}_d$ in terms of the set of $L - d$ linear equations:

$$x_L + c_{j1} x_{L-1} + \ldots + c_{j,L-2} x_2 + c_{j,L-1} x_1 = 0 \tag{5.22}$$

$(1 \leq j \leq L-d)$, where the vectors $(c_{j,L-1}, \ldots, c_{j1}, 1)^\mathrm{T}$ are linearly independent (this is possible in view of the assumption $e_L \notin \mathcal{L}_d$).

Proposition 5.6 *There exists an infinite nonzero series* $\Phi = (\varphi_0, \ldots, \varphi_n, \ldots)$ *such that its L-lagged vectors belong to $\mathcal{L}_d$ if and only if the polynomials*

$$P_j(\lambda) = \lambda^{L-1} + c_{j1} \lambda^{L-2} + \ldots + c_{j,L-2} \lambda + c_{j,L-1} \tag{5.23}$$

$(j = 1, \ldots, L - d)$ *have at least one common root.*

Proof.

If the L-lagged vectors of the series Φ belong to $\mathcal{L}_d$, then this series satisfies $L - d$ linear recurrent formulae

$$\varphi_n = a_{j1} \varphi_{n-1} + \ldots + a_{j,L-2} \varphi_{n-L+1} + a_{j,L-1} \varphi_{n-L}, \quad n \geq L, \tag{5.24}$$

where $a_{jk} = -c_{jk}$. According to Theorem 5.3 the form of the series satisfying (5.24) is fully determined by the roots of the polynomial (5.23). If all the roots are different (this case is the most important in practice), then

$$\varphi_n = b_{1j}\lambda_{1j}^n + \ldots + b_{Lj}\lambda_{Lj}^n,$$

where the λ_{kj} are the roots of the polynomial (5.23). This immediately implies the statement of the proposition for the case when all the roots are different. The case of multiple roots is analogous. □

Corollary 5.2 *Let $d = L - 1$ and $e_L \notin \mathfrak{L}_d$. Then the LRF*

$$\varphi_n = a_{11}\varphi_{n-1} + \ldots + a_{1,L-2}\varphi_{n-L+1} + a_{1,L-1}\varphi_{n-L}, \quad n \geq L,$$

with $a_{1k} = -c_{1k}$ generates all series whose L-lagged vectors belong to $\mathfrak{L}_d$.

5.3 Time series continuation

Let $F_N = (f_0, \ldots, f_{N-1})$ and $d \leq \min(L, K)$, $K = N - L + 1$. Assume that $\text{rank}_L(F_N) = d$, and denote by $\mathfrak{L}_d = \mathfrak{L}^{(L)}(F_N)$ the corresponding d-dimensional L-trajectory space.

We shall say that the series F_N *is continuable in* $\mathfrak{L}_d$ (or simply *L-continuable*) if there is a unique number $\widetilde{f}_N$ such that all the L-dimensional lagged vectors of the series $\widetilde{F}_{N+1} = (f_0, \ldots, f_{N-1}, \widetilde{f}_N)$ still belong to the space $\mathfrak{L}_d$. In this case, the series $\widetilde{F}_{N+1}$ (and also the number $\widetilde{f}_N$) will be called the *one–step L-continuation* of the series F_N.

Note that the condition $e_L \notin \mathfrak{L}_d$ is necessary for L-continuability of a time series. Indeed, if $e_L \in \mathfrak{L}_d$, then the series F_N is not L-continuable since, otherwise, the uniqueness of the continuation would be violated. In particular, if we have $\text{rank}_L(F_N) = L$, then the series F_N is not L-continuable.

On the other hand, if there are two different numbers $\widetilde{f}_N$ such that the L-dimensional lagged vectors of the series $\widetilde{F}_{N+1} = (f_0, \ldots, f_{N-1}, \widetilde{f}_N)$ belong to $\mathfrak{L}_d$, then $e_L \in \mathfrak{L}_d$. This means that the condition $e_L \notin \mathfrak{L}_d$ provides the uniqueness of the L-continuation.

The following theorem gives sufficient conditions for L-continuability in terms of the LRF. Moreover, it shows the close link between series that admit an L-continuation and series that satisfy linear recurrent formulae.

Theorem 5.4

1. *If the series F_N is governed by some LRF*

$$f_{i+d_0} = \sum_{k=1}^{d_0} a_k f_{i+d_0-k}, \quad 0 \leq i \leq N - d_0 - 1, \tag{5.25}$$

with $d_0 \leq \min(K, L-1)$, then it is L-continuable and its L-continuation is achieved by the same LRF (5.25).

2. *We set $L^* = \min(L, K)$. If $d = \text{rank}_{L^*}(F_N) < L^*$ and $e_{L^*} \notin \mathfrak{L}^{(L^*)}(F_N)$,*

then the series F_N is both L-continuable and K-continuable. These continuations coincide and can be achieved by some LRF of dimension $L^* - 1$ or smaller.

Proof.
1. We set
$$\widetilde{f}_N = \sum_{k=1}^{d_0} a_k f_{N-k}, \quad \widetilde{X}_{K+1} = (f_K, \ldots, \widetilde{f}_N)^{\mathrm{T}}.$$

Since $d_0 \leq K$,
$$\widetilde{X}_{K+1} = \sum_{k=1}^{d_0} a_k X_{K+1-k}$$

and therefore $\widetilde{X}_{K+1} \in \mathfrak{L}_d$. The inequality $d_0 < L$ implies the condition $e_L \notin \mathfrak{L}_d$.

2. Let us consider the trajectory space $\mathfrak{L}^{(L^*)}(F_N)$. Since its dimension d is smaller than L^* and $e_{L^*} \notin \mathfrak{L}^{(L^*)}(F_N)$, we obtain (see Theorem 5.2) that the series F_N satisfies an LRF of dimension not exceeding $L^* - 1 < \min(L, K)$. The rest of the proof is similar to the proof of the first assertion of the theorem. □

Remark 5.9
1. If F_N satisfies the LRF (5.25) and $d_0 < \min(L, K)$, then F_N admits both an L-continuation and a K-continuation. The results of these continuations coincide.
2. We see that (under the assumptions of Theorem 5.4) if the series F_N is L-continuable, then so is the series $\widetilde{F}_{N+1}$. Therefore, under these assumptions any L-continuable series F_N of L-rank d can be uniquely continued to an infinite series $\widetilde{F}$ of the same L-rank.
3. By definition, L-continuabity of the series F_N implies the possibility of continuing the series to the right. Analogously, we can define L-continuability of the series F_N to the left. If $d < L^* = \min(L, K)$, then a necessary and sufficient condition for this is $e_1 \notin \mathfrak{L}^{(L^*)}(F_N)$.

Example 5.11 *Right and left continuation*
Among the series of Examples 5.6-5.8, only the series of Example 5.7 is continuable to the right (with all the continued values equal to 1). At the same time, the series of Example 5.6 is continuable to the left, and the series of Example 5.8 cannot be continued either to the right or to the left.

Example 5.12 *Counterexample*
The series of Example 5.9 has L-rank 3 for $L = 5$, with $e_5 \notin \mathfrak{L}_3$. However, it is not L-continuable. Indeed, the condition of orthogonality of the vector $\widetilde{X}_3 = (2, 1, 1, 1, \widetilde{f}_7)^{\mathrm{T}}$ to the basis vectors $e_1 - e_5$, $-4e_1 + e_2 + e_3 + e_4$ of the space $\mathfrak{L}_3^{\perp}$ leads to contradiction. The conditions of Theorem 5.4 (second assertion) are not satisfied since $d = K$.

Despite the fact that in the definition of L-continuability of a series F_N of L-rank d all the terms of this series are formally present, the continued value $\widetilde{f}_N$ is determined (under the fulfillment of the conditions $e_L \notin \mathfrak{L}_d$ and $d < L \leq K$)

only by the last L-lagged vector X_K. More precisely, the lagged vector $\widetilde{X}_{K+1}$ of the L-continued series $\widetilde{F}_{N+1}$ is

$$\widetilde{X}_{K+1} = \mathbf{D}_L X_K, \qquad (5.26)$$

where

$$\mathbf{D}_L = \begin{pmatrix} 0 & 1 & 0 & \cdots & 0 \\ 0 & 0 & 1 & \cdots & 0 \\ \vdots & \vdots & \vdots & \ddots & \vdots \\ 0 & 0 & 0 & \cdots & 1 \\ 0 & a_{L-1} & a_{L-2} & \cdots & a_1 \end{pmatrix} \qquad (5.27)$$

and the vector $\mathcal{R} = (a_{L-1}, \ldots, a_1)$ is defined in (5.16). By definition, $\widetilde{X}_{K+1} \in \mathcal{L}_d$. We also have that the vector $\widetilde{X}^\nabla_{K+1}$ consisting of the first $L-1$ components of the vector $\widetilde{X}_{K+1}$ coincides with the vector $(X_K)_\Delta$ consisting of the last $L-1$ components of the vector X_K.

Consider now a vector $X \in \mathcal{L}_d$, still assuming that $e_L \notin \mathcal{L}_d$. Let us state the problem of finding the vector $Y \in \mathcal{L}_d$ such that the distance between the vectors X_Δ and Y^∇ is minimal. Such a problem arises in forecasting (continuing) a series by the Vector SSA forecasting method discussed in Section 2.3.1.

According to Remark 5.7, $Y = \mathbf{D}_L X$ if and only if $X_\Delta \in \mathcal{L}^\nabla_d$, where the space $\mathcal{L}^\nabla_d$ consists of all the vectors Z^∇ with $Z \in \mathcal{L}_d$. In this case, $Y^\nabla = X_\Delta$, and the last component y of the vector Y has the form $y = (\mathcal{R}, X_\Delta)$. In the general case, the problem stated above has the following solution, which does not require a special proof.

Proposition 5.7 *Denote by Π the operator corresponding to the orthogonal projection $\mathbf{R}^{L-1} \mapsto \mathcal{L}^\nabla_d$. Then $Y^\nabla = \Pi X_\Delta$ and $y = (\mathcal{R}, Y^\nabla)$.*

As discussed above, if $e_L \notin \mathcal{L}_d$, $d < \min(L, K)$ and $X = X_K$ is the last column of the L-trajectory matrix of the series F_N, then the vector X_Δ belongs to $\mathcal{L}^\nabla_d$ and $\Pi X_\Delta = X_\Delta$. Therefore, Proposition 5.7 essentially describes a continuation of the series F_N with the help of the formulae (5.26) and (5.27). Note that for $d = L-1$ the operator Π is the identity operator (the equality $\Pi X_\Delta = X_\Delta$ holds for any X).

Proposition 5.8 *Let $d = L-1$, $e_L \notin \mathcal{L}_{L-1}$ and $X \in \mathcal{L}_{L-1}$. Then $\mathbf{D}_L X \in \mathcal{L}_{L-1}$.*

Proof.
Let us represent the hyperplane $\mathcal{L}_{L-1}$ in the form

$$\mathcal{L}_{L-1} = \{Z \in \mathbf{R}^L \text{ such that } (A, Z) = 0\}.$$

Since $e_L \notin \mathcal{L}_{L-1}$ and the vector $A \in \mathbf{R}^L$ is orthogonal to $\mathcal{L}_{L-1}$, the vector A coincides (up to a multiplier) with the vector Q defined in (5.19). The last component z of the vector X can be expressed in terms of the linear combination of the other components: $z = (\mathcal{R}, X^\nabla)$.

Moreover, we have in this case that $\mathcal{L}_{L-1}^{\triangledown} = \mathbf{R}^{L-1}$. (Indeed, any vector in $\mathbf{R}^{L-1}$ always uniquely determines the point in the hyperplane in $\mathbf{R}^L$, unless this hyperplane is vertical.) This completes the proof. □

Remark 5.10 Proposition 5.7 shows that any time series F_N with $\text{rank}_L(F_N) = L - 1$ and $e_L \notin \mathcal{L}_{L-1}$ admits an L-continuation.

Let us discuss now the properties of the operator Π. Let $P_1, \ldots, P_d$ be an orthonormal basis of the space $\mathcal{L}_d$. We set $\mathbf{V}^{\triangledown} = [P_1^{\triangledown} : \ldots : P_d^{\triangledown}]$ and $W = (\pi_1, \ldots, \pi_d)^{\mathrm{T}}$, where π_i is the last component of the vector P_i. As above, let $\nu^2 = \pi_1^2 + \ldots + \pi_d^2$. Also, we introduce the matrix $\mathbf{A} = (\mathbf{V}^{\triangledown})^{\mathrm{T}} \mathbf{V}^{\triangledown}$.

Proposition 5.9 *If $e_L \notin \mathcal{L}_d$ then we have the following.*
1. *The matrix Π of the linear projection operator Π has the form*

$$\Pi = \mathbf{V}^{\triangledown}(\mathbf{V}^{\triangledown})^{\mathrm{T}} + (1 - \nu^2)\mathcal{R}\mathcal{R}^{\mathrm{T}}. \tag{5.28}$$

2. *In the notation of Proposition 5.7 we have the equalities $y = \mathcal{R}^{\mathrm{T}} X_{\Delta}$ and*

$$Y^{\triangledown} = \Pi X_{\Delta} = \mathbf{V}^{\triangledown}(\mathbf{V}^{\triangledown})^{\mathrm{T}} X_{\Delta} + y(1 - \nu^2)\mathcal{R}.$$

Proof.
1. Let us first prove that the matrix $\mathbf{A}$ has the inverse:

$$\mathbf{A}^{-1} = \mathbf{I}_d + \frac{1}{1 - \nu^2} W W^{\mathrm{T}}, \tag{5.29}$$

where $\mathbf{I}_d$ is the identity $d \times d$ matrix. Indeed, let $\mathbf{V}$ be a matrix with columns $P_1, \ldots, P_d$. Obviously, $\mathbf{V}^{\mathrm{T}} \mathbf{V} = \mathbf{I}_d$. On the other hand,

$$\mathbf{V}^{\mathrm{T}} \mathbf{V} = \mathbf{A} + W W^{\mathrm{T}},$$

and thus $\mathbf{A} = \mathbf{I}_d - W W^{\mathrm{T}}$. Since $W^{\mathrm{T}} W = \nu^2 < 1$, we have

$$\left(\mathbf{I}_d - W W^{\mathrm{T}}\right)\left(\mathbf{I}_d + \frac{1}{1 - \nu^2} W W^{\mathrm{T}}\right) =$$
$$= \mathbf{I}_d - \frac{1}{1 - \nu^2}\left(\nu^2 W W^{\mathrm{T}} - W W^{\mathrm{T}} W W^{\mathrm{T}}\right) = \mathbf{I}_d. \tag{5.30}$$

Since both matrices on the left side of (5.30) are symmetric, the proof of the existence of the inverse to $\mathbf{A}$ is complete. This also implies that the vectors $P_i^{\triangledown}$ ($1 \leq i \leq d$) are linearly independent.

The vectors $P_1^{\triangledown}, \ldots, P_d^{\triangledown}$ constitute a basis of the space $\mathcal{L}_d^{\triangledown}$. This yields that Π has the form

$$\Pi = \mathbf{V}^{\triangledown} \mathbf{A}^{-1} (\mathbf{V}^{\triangledown})^{\mathrm{T}} = \mathbf{V}^{\triangledown}(\mathbf{V}^{\triangledown})^{\mathrm{T}} + \frac{1}{1 - \nu^2} \mathbf{V}^{\triangledown} W W^{\mathrm{T}} (\mathbf{V}^{\triangledown})^{\mathrm{T}}.$$

Using the equality $\mathcal{R} = \mathbf{V}^{\triangledown} W / (1 - \nu^2)$, see (5.16), we immediately obtain the required result.

2. Taking into account the equality $||\mathcal{R}||^2 = \nu^2 / (1 - \nu^2)$, see (5.28) and Proposition 5.5, we obtain

$$\mathcal{R}^{\mathrm{T}} \Pi X_{\Delta} = \mathcal{R}^{\mathrm{T}} \mathbf{V}^{\triangledown}(\mathbf{V}^{\triangledown})^{\mathrm{T}} X_{\Delta} + (1 - \nu^2) \mathcal{R}^{\mathrm{T}} \mathcal{R} \mathcal{R}^{\mathrm{T}} X_{\Delta}$$

$$= \frac{1}{1-\nu^2} W^{\mathrm{T}} \mathbf{A} (\mathbf{V}^\nabla)^{\mathrm{T}} X_\Delta + \nu^2 \mathcal{R}^{\mathrm{T}} X_\Delta =$$
$$= \frac{1}{1-\nu^2} \left(W^{\mathrm{T}} (\mathbf{V}^\nabla)^{\mathrm{T}} X_\Delta - W^T W W^{\mathrm{T}} (\mathbf{V}^\nabla)^{\mathrm{T}} X_\Delta \right) + \nu^2 \mathcal{R}^{\mathrm{T}} X_\Delta$$
$$= \mathcal{R}^{\mathrm{T}} X_\Delta - \nu^2 \mathcal{R}^{\mathrm{T}} X_\Delta + \nu^2 \mathcal{R}^{\mathrm{T}} X_\Delta = \mathcal{R}^{\mathrm{T}} X_\Delta.$$

Using (5.28) we obtain the second statement of the proposition. □

CHAPTER 6
SVD of trajectory matrices

This chapter contains a rigorous mathematical description of four topics used throughout Part 1 of this book: weak separability of time series, Hankelization of matrices, centring in SSA, and specifics of SSA for deterministic stationary sequences.

6.1 Mathematics of separability

The concept of separability is central in the SSA considerations. This concept was thoroughly discussed in Section 1.5 from the standpoints of methodology and practical implementation; here we mainly restrict ourselves to strict definitions and analytical examples.

6.1.1 Definition and examples

Let $F_N^{(1)}$ and $F_N^{(2)}$ be time series of length N and $F_N = F_N^{(1)} + F_N^{(2)}$. Under the choice of window length L, each of the series $F_N^{(1)}$, $F_N^{(2)}$ and F_N generates an L-trajectory matrix: $\mathbf{X}^{(1)}$, $\mathbf{X}^{(2)}$ and $\mathbf{X}$.

Denote by $\mathfrak{L}^{(L,1)}$ and $\mathfrak{L}^{(L,2)}$ the linear spaces spanned by the columns of the trajectory matrices $\mathbf{X}^{(1)}$ and $\mathbf{X}^{(2)}$. Similar notation $\mathfrak{L}^{(K,1)}$ and $\mathfrak{L}^{(K,2)}$ will be used for the spaces spanned by the columns of the transposed matrices $(\mathbf{X}^{(1)})^{\mathrm{T}}$ and $(\mathbf{X}^{(2)})^{\mathrm{T}}$, $K = N - L + 1$.

If $\mathfrak{L}^{(L,1)} \perp \mathfrak{L}^{(L,2)}$ and $\mathfrak{L}^{(K,1)} \perp \mathfrak{L}^{(K,2)}$, then we say that the series $F_N^{(1)}$ and $F_N^{(2)}$ are *weakly L-separable*.

For brevity, we shall use the term 'separability' instead of 'weak L-separability' in cases when no ambiguity occur.

Let us elucidate the last definition. Suppose that the series $F_N^{(1)}$ and $F_N^{(2)}$ are L-separable. Consider certain SVDs of the trajectory matrices $\mathbf{X}^{(1)}$ and $\mathbf{X}^{(2)}$:

$$\mathbf{X}^{(1)} = \sum_k \sqrt{\lambda_{1k}}\, U_{1k} V_{1k}^{\mathrm{T}}, \quad \mathbf{X}^{(2)} = \sum_k \sqrt{\lambda_{2k}}\, U_{2k} V_{2k}^{\mathrm{T}}.$$

Then

$$\mathbf{X} = \mathbf{X}^{(1)} + \mathbf{X}^{(2)} = \sum_k \sqrt{\lambda_{1k}}\, U_{1k} V_{1k}^{\mathrm{T}} + \sum_m \sqrt{\lambda_{2m}}\, U_{2m} V_{2m}^{\mathrm{T}}. \tag{6.1}$$

Proposition 4.2 of Section 4.1 tells us that (6.1) is an SVD of the matrix $\mathbf{X}$. Thus, the representation $F_N = F_N^{(1)} + F_N^{(2)}$ is natural from the viewpoint of the SVD of the matrix $\mathbf{X}$.

If $F_N^{(1)}$ and $F_N^{(2)}$ are weakly L-separable and $\lambda_{1k} \neq \lambda_{2m}$ for all k and m, then we say that g_1 and g_2 are *strongly L-separable*. The difference between separability and strong separability can be expressed as follows. If separability occurs, then an SVD of the matrix $\mathbf{X}$ exists such that we can group its terms in a proper way and obtain $F_N^{(1)}$ and $F_N^{(2)}$ in terms of their trajectory matrices $\mathbf{X}^{(1)}$ and $\mathbf{X}^{(2)}$. In the case of strong separability, we can obtain $F_N^{(1)}$ and $F_N^{(2)}$ for any SVD of the trajectory matrix $\mathbf{X}$. In this section we study features of weak separability.

Remark 6.1 Suppose that nonzero series $F_N^{(1)}$ and $F_N^{(2)}$ are weakly L-separable. Denote by d_1, d_2 the ranks of the trajectory matrices $\mathbf{X}^{(1)}$ and $\mathbf{X}^{(2)}$. Since

$$d_1 + d_2 = \operatorname{rank} \mathbf{X} \leq L,$$

both d_1 and d_2 do not exceed $L - 1$. Therefore, the time series $F_N^{(1)}$ and $F_N^{(2)}$ have L-ranks smaller than L and can be studied by the methods of Chapter 5.

In particular, Theorem 5.2 shows that if the vector $e_L = (0, 0, \ldots, 1)^\mathrm{T}$ does not belong to the L-trajectory space of the series F_N, then both $F_N^{(1)}$ and $F_N^{(2)}$ satisfy certain LRFs of dimension $L - 1$.

We set $F_N^{(1)} = (f_0^{(1)}, \ldots, f_{N-1}^{(1)})$ and $F_N^{(2)} = (f_0^{(2)}, \ldots, f_{N-1}^{(2)})$.

Proposition 6.1 *Let $K = N - L + 1$. Time series $F_N^{(1)}$ and $F_N^{(2)}$ are weakly L-separable if and only if*
1. for any $0 \leq k, m < K - 1$

$$f_k^{(1)} f_m^{(2)} = f_{k+L}^{(1)} f_{m+L}^{(2)}; \tag{6.2}$$

2. for any $0 \leq m \leq K - 1$

$$f_m^{(1)} f_0^{(2)} + \ldots + f_{m+L-1}^{(1)} f_{L-1}^{(2)} = 0; \tag{6.3}$$

3. for any $0 \leq k, m < L - 1$

$$f_k^{(1)} f_m^{(2)} = f_{k+K}^{(1)} f_{m+K}^{(2)}; \tag{6.4}$$

4. for any $0 \leq m \leq L - 1$

$$f_m^{(1)} f_0^{(2)} + \ldots + f_{m+K-1}^{(1)} f_{K-1}^{(2)} = 0. \tag{6.5}$$

Proof.
By definition, weak L-separability is equivalent to the matrix equalities

$$(\mathbf{X}^{(1)})^\mathrm{T} \mathbf{X}^{(2)} = \mathbf{0}_{KK} \quad \text{and} \quad \mathbf{X}^{(1)} (\mathbf{X}^{(2)})^\mathrm{T} = \mathbf{0}_{LL}. \tag{6.6}$$

Taking the first equality in (6.6) we obtain the condition

$$f_k^{(1)} f_m^{(2)} + \ldots + f_{k+L-1}^{(1)} f_{m+L-1}^{(2)} = 0, \quad 0 \leq k, m \leq K - 1,$$

MATHEMATICS OF SEPARABILITY 259

which is equivalent to (6.2), (6.3). The second equality in (6.6) is equivalent to (6.4), (6.5). □

Let us consider several examples of separation taking some simple test series $F_N^{(1)}$ and finding conditions for their separation from some nonzero series $F_N^{(2)} = (f_0^{(2)}, \ldots, f_{N-1}^{(2)})$.

Example 6.1 *Separation of a nonzero constant*
Consider the constant series $F_N^{(1)}$ with $f_n^{(1)} \equiv c \neq 0, 0 \leq n \leq N - 1$. In this case the conditions for separability are rather simple. Equalities (6.2) and (6.4) show that the series F_N have both periods L and K. Therefore, there exists an integer M such that L/M and $(N+1)/M$ are integers and the series $F_N^{(2)}$ has period M. Equalities (6.3) and (6.5) imply

$$f_0^{(2)} + \ldots + f_{M-1}^{(2)} = 0.$$

For example, if $f_n^{(2)} = \cos(2\pi n/T)$, where T and $(N+1)/T$ are integers, then the choice $L = kT < N$ with k an integer implies separability.

Example 6.2 *Separation of an exponential series*
Proposition 6.1 implies that the conditions for separability of the exponential series $f_n^{(1)} = e^{\alpha n}$ from a series $F_N^{(2)}$ are exactly the same as the conditions for separability of the constant series $\widetilde{f}_n^1 \equiv c = 1$ from a series $\widetilde{F}_N^{(2)}$ with $\widetilde{f}_n^{(2)} = e^{\alpha n} f_n^{(2)}$. Therefore, the exponential series is separable from a nonzero series $F_N^{(2)}$ if and only if there exists integer $T > 1$ such that L/T and $(N+1)/T$ are integers, T is a period of the series $e^{\alpha n} f_n^{(2)}$, and

$$\sum_{m=0}^{T-1} e^{\alpha m} f_m^{(2)} = 0.$$

For example, if $(N+1)/L$ is an integer, then the choice of window length L leads to separation of the series $f_n^{(1)} = e^{\alpha n}$ and $f_n^{(2)} = e^{-\alpha n} \cos(2\pi n/L)$.

Example 6.3 *Separation of a harmonic series*
Here we deal with the series $f_n^{(1)} = \cos(2\pi\omega n + \phi)$ assuming that $0 < \omega < 1/2$ and $L, K > 2$. We set $T = 1/\omega$. The equality (6.2) leads to two equalities:

$$\cos(2\pi\omega L) f_{m+L}^{(2)} = f_m^{(2)}, \quad \sin(2\pi\omega L) f_{m+L}^{(2)} = 0,$$

$0 \leq m < K - 1$. Since $F_N^{(2)}$ is a nonzero series, it follows that $2L/T$ is an integer. Therefore, two general situations can occur: if L/T is an integer, then the series $F_N^{(2)}$ has period L; if L/T is not an integer but $2L/T$ is an integer, then $f_{m+L}^{(2)} = -f_m^{(2)}$. In the same manner, (6.4) implies that $2K/T$ is an integer and similar situation holds with K substituted for L.

The equalities (6.3) and (6.5) yield a pair of conditions:

$$\sum_{k=0}^{L-1} e^{i2\pi\omega k} f_k^{(2)} = 0,$$

$$\sum_{k=0}^{K-1} e^{i2\pi\omega k} f_k^{(2)} = 0,$$

where $i^2 = -1$.

Let us give a typical example of separation. For an integer T_1 consider $f_n^{(2)} = \cos(2\pi n/T_1)$ assuming that L/T, L/T_1, K/T and K/T_1 are integers. Then $F_N^{(1)}$ and $F_N^{(2)}$ are separable.

Example 6.4 *Separation of an exponential-cosine series*
Assume that $f_n^{(1)} = e^{\alpha n}\cos(2\pi\omega n + \phi)$. It can be easily seen that the conditions for separability of this series from a nonzero series F_N coincide with the conditions for separability of the harmonic series $\widetilde{f}_n^{(1)} = \cos(2\pi\omega n + \phi)$ from the series $\widetilde{F}_N^{(2)}$ with $\widetilde{f}_n^{(2)} = e^{\alpha n} f_n^{(2)}$.

Example 6.5 *Separation of a linear series*
Let $f_n^{(1)} = an + b$ with $a \neq 0$. Then (6.2) has the form

$$ak\, f_m^{(2)} + b f_m^{(2)} = ak\, f_{m+L}^{(2)} + (aL + b) f_{m+L}^{(2)}, \quad a \neq 0,$$

and since $0 \le k < K - 1$ we have $f_m^{(2)} = f_{m+L}^{(2)}$ and $b f_m^{(2)} = (aL + b) f_{m+L}^{(2)}$. Therefore, $f_m^{(2)} = 0$ for $0 \le m \le N-1$. As a result, a nonconstant linear series is not separable from any nonzero series. A similar result holds for the polynomials of higher order.

6.1.2 Approximate and asymptotic separability

For a fixed window length L, the definition of weak separability of series $F_N^{(1)}$ and $F_N^{(2)}$ is formulated in terms of orthogonality for their subseries. This leads to the natural concept of *approximate separability* of two time series. For any series $F_N = (f_0, \ldots, f_{N-1})$ we set

$$F_{i,j} = (f_{i-1}, \ldots, f_{j-1}), \quad 1 \le i \le j < N.$$

Let $F_N^{(1)} = (f_0^{(1)}, \ldots, f_{N-1}^{(1)})$, $F_N^{(2)} = (f_0^{(2)}, \ldots, f_{N-1}^{(2)})$. For $i, j \ge 1$ and $M \le N - 1 - \max(i, j)$ we set

$$\rho_{i,j}^{(M)} = \frac{\left(F_{i,i+M-1}^{(1)}, F_{j,j+M-1}^{(2)}\right)}{\|F_{i,i+M-1}^{(1)}\|\, \|F_{j,j+M-1}^{(2)}\|} \tag{6.7}$$

under the assumption that the denominator is positive.

MATHEMATICS OF SEPARABILITY

The notation $(\,\cdot\,,\,\cdot\,)$ stands for the usual inner product of Euclidean vectors and $\|\cdot\|$ is the Euclidean norm. If the denominator in (6.7) is equal to zero, then we assume that $\rho_{i,j}^{(M)} = 0$.

The number $\rho_{i,j}^{(M)}$ has the sense of the cosine of the angle between the vectors $F_{i,i+M-1}^{(1)}$ and $F_{j,j+M-1}^{(2)}$. Using the statistical terminology, we can call $\rho_{i,j}^{(M)}$ the *correlation coefficient* between $F_{i,i+M-1}^{(1)}$ and $F_{j,j+M-1}^{(2)}$.

Time series $F_N^{(1)}, F_N^{(2)}$ are (weakly) *ε-separable for the window length L* if

$$\rho^{(L,K)} \stackrel{\text{def}}{=} \max\left(\max_{1\leq i,j\leq K}|\rho_{i,j}^{(L)}|,\max_{1\leq i,j\leq L}|\rho_{i,j}^{(K)}|\right) < \varepsilon, \qquad (6.8)$$

$K = N - L + 1$.

If the number ε is small, then the series are *approximately separable*. Of course, if separable time series $F_N^{(1)}$ and $F_N^{(2)}$ are slightly perturbed, they become ε-separable with some small ε. Suppose that the parameters L and N provide weak separability of the series $F_N^{(1)}, F_N^{(2)}$. Then another way from separability to approximate separability is in a small perturbation of the parameters L and N.

The concept of approximate separability has its asymptotic variant. Consider infinite time series $F^{(1)} = (f_0^{(1)}, \ldots, f_n^{(1)}, \ldots)$ and $F^{(2)} = (f_0^{(2)}, \ldots, f_n^{(2)}, \ldots)$. For each $N > 2$ let the series $F_N^{(1)}$ and $F_N^{(2)}$ consist of the first N terms of the series $F^{(1)}$ and $F^{(2)}$, respectively. Choosing a sequence of window lengths $1 < L = L(N) < N$, we obtain the related sequence of the *maximum correlation coefficients* $\rho_N = \rho^{(L,K)}$ defined by (6.8).

If there exists a sequence $L = L(N)$ such that $\rho_N \to 0$ as $N \to \infty$, then the time series $F^{(1)}$ and $F^{(2)}$ are called *asymptotically separable*. If $F^{(1)}$ and $F^{(2)}$ are asymptotically separable for any choice of L such that $L \to \infty$ and $K \to \infty$, then they are called *regularly asymptotically separable*. Conditions for regular asymptotic separability can be written as follows:

$$\rho(N_1, N_2) \stackrel{\text{def}}{=} \max_{i,j<N_1} \frac{\left|\sum_{k=0}^{N_2-1} f_{i+k}^{(1)} f_{j+k}^{(2)}\right|}{\sqrt{\sum_{k=0}^{N_2-1}\left(f_{i+k}^{(1)}\right)^2}\sqrt{\sum_{k=0}^{N_2-1}\left(f_{j+k}^{(2)}\right)^2}} \xrightarrow[N_1,N_2\to\infty]{} 0. \qquad (6.9)$$

Now let us consider examples of asymptotic separability taking into consideration the rate of convergence in (6.9). Since we will deal only with regular separability, the term 'regular' will be omitted for brevity.

Example 6.6 *Asymptotic separation of a constant series*

1. Consider an oscillatory series of the form

$$f_n^{(2)} = \sum_{k=0}^{m} c_k \cos(2\pi\omega_k n + \phi_k) \qquad (6.10)$$

with different frequencies $\omega_k \in (0, 1/2]$ and some phases $\phi_k \in [0, 2\pi)$. We assume that $\phi_k \neq \pi/2, 3\pi/2$ if $\omega_k = 1/2$. Then the series (6.10) is regularly asymptotically separable from a constant series $f_n^{(1)} \equiv c \neq 0$.

Indeed, uniformly in i, j and N_1, the numerator in (6.9) is $O(1)$, while both factors in the denominator are of order $N_2^{-1/2}$. Therefore, the rate of the asymptotic separation of the series (6.10) from a constant is $1/\min(L, K)$.

2. The exponential series $f_n^{(2)} = e^{\alpha n}$ with $\alpha \neq 0$ is asymptotically separable from a constant series.

Note that in this case the ratio in (6.9) does not depend on i or j. If $\alpha > 0$, then the numerator in (6.9) is of order $e^{\alpha N_2}$, while the denominator tends to infinity as $e^{\alpha N_2}\sqrt{N_2}$, $N_2 \to \infty$. If $\alpha < 0$, then the numerator tends to a constant, while the denominator is of order $N_2^{1/2}$. In both cases, the rate of the asymptotic separation is $1/\min(\sqrt{L}, \sqrt{K})$, which is rather slow.

If $|\alpha|$ is close to zero, then the true order of (6.9) is $(|\alpha|N_2)^{-1/2}$. Therefore, a small α produces strong requirements on the size of L and K.

3. Consider an exponential-cosine series

$$f_n^{(2)} = e^{\alpha n} \cos(2\pi\omega n + \phi), \qquad (6.11)$$

$\alpha \neq 0$. If $\alpha > 0$, then the numerator in (6.9) has the form $O(e^{\alpha N_2})$, while the denominator is equivalent to $cN_2^{1/2}e^{\alpha N_2}$. For $\alpha < 0$ the numerator tends to a constant and the denominator is of order $N_2^{1/2}$. Since the convergence is uniform in i, j, the rate of the asymptotic separation is $1/\min(\sqrt{L}, \sqrt{K})$.

4. Any polynomial series

$$f_n^{(2)} = \sum_{k=0}^{m} c_k n^k, \qquad c_m = 1, \qquad (6.12)$$

is not asymptotically separable from a constant nonzero series. Indeed, in this case for fixed i and j, both the numerator and the denominator in (6.9) is of order N_2^{m+1} as $N_2 \to \infty$.

Example 6.7 *Asymptotic separation of an oscillatory series*

1. Consider two oscillatory sequences of the form (6.10) with disjoint sets of frequencies. Simple calculations show that this case is similar to that of separation of a constant from an oscillatory series, and the rate of separation is also $1/\min(L, K)$.

Note that the actual separation depends on the distance between the two sets of frequencies. For example, if we consider two harmonic time series with frequencies $0 < \omega_1 < \omega_2 < 1/2$, then there exists a subsequence of $N_2 \to \infty$ such that

$$\rho(N_1, N_2) \sim \frac{C}{N_2(\omega_2 - \omega_1)}.$$

In other words, if ω_1 and ω_2 are close to each other, then both L and $K = N - L + 1$ have to be essentially greater than $1/|\omega_1 - \omega_2|$.

MATHEMATICS OF SEPARABILITY

2. Any polynomial (6.12) is asymptotically separable from an oscillatory series (6.10). In this case, the numerator in (6.9) has the form $O(N_2^m)$, while the denominator is of order N_2^{m+1}. Since the convergence is uniform in i and j, the rate of separation is $1/\min(L, K)$.

3. Consider the exponential series $F_N^{(2)}$ with $f_n^{(2)} = e^{\alpha n}$.

In the same manner as in the case of an exponential and a constant series, for $\alpha > 0$ the numerator in (6.9) has the form $O(e^{\alpha N_2})$, while the denominator is equivalent to $cN_2^{1/2} e^{\alpha N_2}$, $c > 0$. If $\alpha < 0$, then the correlation coefficient is of order $N_2^{-1/2}$ uniformly in i, j. The rate of separation is $1/\min(\sqrt{L}, \sqrt{K})$.

4. The result concerning separation of the series (6.10) from the exponential-cosine series (6.11) is similar to the result obtained for (6.10) and the pure exponential series; the rate of separation is also the same.

Example 6.8 *Asymptotic separation of a polynomial*

1. Any exponential series $f_n^{(2)} = e^{\alpha n}$ with $\alpha \neq 0$ is asymptotically separable from a polynomial (6.12).

The order of decrease of the correlation coefficient (6.9) is $O(N_2^{-1/2})$, since the numerator is of order $O(N_2^m e^{\alpha N_2})$, while the order of the denominator is $N^{m+1/2} e^{\alpha N_2}$. The convergence is uniform in i and j, and the rate of asymptotic separation is $1/\min(\sqrt{L}, \sqrt{K})$.

2. Two polynomials are not asymptotically separable in the sense of convergence in (6.9). This case is analogous to the case of a constant and a polynomial.

Example 6.9 *Asymptotic separation of an exponential series*

1. Two exponential series $f_n^{(1)} = e^{\alpha n}$ and $f_n^{(2)} = e^{\beta n}$ with nonzero $\alpha \geq \beta$ are asymptotically separable from each other in the sense of convergence (6.9) if and only if $\beta < 0 < \alpha$.

It can be easily seen that if α and β are either both positive or both negative, then the expression (6.9) tends to a constant depending on α and β. If the signs of α and β are different, the order of convergence to zero in (6.9) is $O(e^{\min(-\alpha, \beta) N_2})$, $N_2 \to \infty$.

For stationary sequences (see Section 6.4) we use another definition of asymptotic separability. We say that two infinite time series $F^{(1)}, F^{(2)}$ are *weakly pointwise regularly asymptotically separable* if both the window length L and $K = N - L + 1$ tend to infinity and

$$\frac{\sum_{k=0}^{M-1} f_{n+k}^{(1)} f_{m+k}^{(2)}}{\sqrt{\sum_{k=0}^{M-1} \left(f_{n+k}^{(1)}\right)^2} \sqrt{\sum_{k=0}^{M-1} \left(f_{m+k}^{(2)}\right)^2}} \xrightarrow[M \to \infty]{} 0 \qquad (6.13)$$

for any $m, n \geq 0$. Evidently, (6.13) follows from (6.9).

6.1.3 Separation of a signal from noise

According to our definition (see Section 1.4.1 for the related discussions and Section 6.4 for the mathematical results), we consider the stationary noise series as having a deterministic rather than random nature. Within this approach, it can be demonstrated that, roughly speaking, any stationary signal is pointwise asymptotically separable from any stationary noise (see Section 6.4.4).

However, the theory fails for nonstationary signals. We thus consider the problem of separating a signal from a random noise separately. Note that in this case we shall deal with uniform separability of the kind (6.9) rather than with pointwise separability defined in (6.13), though the latter is much simpler for the analysis and more natural for deterministic stationary sequences.

Let $(\Omega, \mathcal{F}, \mathbf{P})$ be a probability space and let $F^{(1)}$, $F^{(2)}$ be two random infinite sequences. If there exists a sequence $L = L(N)$ such that the sequence of random variables $\rho^{(L,K)}$, defined in (6.8), (6.7) tends to zero in probability as $N \to \infty$, then we call the random sequences $F^{(1)}$ and $F^{(2)}$ *stochastically separable*.

When $F^{(1)}$ is a nonrandom sequence ('signal'), and $F^{(2)}$ is a 'purely random' sequence ('noise'), it is convenient to introduce another notation.

Let $F = (f_0, \ldots, f_n, \ldots)$ be some nonrandom infinite time series, while $\Xi = (\xi_0, \ldots, \xi_n, \ldots)$ is a random sequence with zero mean. The following conditions are sufficient for stochastic separability of a signal F from a noise Ξ.

For any $\delta > 0$ let us introduce the random event

$$A_{L,K}(\delta) = \left\{ \omega \in \Omega : \min_{0 \leq j \leq L-1} \frac{1}{K} \sum_{m=0}^{K-1} \xi_{m+j}^2 < \delta \right\}$$

and set $P_{L,K}(\delta) = \mathbf{P}(A_{L,K}(\delta))$. Let

$$\widetilde{f}_{i,k} \stackrel{\text{def}}{=} \frac{f_{i+k}}{\sqrt{\sum_{k=0}^{K-1} f_{i+k}^2}}$$

and

$$\Xi_{L,K} \stackrel{\text{def}}{=} \max_{i,j \leq L-1} \left| K^{-1/2} \sum_{k=0}^{K-1} \widetilde{f}_{i+k} \xi_{j+k} \right|.$$

Proposition 6.2 *If there exist $\delta > 0$, $L = L(N)$ and $K = K(N)$ such that*

$$\max(P_{L,K}(\delta), P_{K,L}(\delta)) \to 0, \quad \max(\Xi_{L,K}, \Xi_{K,L}) \to 0$$

in probability as $N \to \infty$, then F and Ξ are stochastically separable.

MATHEMATICS OF SEPARABILITY

Proof.
We shall prove that there exist $L, K \to \infty$ such that

$$\rho_L^{(K)} \stackrel{\text{def}}{=} \max_{i,j \leq L-1} \frac{\left| K^{-1/2} \sum_{k=0}^{K-1} f_{i+k} \xi_{j+k} \right|}{\sqrt{\sum_{k=0}^{K-1} f_{i+k}^2} \sqrt{\frac{1}{K} \sum_{k=0}^{K-1} \xi_{j+k}^2}} \xrightarrow[L,K \to \infty]{} 0$$

in probability, and that the analogous convergence holds for $\rho_K^{(L)}$. Since

$$\mathbf{P}\left(\rho_L^{(K)} > \varepsilon\right) = \mathbf{P}\left(\rho_L^{(K)} > \varepsilon, A_{L,K}(\delta)\right) + \mathbf{P}\left(\rho_L^{(K)} > \varepsilon, A_{L,K}^C(\delta)\right) \quad (6.14)$$

and the first term on the right side of (6.14) tends to zero as $L, K \to \infty$, we need to check the convergence of the second term. Inequalities

$$\mathbf{P}\left(\rho_L^{(K)} > \varepsilon, A_{L,K}^C(\delta)\right) \leq \mathbf{P}\left(\Xi_{L,K} > \varepsilon_1, A_{L,K}^C(\delta)\right) \leq \mathbf{P}(\Xi_{L,K} > \varepsilon_1)$$

with $\varepsilon_1 = \varepsilon\sqrt{\delta}$ lead to the convergence of $\rho_L^{(K)}$. The case of $\rho_K^{(L)}$ is similar. □

Corollary 6.1 *Any infinite time series F is stochastically separable from Gaussian white noise Ξ if $L, K \to \infty$ and $L/K \to a > 0$.*

Proof.
We set $\delta = 1 - \Delta$, $0 < \Delta < 1$. Applying the analogue of the Chebyshev inequality for the fourth moment we obtain that

$$\mathbf{P}\left(\frac{1}{K}\sum_{m=0}^{K-1} \xi_{m+j}^2 < \delta\right) = \mathbf{P}\left(\frac{1}{K}\sum_{m=0}^{K-1}(\xi_{m+j}^2 - 1) < -\Delta\right) = O(K^{-2})$$

uniformly in j. Then

$$P_{L,K}(\delta) \leq L\,\mathbf{P}\left(\frac{1}{K}\sum_{m=0}^{K-1} \xi_m^2 < \delta\right) = O(L/K^2).$$

Since $\widetilde{f}_{i,0}^2 + \ldots + \widetilde{f}_{i,K-1}^2 = 1$ for any i, it follows that

$$G_{ij}^{(K)} \stackrel{\text{def}}{=} \sum_{k=0}^{K-1} \widetilde{f}_{i+k} \xi_{j+k} \in N(0,1)$$

with $N(0, 1)$ standing for the normal distribution with parameters 0 and 1.

Therefore, $\mathbf{P}\left(K^{-1/2}|G_{ij}^{(K)}| > \varepsilon\right) = 2(1 - \Phi(K^{1/2}\varepsilon))$ and

$$\mathbf{P}(|\Xi_{L,K}| > \varepsilon) \leq 2L(1 - \Phi(K^{1/2}\varepsilon)),$$

where Φ stands for the cumulative distribution function of the $\mathrm{N}(0,1)$ distribution. Since

$$1 - \Phi(x) \sim \frac{1}{\sqrt{2\pi}x} e^{-x^2}$$

as $x \to +\infty$, the assertion is proved. $\square$

6.2 Hankelization

By definition, an $L \times K$ matrix $\mathbf{X}$ is a *Hankel* matrix if its elements x_{ij} coincide on the 'matrix diagonals' $i + j = s$ for any $2 \leq s \leq L + K$. Our interest in Hankel matrices is based on the fact that any Hankel matrix is the trajectory matrix of some time series F_N with $N = K + L - 1$ and vice versa.

Assume that we have some $L \times K$ matrix $\mathbf{Y}$ with elements y_{ij}. Our aim is to find a Hankel matrix $\mathbf{Z} = \mathcal{H}\mathbf{Y}$ of the same dimension such that the difference $\mathbf{Y} - \mathbf{Z}$ has minimal Frobenius norm (see Section 4.2.1). The linear operator $\mathcal{H}$ is the *Hankelization* operator.

We set $L^* = \min(L, K)$, $K^* = \max(L, K)$ and $N = L + K - 1$. Let $y_{ij}^* = y_{ij}$ if $L < K$ and $y_{ij}^* = y_{ji}$ otherwise.

The following assertion can be found in Buchstaber (1994).

Proposition 6.3 *Let $s = i + j$. Then the element $\widetilde{y}_{ij}$ of the matrix $\mathcal{H}\mathbf{Y}$ is*

$$\widetilde{y}_{ij} = \begin{cases} \dfrac{1}{s-1} \sum_{l=1}^{s-1} y_{l,s-l}^* & \text{for } 2 \leq s \leq L^* - 1, \\ \dfrac{1}{L^*} \sum_{l=1}^{L^*} y_{l,s-l}^* & \text{for } L^* \leq s \leq K^* + 1, \\ \dfrac{1}{N-s+2} \sum_{l=s-K^*}^{L^*} y_{l,s-l}^* & \text{for } K^* + 2 \leq s \leq N + 1. \end{cases} \quad (6.15)$$

Proof.
By definition, a Hankel matrix $\mathbf{Z}$ with elements z_{ij} satisfies the conditions $z_{ij} = g_s$ for $i + j = s$ and some numbers g_s. Since the square of the Frobenius norm of a matrix is the sum of squares of all its elements, we obtain

$$\|\mathbf{Y} - \mathbf{Z}\|_\mathcal{M}^2 = \sum_{i,j} |y_{ij} - z_{ij}|^2 = \sum_{s=2}^{L+K} \sum_{i+j=s} |y_{ij} - g_s|^2. \quad (6.16)$$

HANKELIZATION 267

Therefore, we must find the numbers g_s such that the right side of (6.16) attains its minimum. The result is well-known:

$$g_s = \frac{1}{n_s} \sum_{i+j=s} y_{ij}, \qquad (6.17)$$

where n_s stands for the number of (i,j) such that $1 \leq i \leq L$, $1 \leq j \leq K$ and $i+j = s$. Formula (6.15) is just another form of (6.17) and gives exact expressions for the g_s. □

Remark 6.2
1. If we fix L and K and consider the linear space $\mathcal{M}_{LK}$ of $L \times K$ real matrices, then the set $\mathcal{M}_{LK}^{(H)}$ of Hankel $L \times K$ matrices is a linear subset of $\mathcal{M}_{LK}$. If we equip $\mathcal{M}_{LK}$ with the (Frobenius) inner product (4.7), then the optimal Hankel matrix of Proposition 6.3 is the orthogonal projection of the matrix $\mathbf{Y}$ onto the linear subspace $\mathcal{M}_{LK}^{(H)}$ of Hankel matrices. Therefore, the linear operator $\mathcal{H}: \mathcal{M}_{LK} \mapsto \mathcal{M}_{LK}^{(H)}$ is an orthogonal projection operator.
2. Any Hankel matrix can be represented as the trajectory matrix of some time series $F_N = (f_0, \ldots, f_{N-1})$ with $N = L + K - 1$. Formula (6.17) determines the series F_N with the trajectory matrix that is nearest to the matrix $\mathbf{Y}$: $f_n = g_{n+2}$. In terms of Section 1.1 the corresponding operator $\mathfrak{P}$ performs diagonal averaging of the matrix $\mathbf{Y}$.

The restriction of the Frobenius inner product (4.7) from $\mathcal{M}_{LK}$ to $\mathcal{M}_{LK}^{(H)}$ determines the corresponding inner product on the linear space $\mathfrak{F}_N$ of time series of length $N = K + L - 1$.

Let us fix integers $1 \leq L \leq N$ and set $K = N - L + 1$, $L^* = \min(L, K)$ and $K^* = \max(L, K)$. For the weights

$$w_i = \begin{cases} i+1 & \text{for } 0 \leq i < L^*, \\ L^* & \text{for } L^* \leq i < K^*, \\ N-i & \text{for } K^* \leq i < N, \end{cases} \qquad (6.18)$$

we define the *inner product* of series $F_N^{(1)}, F_N^{(2)} \in \mathfrak{F}_N$ as

$$\left(F_N^{(1)}, F_N^{(2)} \right)_w \stackrel{\text{def}}{=} \sum_{i=0}^{N-1} w_i f_i^{(1)} f_i^{(2)} \qquad (6.19)$$

where $f_i^{(1)}$ and $f_i^{(2)}$ are the terms of the series $F_N^{(1)}$ and $F_N^{(2)}$.

Proposition 6.4 *Let $\mathbf{X}^{(1)}$ and $\mathbf{X}^{(2)}$ be the L-trajectory matrices of the series $F_N^{(1)}$ and $F_N^{(2)}$. Then $\left(F_N^{(1)}, F_N^{(2)} \right)_w = \left\langle \mathbf{X}^{(1)}, \mathbf{X}^{(2)} \right\rangle_{\mathcal{M}}$.*

Proof.
Denote by $x_{ij}^{(1)}$ and $x_{ij}^{(2)}$ the elements of the trajectory matrices $\mathbf{X}^{(1)}$ and $\mathbf{X}^{(2)}$. Then, analogous to (6.16), we obtain

$$\left\langle \mathbf{X}^{(1)}, \mathbf{X}^{(2)} \right\rangle_{\mathcal{M}} = \sum_{i,j} x_{ij}^{(1)} x_{ij}^{(2)}$$

$$= \sum_{s=2}^{L+K} \sum_{i+j=s} f_{s-2}^{(1)} f_{s-2}^{(2)} = \sum_{i=0}^{N-1} n_{i+2} f_i^{(1)} f_i^{(2)}$$

$$= \left(F_N^{(1)}, F_N^{(2)}\right)_w$$

where we have used the equality $w_i = n_{i+2}$. □

If $\left(F_N^{(1)}, F_N^{(2)}\right)_w = 0$, then we shall call the series $F_N^{(1)}$ and $F_N^{(2)}$ w-*orthogonal*.

Corollary 6.2 *If the series* $F_N^{(1)}$ *and* $F_N^{(2)}$ *are weakly L-separable, then they are* w-*orthogonal*.

Proof.
If $F_N^{(1)}$ and $F_N^{(2)}$ are weakly separable, then their trajectory matrices are biorthogonal in the sense of equalities (6.6). Therefore (see Section 4.2.1), these matrices are orthogonal with respect to the Frobenius inner product. An application of Proposition 6.4 now gives us the result. □

Remark 6.3 The w-orthogonality of the series follows from pairwise orthogonality of the columns (or rows) of their L-trajectory matrices. Hence the notion of w-orthogonality is useful not only for the SVD of the trajectory matrices but also for certain other orthogonal decompositions (for example, for the Toeplitz and Centring SVDs).

6.3 Centring in SSA

Some features of Centring decompositions of general matrices were discussed in Section 4.4. The case of trajectory matrices has its own peculiarities related to the problems of separability and special tasks of centring.

6.3.1 Single centring SSA and the constant series

Let us consider a time series F_N, some window length L, and let, as usual, $K = N - L + 1$. As described in Section 4.4, the *single centring SVD* of the trajectory matrix $\mathbf{X}$ of the series F_N has the form

$$\mathbf{X} = \mathcal{A}_1(\mathbf{X}) + \sum_{i=1}^{d} \sqrt{\lambda_i} U_i V_i^{\mathrm{T}} = \mathcal{A}_1(\mathbf{X}) + \sum_{i=1}^{d} \mathbf{X}_i \qquad (6.20)$$

with

$$\mathcal{A}_1(\mathbf{X}) = \mathcal{E}_1(\mathbf{X}) \mathbf{1}_K^{\mathrm{T}}, \quad \mathcal{E}_1(\mathbf{X}) = \frac{1}{K} \mathbf{X} \mathbf{1}_K \in \mathbf{R}^L, \qquad (6.21)$$

$\mathbf{1}_K = (1, \ldots, 1)^{\mathrm{T}} \in \mathbf{R}^K$, and $(\sqrt{\lambda_i}, U_i, V_i)$ standing for the eigentriples of the SVD of the matrix $\mathbf{X} - \mathcal{A}_1(\mathbf{X})$. Evidently, $\mathcal{A}_1(\mathbf{X})$ is an elementary (unit-rank) matrix.

Single centring is a standard transformation in principal component analysis. It can easily be proved (see Proposition 4.8 of Section 4.4) that all the factor vectors V_i are orthogonal to the constant vector $\mathbf{1}_K$. This property can be written as $\mathcal{E}_1(\mathbf{X}_i) = \mathbf{0}_L$.

If we apply the decomposition (6.20) for the reconstruction of the series component, then we can include the (first) average triple $(\sqrt{\lambda_{0(1)}}, U_{0(1)}, V_{0(1)})$ defined by

$$U_{0(1)} = \mathcal{E}_1(\mathbf{X})/\|\mathcal{E}_1(\mathbf{X})\|, \quad V_{0(1)} = \mathbf{1}_K/\sqrt{K}$$

and $\sqrt{\lambda_{0(1)}} = \|\mathcal{E}_1(\mathbf{X})\|\sqrt{K}$ to the list of eigentriples selected for reconstruction.

Thus, the definition of weak L-separability (see Section 6.1.1) has to be slightly modified. Indeed, let $F_N = F_N^{(1)} + F_N^{(2)}$ and $\mathbf{X}$, $\mathbf{X}^{(1)}$ and $\mathbf{X}^{(2)}$ stand for the L-trajectory matrices of the series F_N, $F_N^{(1)}$ and $F_N^{(2)}$. If we assume that the trajectory matrix $\mathbf{X}^{(1)}$ can be expressed as the sum of some terms of the decomposition (6.20), then the term $\mathcal{A}_1(\mathbf{X})$ either belongs to the set of these terms or does not.

We can now give the following natural definition. The series $F_N^{(1)}$ and $F_N^{(2)}$ are *weakly L-separable under single centring* if either $\mathcal{E}_1(\mathbf{X}^{(1)}) = \mathbf{0}_L$ and the matrices $\mathbf{X}^{(1)}$, $\mathbf{X}^{(2)} - \mathcal{A}_1(\mathbf{X}^{(2)})$ are biorthogonal, or $\mathcal{E}_1(\mathbf{X}^{(2)}) = \mathbf{0}_L$ and $\mathbf{X}^{(2)}$ is biorthogonal to $\mathbf{X}^{(1)} - \mathcal{A}_1(\mathbf{X}^{(1)})$.

This definition of weak separability gives rise to a new series of examples, which are analogous to those provided in Section 6.1.1. Most of the results are similar, and we do not discuss them here. We consider only the simplest example of separation of a constant series where the difference from the previous case is apparent.

(a) Separation of a constant series

A specific interest in centring in SSA is in connection with the fact that for any L the transformation $\mathbf{X} \mapsto \mathbf{X} - \mathcal{A}_1(\mathbf{X})$ transfers the trajectory matrix of a constant time series to the zero matrix. In other words, if the series F_N is a constant series, and $\mathbf{X}$ stands for its trajectory matrix, then $\mathcal{A}_1(\mathbf{X}) = \mathbf{X}$.

To extract a constant series $F_N^{(1)}$ from the sum $F_N = F_N^{(1)} + F_N^{(2)}$, we need not take into consideration the eigentriples in the SVD of the matrix $\mathbf{X} - \mathcal{A}_1(\mathbf{X})$. All we need is to consider the average term $\mathcal{A}_1(\mathbf{X})$. In view of the equality

$$\mathcal{A}_1(\mathbf{X}) = \mathcal{A}_1(\mathbf{X}^{(1)}) + \mathcal{A}_1(\mathbf{X}^{(2)}) = \mathbf{X}^{(1)} + \mathcal{A}_1(\mathbf{X}^{(2)}),$$

the conditions for separability of the constant series $F_N^{(1)}$ from a series $F_N^{(2)}$ is the equality $\mathcal{A}_1(\mathbf{X}^{(2)}) = \mathbf{0}_{LK}$. This means that $\mathcal{E}_1(\mathbf{X}^{(2)}) = \mathbf{0}_L$ or, in terms of elements $f_n^{(2)}$ of the series $F_N^{(2)}$,

$$f_i^{(2)} + \ldots + f_{i+K-1}^{(2)} = 0, \quad i = 0, \ldots, L-1. \tag{6.22}$$

Therefore, the series $F_N^{(2)}$ is separable from a constant with the help of the first average triple if it has period K and its average with respect to this period is zero. Note that these conditions are weaker than those given in Example 6.1 of Section 6.1.1.

(b) Asymptotic separation of a constant series

Now let us turn to asymptotic separability of a constant series in case of single centring. For fixed L, K we denote by $\mathfrak{P} = \mathfrak{P}^{(L,K)}$ the diagonal averaging operator. It transfers any $L \times K$ matrix $\mathbf{Y}$ into the time series $F = F(\mathbf{Y})$ by diagonal averaging of the matrix $\mathbf{Y}$ (see Section 6.2). In terms of Section 1.1, if $\mathbf{Y}$ is the resultant matrix, then $F = \mathfrak{P}\mathbf{Y}$ is the corresponding reconstructed series.

Suppose $F^{(1)}$ is an infinite constant series, $F^{(2)}$ is some infinite series, and denote by $F_N^{(1)}$ and $F_N^{(2)}$ the series consisting of the first N terms of the series $F^{(1)}$ and $F^{(2)}$. We set $F_N = F_N^{(1)} + F_N^{(2)}$ and denote by $\mathbf{X}^{(1)}$ and $\mathbf{X}^{(2)}$ the corresponding trajectory matrices for some window length L. Note that $\mathcal{A}_1(\mathbf{X}^{(1)}) = \mathbf{X}^{(1)}$ and $\mathfrak{P}\mathcal{A}_1(\mathbf{X}^{(1)}) = F_N^{(1)}$.

Thus, if we reconstruct $F_N^{(1)}$ with the help of the average triple, then (using the linearity of the operators $\mathcal{A}_1$ and $\mathfrak{P}$) we obtain the series $F_N^{(1)} + \mathfrak{P}\mathcal{A}_1(\mathbf{X}^{(2)})$. Therefore, the series $\Delta_N \stackrel{\text{def}}{=} \mathfrak{P}\mathcal{A}_1(\mathbf{X}^{(2)})$ can be regarded as an error series.

Let us assume now that $N \to \infty$ and $L = L(N)$. Our aim is to find conditions guaranteeing that each term δ_k of the error series Δ_N tends to zero as $N \to \infty$. In this case we shall say that the constant series $F_N^{(1)}$ is *asymptotically separable from* $F_N^{(2)}$ *with the help of the first average triple*.

This definition corresponds to pointwise asymptotic separability of the kind (6.13). To obtain an analogue of the asymptotic separation in the sense of (6.9), we must require the convergence $\delta_k \to 0$ be uniform with respect to k.

Proposition 6.5 *Assume that one of the following conditions is valid*:
1. $\min(L, K) \to \infty$ *and*

$$\frac{1}{K}\sum_{m=0}^{K-1} f_{i+m}^{(2)} \xrightarrow[K \to \infty]{} 0, \quad i \geq 0; \tag{6.23}$$

2. *L is bounded and* (6.23) *holds uniformly with respect to i.*

Then the constant series $F_N^{(1)}$ is asymptotically separable from $F_N^{(2)}$ with the help of the first average triple.

Proof.
We demonstrate the validity of both assertions at the same time assuming for brevity that $L \leq (N+1)/2$. By (6.15) and (6.21), we have

$$\delta_k = \begin{cases} \dfrac{1}{k+1}\sum_{i=1}^{k+1} y_i & \text{for } 0 \leq k \leq L-2, \\[1ex] \dfrac{1}{L}\sum_{i=1}^{L} y_i & \text{for } L-1 \leq k \leq K-1, \\[1ex] \dfrac{1}{N-k}\sum_{i=k-K+2}^{L} y_i & \text{for } K \leq k \leq N-1, \end{cases} \tag{6.24}$$

where y_i stands for the ith component of the vector $\mathcal{E}_1(\mathbf{X}^{(2)})$. More precisely, if we denote by $f_n^{(2)}$ the terms of the series $F_N^{(2)}$, then

$$y_i = \frac{1}{K}\left(f_{i-1}^{(2)} + \ldots + f_{i+K-1}^{(2)}\right). \tag{6.25}$$

Since $K \to \infty$, it follows that $y_i \to 0$ for any i. Therefore, $\delta_k \to 0$ for $k \leq L-2$. If $y_i \to 0$ uniformly in i, then $\delta_k \to 0$ uniformly in k. $\square$

We observe that the conditions for asymptotic separability of a constant series under centring differ from the related conditions in the case with no centring; see formulae (6.9) and (6.13).

(c) Stochastic separability

Let us take a random sequence $\Xi = (\xi_0 \ldots, \xi_n, \ldots)$ with zero mean and a covariance function $R_\xi(i,j)$ and consider the problem of the asymptotic stochastic separation of a constant signal F from a noise Ξ in the sum $F + \Xi$. For fixed N and window length L, we denote the truncated series F_N and Ξ_N in the same manner as in the previous section.

Let us fix the window length L and denote by $\mathbf{X}$ and $\mathbf{H}$ the L-trajectory matrices of the series F_N and Ξ_N. Then the random process

$$\widehat{F}_N = \mathfrak{P}\mathcal{A}_1(\mathbf{X} + \mathbf{H}) = F_N + \mathfrak{P}\mathcal{A}_1(\mathbf{H})$$

is an unbiased estimator of the constant series F_N. The general term η_k of the process $\mathfrak{E}_N \stackrel{\text{def}}{=} \mathfrak{P}\mathcal{A}_1(\mathbf{H})$ is obtained by (6.24), (6.25) with the replacement of $f_i^{(2)}$ by ξ_i in (6.25). This process has the meaning of a (random) estimation error.

The covariance function of the process $\mathfrak{E}_N$ can easily be written. Let us restrict ourselves to the variance of the process η_k under the assumptions $N \to \infty$ and $L \leq K = N - L + 1$. Evidently, $\mathbf{E}\eta_k = 0$.

To be in accordance with (6.25), we set $\zeta_i = (\xi_{i-1} + \ldots + \xi_{i+K-1})/K$. Then

$$R_\zeta(i,j) \stackrel{\text{def}}{=} \mathbf{E}\zeta_i\zeta_j = \frac{1}{K^2}\sum_{m,l=0}^{K-1} R_\xi(i+m, j+l).$$

Thus,

$$\mathbf{D}\eta_k = \begin{cases} \dfrac{1}{(k+1)^2}\displaystyle\sum_{i,j=1}^{k+1} R_\zeta(i,j) & \text{for } 0 \leq k \leq L-2, \\[2mm] \dfrac{1}{L^2}\displaystyle\sum_{i,j=1}^{L} R_\zeta(i,j) & \text{for } L-1 \leq k \leq K-1, \\[2mm] \dfrac{1}{(N-k)^2}\displaystyle\sum_{i,j=k-K+2}^{L} R_\zeta(i,j) & \text{for } K \leq k \leq N-1. \end{cases}$$

Suppose now that $|R_\varepsilon(k,m)| \leq ce^{-\beta|k-m|}$ with some positive β. Then it can be easily checked that $|R_\zeta(i,j)| = O(K^{-1})$ for $i,j \leq K$. Thus, if $K \to \infty$ and $L \leq K$, then

$$\sup_{0 \leq k \leq N-1} \mathbf{D}\eta_k = O(N^{-1})$$

as $N \to \infty$.

6.3.2 Double centring and linear series

The double centring version of SVD of an $L \times K$ matrix $\mathbf{X}$ was described in Section 4.4.2. The decomposition has the form

$$\mathbf{X} = \mathcal{A}^{(12)}(\mathbf{X}) + \sum_{i \geq 1} \sqrt{\lambda_i} U_i V_i^\mathrm{T}, \qquad (6.26)$$

where

$$\mathcal{A}^{(12)}(\mathbf{X}) = \mathcal{A}_1(\mathbf{X}) + \mathcal{A}_2(\mathbf{X}) - \mathcal{A}_1(\mathcal{A}_2(\mathbf{X})),$$

$\mathcal{A}_1(\mathbf{X})$ is defined in (6.21) and the expression for $\mathcal{A}_2(\mathbf{X})$ can be found in (4.33), (4.34).

Whereas $\mathcal{A}_1(\mathbf{X})$ is a matrix with equal columns $\mathcal{E}_1(\mathbf{X})$ resulting from the row centring of the matrix $\mathbf{X}$, the matrix $\mathcal{A}_2(\mathbf{X})$ has equal rows $(\mathcal{E}_2(\mathbf{X}))^\mathrm{T}$ corresponding to the column centring of $\mathbf{X}$.

The matrix $\mathcal{A}_{12}(\mathbf{X}) \stackrel{\text{def}}{=} \mathcal{A}_1(\mathcal{A}_2(\mathbf{X}))$ is a matrix with the identical elements. They are equal to the average value of all the elements of the matrix $\mathbf{X}$. The sum on the right side of (6.26) is the SVD of the matrix $\mathbf{X}'' \stackrel{\text{def}}{=} \mathbf{X} - \mathcal{A}^{(12)}(\mathbf{X})$. The decomposition (6.26) can be rewritten as

$$\mathbf{X} = \mathcal{A}_1(\mathbf{X}) + \mathcal{A}_2(\mathbf{X}') + \sum_{i \geq 1} \mathbf{X}_i \qquad (6.27)$$

where $\mathbf{X}' = \mathbf{X} - \mathcal{A}_1(\mathbf{X})$ and $\mathbf{X}_i = \sqrt{\lambda_i} U_i V_i^\mathrm{T}$.

According to Section 4.4.2, all the terms on the right side of (6.27) are elementary pairwise orthogonal matrices. Both matrices $\mathcal{A}_1(\mathbf{X})$ and $\mathcal{A}_2(\mathbf{X}')$ can be expressed in terms of the first and second average triples (see Section 1.7.1).

Double centring is used in multivariate statistics (Jolliffe, 1986, Chapter 12.3) but it seems to be a little exotic there.

In SSA, the matrix $\mathbf{X}$ is the trajectory matrix of some time series F_N. If the decomposition (6.26) is performed, then the reconstruction is obtained in the usual manner: after grouping the terms of the double centring decomposition (6.26), we apply the diagonal averaging operator $\mathfrak{P}$ to each resultant matrix and arrive at the decomposition of the initial series F_N.

CENTRING IN SSA

(a) Separation of a linear series

The following proposition shows that a linear time series plays the same role for double centring as a constant series for single centring.

Proposition 6.6 *If* $\mathbf{X}$ *is an L-trajectory matrix of a linear series* F_N, *then*
$$\mathcal{A}^{(12)}(\mathbf{X}) = \mathbf{X}.$$

Proof.
Let $f_n = an + b$ be the general term of the series F_N. Then the L-trajectory matrix $\mathbf{X}$ of the series F_N has the elements
$$x_{ij} = a(i+j-2) + b = f_{i-1} + f_{j-1} - b,$$
($1 \le i \le L, 1 \le j \le K$). Denote by $a_{ij}^{(12)}$ the elements of the matrix $\mathcal{A}^{(12)}(\mathbf{X})$. Then
$$a_{ij}^{(12)} = \frac{1}{K}\sum_{m=1}^{K} x_{im} + \frac{1}{L}\sum_{k=1}^{L} x_{kj} - \frac{1}{LK}\sum_{k=1}^{L}\sum_{m=1}^{K} x_{km}. \quad (6.28)$$

The three terms on the right side of (6.28) are equal to $f_{i-1} + 0.5 f_{K-1} - b/2$, $f_{j-1} + 0.5 f_{L-1} - b/2$ and $0.5(f_{L-1} + f_{K-1})$, respectively. This clearly implies the required result. $\square$

We have thus arrived at the conditions for extracting a linear series with the help of double centring. Let $F_N = F_N^{(1)} + F_N^{(2)}$ and suppose that $F_N^{(1)}$ is a linear time series. For a given window length L we have the following representation of the corresponding trajectory matrices:
$$\mathbf{X} = \mathbf{X}^{(1)} + \mathbf{X}^{(2)}. \quad (6.29)$$

Applying the linear double centring operator $\mathcal{A}^{(12)}$ to (6.29) we obtain that
$$\mathcal{A}^{(12)}(\mathbf{X}) = \mathbf{X}^{(1)} + \mathcal{A}^{(12)}(\mathbf{X}^{(2)}).$$

If the matrix $\mathbf{Z} \stackrel{\text{def}}{=} \mathcal{A}^{(12)}(\mathbf{X}^{(2)})$ is the zero matrix, then double centring will separate the linear series $F_N^{(1)}$ from the series $F_N^{(2)}$ with the help of both average triples.

Denote by $f_n^{(2)}$ the terms of the series $F_N^{(2)}$. The condition $\mathbf{Z} = \mathbf{0}_{LK}$ can be rewritten in the form
$$\frac{1}{K}\sum_{m=1}^{K} f_{i+m-2}^{(2)} + \frac{1}{L}\sum_{k=1}^{L} f_{k+j-2}^{(2)} = \frac{1}{LK}\sum_{k=1}^{L}\sum_{m=1}^{K} f_{k+m-2}^{(2)}, \quad (6.30)$$
($1 \le i \le L, 1 \le j \le K$). If we sum the equalities (6.30) over $1 \le j \le K$ for fixed i, then we arrive at the equality
$$f_i^{(2)} + \ldots + f_{i+K-1}^{(2)} = 0 \quad (6.31)$$
which is valid for any $0 \le i \le L-1$. In similar fashion,
$$f_j^{(2)} + \ldots + f_{j+L-1}^{(2)} = 0 \quad (6.32)$$

for any $0 \leq j \leq K-1$. The conditions (6.31) and (6.32) are the conditions of Example 6.1 (see Section 6.1.1), where separation of a constant series from another series was discussed.

As a result, if there exists $M > 1$ such that M is a divisor of both L and $N-1$, the series $F_N^{(2)}$ has period M and

$$f_0^{(2)} + \ldots + f_{M-1}^{(2)} = 0,$$

and then double centring separates the linear series $F_N^{(1)}$ from $F_N^{(2)}$ under the choice of the window length L and with the help of both average triples.

(b) Asymptotic separability

The concept of asymptotic extraction of a linear series under double centring is similar to the case of single centring and a constant series.

Suppose that $F^{(1)}$ is an infinite linear series, $F^{(2)}$ is an infinite series, and denote by $F_N^{(1)}$ and $F_N^{(2)}$ the series consisting of the first N terms of the series $F^{(1)}$ and $F^{(2)}$.

We set $F_N = F_N^{(1)} + F_N^{(2)}$ and denote by $\mathbf{X}^{(1)}$ and $\mathbf{X}^{(2)}$ the corresponding L-trajectory matrices for some window length L. Note that in view of Proposition 6.6, we have $\mathcal{A}^{(12)}(\mathbf{X}^{(1)}) = \mathbf{X}^{(1)}$ and therefore $\mathfrak{P}\mathcal{A}^{(12)}(\mathbf{X}^{(1)}) = F_N^{(1)}$.

Thus, if we reconstruct $F_N^{(1)}$ with the help of both average triples, then using the linearity of the operators $\mathcal{A}^{(12)}$ and $\mathfrak{P}$ we have an error $\Delta_N^{(12)} \stackrel{\text{def}}{=} \mathfrak{P}\mathcal{A}^{(12)}(\mathbf{X}^{(2)})$. As before, $\mathfrak{P}$ stands for the diagonal averaging operator.

Now let us suppose that $N \to \infty$ and $L = L(N)$. Our aim is to obtain conditions guaranteeing that all the terms $\delta_n^{(12)}$ of the error series $\Delta_N^{(12)}$ tend to zero as $N \to \infty$.

In view of (6.28) the element $a_{ij}^{(N)}$ of the matrix $\mathcal{A}^{(12)}(\mathbf{X}^{(2)})$ is

$$a_{ij}^{(N)} = \frac{1}{K} \sum_{m=0}^{K-1} f_{i+m}^{(2)} + \frac{1}{L} \sum_{k=0}^{L-1} f_{k+j}^{(2)} - \frac{1}{LK} \sum_{k=0}^{L-1} \sum_{m=0}^{K-1} f_{k+m}^{(2)}, \quad (6.33)$$

where the $f_n^{(2)}$ are the terms of the series $F^{(2)}$.

Since $\mathfrak{P}$ is a linear averaging operator, we can easily obtain $\delta_n^{(12)} \to 0$ for any n, if $\min(L, K) \to \infty$ and

$$\frac{1}{M} \sum_{j=0}^{M-1} f_{i+j}^{(2)} \xrightarrow[M \to \infty]{} 0$$

for any i (see Proposition 6.5).

(c) Stochastic separability

The conditions for separability of a linear signal from a general random noise under double centring can be derived in the manner of Section 6.3.1. However, the corresponding covariance function is rather cumbersome, and we thus restrict ourselves to the simplest (and, in a way, the main) particular case of a random white noise series.

Suppose that we have an infinite linear series $F^{(1)}$ with $f_n^{(1)} = an + b$ and a white noise series $\Xi = (\xi_0, \ldots, \xi_n, \ldots)$. Assume that $\mathbf{E}\xi_n = 0$, $\mathbf{D}\xi_n = \sigma^2$ and $\mathbf{E}\xi_n\xi_m = 0$ for $n \neq m$.

If we cut both series at the moment of time $N - 1$, we then obtain a finite linear series $F_N^{(1)}$ and the corresponding noise series Ξ_N. We set $F_N = F_N^{(1)} + \Xi_N$.

Under the choice of window length $L = L(N)$, we obtain the L-trajectory matrices $\mathbf{X}$, $\mathbf{X}^{(1)}$ and $\mathbf{H}$ of the series F_N, $F_N^{(1)}$ and Ξ_N, respectively. We then apply double centring to the matrix $\mathbf{X}$. If we select both average triples in the double centring SSA for reconstruction, we obtain the series $\widetilde{F}_N^{(1)}$ with $\widetilde{f}_n^{(1)} = an + b + \delta_n$, where δ_n can be considered as the general term of a random error series Δ_N. Evidently, $\mathbf{E}\delta_n = 0$. Our aim is to investigate the variance of the series δ_n.

To simplify all the expressions, we consider the case of odd N and $L = K = (N+1)/2$. Then it is sufficient to consider $0 \le n \le L - 1$.

The series Δ_N has the form

$$\Delta_N = \mathfrak{P}\mathcal{A}^{(12)}(\mathbf{H}) = \mathfrak{P}\mathcal{A}_1(\mathbf{H}) + \mathfrak{P}\mathcal{A}_2(\mathbf{H}) - \mathfrak{P}\mathcal{A}_{12}(\mathbf{H}).$$

We set $\eta_i = (\xi_i + \ldots + \xi_{i+L-1})/L$. Then the matrix $\mathfrak{P}\mathcal{A}_{12}(\mathbf{H})$ has identical elements that are equal to $(\eta_0 + \ldots + \eta_{L-1})/L$ and, since $L = K$, we have

$$\mathfrak{P}\mathcal{A}_1(\mathbf{H}) = \mathfrak{P}\mathcal{A}_2(\mathbf{H}).$$

The kth term ($1 \le k \le L$) of the series $\mathfrak{P}\mathcal{A}_1(\mathbf{H})$ is equal to $(\eta_0 + \ldots + \eta_{k-1})/k$. Therefore, for $0 \le n \le L - 1$, the term δ_n of the series Δ_N has the form

$$\delta_n = \frac{2}{n+1} \sum_{i=0}^{n} \eta_i - \frac{1}{L} \sum_{i=0}^{L-1} \eta_i$$

$$= \frac{2L - n - 1}{(n+1)L} \sum_{i=0}^{n} \eta_i - \frac{1}{L} \sum_{i=n+1}^{L-1} \eta_i. \qquad (6.34)$$

It is easy to check that $\mathbf{E}\eta_i\eta_j = \sigma^2(L - |i - j|)/L^2$, $0 \le i, j \le L - 1$. We set $\tau = (n+1)/L$. In this notation

$$\mathbf{D}\left(\sum_{i=0}^{n} \eta_i\right) = \sigma^2 \tau^2 L \left(1 - \tau/3\right) + O(1), \qquad (6.35)$$

$$\mathbf{D}\left(\sum_{i=n+1}^{L-1} \eta_i\right) = \sigma^2 (1 - \tau)^2 L \left(1 - (1 - \tau)/3\right) + O(1) \qquad (6.36)$$

and

$$\mathbf{E}\left(\sum_{i=0}^{n} \eta_i \sum_{j=n+1}^{L-1} \eta_j\right) = 0.5\sigma^2 \tau(1-\tau)L + O(1). \tag{6.37}$$

Since
$$\frac{2}{n+1} - \frac{1}{L} = \frac{2-\tau}{\tau L},$$
we obtain from (6.34), (6.35), (6.36) and (6.37) that

$$\mathbf{D}\delta_n = \frac{2\sigma^2}{3L}(4 - 5\tau + 2\tau^2) + O(1/L^2). \tag{6.38}$$

For $n > L - 1$ we have to use $2 - \tau$ in place of τ in (6.38).

The result (6.38) is slightly worse than that obtained for the standard maximum likelihood estimator. More precisely, formula (1.4.9) in Draper and Smith (1998, Section 1.4) gives us $0.5\sigma^2(4 - 6\tau + 3\tau^2)/L$ for the order of the variance (for $n \leq L - 1$); this result should be compared with (6.38).

Nevertheless, the examples of Section 1.7.1 show that for relatively small N, double centring should be preferred to linear regression in the problem of extracting a linear tendency from a time series.

6.4 SSA for stationary series

This section is devoted to the study of the properties of SSA applied to deterministic stationary sequences.

First, we describe the notion of a deterministic stationary sequence and write the spectral representation for such a sequence. This representation is analogous to the spectral representation for a random stationary sequence; this material is in accordance with Brillinger (1975, Chapter 3.9).

Second, Section 6.4.2 is devoted to the classification of stationary series. With respect to the structure of their spectral measures, these series are divided into three classes: periodic, quasi-periodic and aperiodic (or chaotic). This division is useful in the theory of chaotic dynamical systems (see, for example, Schuster, 1995), where the transition from periodic to quasi-periodic and then to chaotic motion indicates an increase of the level of the system complexity.

On the other hand, the class of (Bohr) almost-periodic series and functions, connected with problems of wave mechanics and thoroughly studied in the literature (see the bibliography in Corduneanu, 1968), is a good example of deterministic stationary sequences with a discrete spectral measure.

Third, we briefly consider the question of the asymptotic behaviour of the SVD of Hankel matrices associated with stationary series. We discuss several results concerning the asymptotic distribution of the eigenvalues and eigenvectors of Toeplitz matrices (see Grenander and Szegö, 1984). These results are useful for

SSA FOR STATIONARY SERIES

SSA since they help to identify the eigentriples of series with a complex structure and explain some phenomena discussed in Chapter 1.

The last two sections are devoted to weak separability of stationary series and the role of periodogram analysis in their study.

6.4.1 Spectral representation of stationary series

Let us start with the following definition (see Brillinger, 1975, Chapter 3.9).

Definition 6.1 Let $F = (f_0, \ldots, f_n, \ldots)$ be a time series. The series F is called *stationary* if there exists a function $R_f(k)$ ($-\infty < k < +\infty$) such that for any $k, l \geq 0$

$$R_f^{(N)}(k, l) \stackrel{\text{def}}{=} \frac{1}{N} \sum_{m=0}^{N-1} f_{k+m} f_{l+m} \xrightarrow[N \to \infty]{} R_f(k - l). \tag{6.39}$$

If (6.39) is valid, then R_f is called the *covariance function* of the stationary series F.

The following proposition shows that if the limit (6.39) exists, then it depends on the pair (k, l) only via $k - l$.

Proposition 6.7 *Suppose that for any $k, l \geq 0$ there exists a finite limit*

$$R_f^*(k, l) = \lim_{N \to \infty} \frac{1}{N} \sum_{m=0}^{N-1} f_{k+m} f_{l+m}.$$

Then $R_f^*(k + n, l + n) = R_f^*(k, l)$ for any $k, l, n \geq 0$.

Proof.
If $k \geq l$, then

$$\frac{1}{N} \sum_{m=0}^{N-1} f_{k+m} f_{l+m} = \frac{1}{N} \sum_{j=l}^{N-1+l} f_{k-l+j} f_j$$

$$= \frac{1}{N} \sum_{j=0}^{N-1+l} f_{k-l+j} f_j - \frac{1}{N} \sum_{j=0}^{l} f_{k-l+j} f_j. \tag{6.40}$$

Since the first term on the right side of (6.40) tends to $R_f^*(k - l, 0)$ and the second is $O(N^{-1})$, the proof is complete. □

Proposition 6.8 *Let R_f be the covariance function of a stationary series F. Then there exists a finite measure m_f defined on the Borel subsets of $(-1/2, 1/2]$ such that*

$$R_f(k) = \int_{(-1/2, 1/2]} e^{i2\pi k\omega} m_f(d\omega).$$

Proof.
For any $n \geq 1$, any integer k_p and any complex z_p ($1 \leq p \leq n$) we have

$$\sum_{j,m=1}^{n} z_j \bar{z}_m \frac{1}{N} \sum_{l=0}^{N-1} f_{k_j+l} f_{k_m+l}$$

$$= \frac{1}{N} \sum_{l=0}^{N-1} \left| \sum_{j=1}^{n} f_{k_j+l} z_j \right|^2 \geq 0.$$

Therefore, the function $R_f^{(N)}$ is positive semidefinite. Since the class of positive semidefinite functions is closed under pointwise limit transition, the covariance function R_f is also positive semidefinite. Hence, in view of Herglotz's theorem (Loève, 1963, Chapter IV), we obtain the required result. □

The measure m_f is called the *spectral measure* of the series F. If m_f is absolutely continuous with respect to Lebesgue measure with the Radon-Nikodým derivative p_f, then p_f is called the *spectral density* of F. Accordingly, the function $\Phi_f(\omega) \stackrel{\text{def}}{=} m_f((-1/2, \omega])$ is the *spectral function*.

Since F is a real-valued series, $R_f(-k) = R_f(k)$, and the spectral measure m_f has the following property: if A is a Borel subset of $(-1/2, 1/2)$, then $m_f(-A) = m_f(A)$. Therefore,

$$R_f(k) = 2 \int_{[0,1/2)} \cos(2\pi k w) m_f(dw) + (-1)^k m_f(\{1/2\}).$$

The spectral density can be considered as an even function: $p_f(-\omega) = p_f(\omega)$.

The following proposition shows that we can extend the domain of stationary sequences to the set $\mathcal{Z} = \{\pm n, n = 0, 1, \ldots\}$.

Proposition 6.9 *Let F be a stationary series with a covariance function R_f. Let $g_n = f_{|n|}$ for $n \in \mathcal{Z}$. Then for all $k, l \in \mathcal{Z}$ we have*

$$\frac{1}{2N-1} \sum_{j=-N+1}^{N-1} g_{k+j} g_{l+j} \xrightarrow[N \to \infty]{} R_f(k-l).$$

Proof.
Suppose that $k \geq l \geq 0$. Then

$$\frac{1}{2N-1} \sum_{j=-N+1}^{N-1} g_{k+j} g_{l+j}$$

$$= \frac{1}{2N-1} \sum_{j=0}^{N-1} f_{k+j} f_{l+j} + \frac{1}{2N-1} \sum_{j=k}^{N-1} f_{j-k} f_{j-l} + O(N^{-1}).$$

Since both terms on the right side of the above equality tend to $0.5 R_f(k-l)$, the assertion is verified. Other cases for k, l are analogous. □

SSA FOR STATIONARY SERIES

From now on, if necessary we shall assume that a stationary sequence has the domain $\mathcal{Z}$.

Let us introduce some more notation. For fixed $k \in \mathcal{Z}$, we write $e_k(\omega)$ for the imaginary exponential $e^{i2\pi k\omega}$, $\omega \in (-1/2, 1/2]$. In the same manner, for fixed $k \in \mathcal{Z}$ and a time series $F = (\ldots, f_{-2}, f_{-1}, f_0, f_1, f_2, \ldots)$, we denote by F_k the series G with $g_n = f_{k+n}$.

Let F be a stationary series with covariance function R_f. Denote by $\mathcal{L}_f$ the linear space spanned by the series F_k, $k \in \mathcal{Z}$, with complex coefficients. The formula

$$\left[F_k, F_l\right] \stackrel{\text{def}}{=} \lim_{N \to \infty} \frac{1}{2N-1} \sum_{j=-N+1}^{N-1} f_{k+j} \overline{f}_{l+j} = R_f(k-l), \qquad (6.41)$$

extended to $\mathcal{L}_f$ by bilinearity, defines an inner product in $\mathcal{L}_f$. The corresponding norm is denoted by $|||\cdot|||$.

Moreover, if we consider the Hilbert space $\mathbf{L}_f^2$ of complex functions that are square integrable relative to the spectral measure m_f, and we denote by $(\cdot, \cdot)_f$ and $||\cdot||_f$ the corresponding inner product and the norm, then

$$\left[F_k, F_l\right] = \int_{(-1/2, 1/2]} e^{i2\pi k\omega} e^{-i2\pi l\omega} m_f(d\omega) = (e_k, e_l)_f$$

and for finite linear combinations $H_1 = \sum_j \alpha_j F_{k_j}$ and $H_2 = \sum_l \beta_l F_{m_l}$

$$\left[H_1, H_2\right] = \left(\sum_j \alpha_j e_{k_j}, \sum_l \beta_l e_{m_l}\right)_f. \qquad (6.42)$$

Denote by $\mathcal{L}^{(e)}$ the linear subspace of $\mathbf{L}_f^2$ spanned by the imaginary exponents $e_k \in \mathcal{Z}$. Then (6.42) shows that the linear spaces $\mathcal{L}_f$ and $\mathcal{L}^{(e)}$ are in natural one-to-one isometric correspondence:

$$\mathcal{L}_f \ni H = \sum_j \alpha_j F_{k_j} \longleftrightarrow \sum_j \alpha_j e_{k_j} = \psi \in \mathcal{L}^{(e)}.$$

Evidently, this isometric correspondence of $\mathcal{L}_f$ and $\mathcal{L}^{(e)}$ can be extended to their closures $\mathcal{H}_f = \overline{\mathcal{L}_f}$ and $\mathbf{L}_f^2 = \overline{\mathcal{L}^{(e)}}$ in the corresponding agreed norms $|||\cdot|||$ and $||\cdot||_f$. Therefore, we obtain the Hilbert space $\mathcal{H}_f$ isometric to $\mathbf{L}_f^2$ and equipped with the inner product $[\cdot, \cdot]$ and the norm $|||\cdot|||$.

To point out a difference between the stationary time series and the elements of $\mathcal{H}_f$, we write $H^{(1)} \simeq H^{(2)}$ if $H^{(1)}$ and $H^{(2)}$ are identical elements of $\mathcal{H}_f$. For two time series $H^{(1)}, H^{(2)} \in \mathcal{H}_f$ we have $H^{(1)} \simeq H^{(2)}$ if and only if

$$\frac{1}{2N-1} \sum_{n=-N+1}^{N-1} \left|h_n^{(1)} - h_n^{(2)}\right|^2 \xrightarrow[N \to \infty]{} 0. \qquad (6.43)$$

Therefore, the stationary series F coincides with the zero series $\mathbf{0}$ in $\mathcal{H}_f$ if and only if
$$\frac{1}{2N-1} \sum_{n=-N+1}^{N-1} |f_n|^2 \xrightarrow[N\to\infty]{} 0.$$

Analogously, the sequence of stationary series
$$F^{(m)} = (\ldots, f_{-1}^{(m)}, f_0^{(m)}, f_1^{(m)}, \ldots) \in \mathcal{H}_f$$
tends to $\mathbf{0}$ in $\mathcal{H}_f$ if and only if
$$\lim_{m\to\infty} \left|\left|\left| F^{(m)} \right|\right|\right|^2 = \lim_{m\to\infty} \lim_{N\to\infty} \frac{1}{2N-1} \sum_{n=-N+1}^{N-1} \left| f_n^{(m)} \right|^2 = 0.$$

The following theorem is similar to the Cramér spectral theorem in the theory of stationary processes (for example, Rozanov, 1967, Chapter 1, §4). The proof, in the main, corresponds to Brillinger (1975, Chapter 3.9). We set $I = (-1/2, 1/2]$.

Theorem 6.1 *Let F be a stationary series with covariance function R_f and spectral measure m_f. Then*

1. *There exists a function $M_f : I \mapsto \mathcal{H}_f$ such that M_f has finite variation in $\mathcal{H}_f$ and for all $k \in \mathcal{Z}$*
$$F_k \simeq \int_I e_k \, dM_f. \tag{6.44}$$

2. *For all $\omega_1, \omega_2 \in I$*
$$\left[M_f(\omega_1), M_f(\omega_2) \right] = \Phi_f(\min(\omega_1, \omega_2)). \tag{6.45}$$

Proof.
Let $H \in \mathcal{H}_f$ and consider $\psi \in \mathbf{L}_f^2$ such that $H \leftrightarrow \psi$. As $\psi_1 \stackrel{\text{def}}{=} e_1\psi \in \mathbf{L}_f^2$, there exists a unique $H_1 \in \mathcal{H}_f$ such that $H_1 \leftrightarrow \psi_1$. Setting $\mathbf{U}H \stackrel{\text{def}}{=} H_1$ we obtain the operator $\mathbf{U} : \mathcal{H}_f \mapsto \mathcal{H}_f$.

In view of the isometry $\mathcal{H}_f \leftrightarrow \mathbf{L}_f^2$, it is easy to check that $\mathbf{U}$ is a unitary operator. Indeed, for any $H \in \mathcal{H}_f$ with $H \leftrightarrow \psi$ we have
$$\mathbf{U}^{-1}H \leftrightarrow e_{-1}\psi$$
and for $H^{(1)} \leftrightarrow \psi^{(1)}$, $H^{(2)} \leftrightarrow \psi^{(2)}$,
$$\left[\mathbf{U}H^{(1)}, \mathbf{U}H^{(2)} \right] = (e_1\psi^{(1)}, e_1\psi^{(2)})_f$$
$$= \int_I e^{i2\pi\omega}\psi^{(1)}(\omega) e^{-i2\pi\omega}\overline{\psi^{(2)}}(\omega) m_f(d\omega)$$
$$= (\psi^{(1)}, \psi^{(2)})_f = \left[H^{(1)}, H^{(2)} \right]. \tag{6.46}$$

Moreover, $F_k \leftrightarrow e_k$ implies $\mathbf{U}F_k \leftrightarrow e_{k+1} \leftrightarrow F_{k+1}$, and therefore $\mathbf{U}$ has the meaning of the trajectory shift operator.

According to the theory of the spectral representation of unitary operators (see, for example, Riesz and Sz.-Nagy, 1990, Chapter VIII, §2), there exists a unique *spectral family* $\{\mathfrak{S}_\omega, \omega \in I\}$ of operators $\mathfrak{S}_\omega : \mathcal{H}_f \mapsto \mathcal{H}_f$ such that

$$\mathbf{U} = \int_I e^{i2\pi\omega} d\mathfrak{S}_\omega. \tag{6.47}$$

Recall that a family of operators $\{\mathfrak{S}_\omega, \omega \in I\}$ is called spectral if

1. $\mathfrak{S}_{\omega_1} \circ \mathfrak{S}_{\omega_2} = \mathfrak{S}_{\min(\omega_1,\omega_2)}$ (and thus $\mathfrak{S}_\omega$ is a projection operator for any ω);
2. $\mathfrak{S}_{\omega+0} = \mathfrak{S}_\omega$, where $\mathfrak{S}_{\omega+0}$ stands for the strong limit $\lim_{\omega_1 \downarrow \omega} \mathfrak{S}_{\omega_1}$;
3. $\mathfrak{S}_{-1/2+0} = \mathbf{0}$, $\mathfrak{S}_{1/2} = \mathbf{1}$, where $\mathbf{0}$ stands for the zero operator and $\mathbf{1}$ denotes the identity operator.

To understand the nature of the spectral operators $\mathfrak{S}_\omega$, we observe that if $\mathcal{H}_f \ni H \leftrightarrow \psi \in \mathbf{L}_f^2$, then

$$\mathcal{H}_f \ni \mathfrak{S}_\omega H \leftrightarrow \mathbf{1}_\omega \psi \in \mathbf{L}_f^2 \tag{6.48}$$

where $\mathbf{1}_\omega$ is the indicator of the set $(-1/2, \omega]$:

$$\mathbf{1}_\omega(\omega_0) = \begin{cases} 1 & \text{for } \omega_0 \leq \omega, \\ 0 & \text{for } \omega_0 > \omega. \end{cases}$$

Indeed, if we define the operator family $\{\mathfrak{S}_\omega, \omega \in I\}$ by (6.48) and set $\mathfrak{S}_\Delta = \mathfrak{S}_{\omega_1} - \mathfrak{S}_{\omega_2}$ for $\Delta = (\omega_1, \omega_2]$, then for any $H \leftrightarrow \psi$ and any elementary complex function

$$g \stackrel{\text{def}}{=} \sum_m c_m \mathbf{1}_{\Delta_k}$$

we can define the $\mathcal{H}_f$-valued integral

$$J_H(g) = \int_I g(\omega) d(\mathfrak{S}_\omega H) \stackrel{\text{def}}{=} \sum_m c_m \mathfrak{S}_{\Delta_m} H$$

and check that

$$\left[J_{H^{(1)}}(g_1), J_{H^{(2)}}(g_2) \right] = \int_I g_1 \bar{g}_2 \psi_1 \bar{\psi}_2 dm_f = (g_1\psi_1, g_2\psi_2)_f \tag{6.49}$$

for elementary functions g_1, g_2 and $H^{(1)}, H^{(2)} \in \mathcal{H}_f$ such that $H^{(1)} \leftrightarrow \psi_1$ and $H^{(2)} \leftrightarrow \psi_2$.

Now, taking $\mathcal{H}_f \ni H \leftrightarrow \psi \in \mathbf{L}_f^2$ and $g \in \mathbf{L}^2(|\psi|^2 dm_f)$, we get the $\mathcal{H}_f$-valued integral

$$\mathcal{H}_f \ni J_H(g) = \int_I g(\omega) d(\mathfrak{S}_\omega H) \leftrightarrow g\psi \in \mathbf{L}_f^2,$$

which is formally defined via the convergence $g_n \to g$ where the g_n are elementary functions approximating g in $\mathbf{L}^2(|\psi|^2 dm_f)$. The equality (6.49) holds also under the assumption that $g_1\psi_1, g_2\psi_2 \in \mathbf{L}_f^2$.

If we take $g = e_1$, then we obtain

$$J_H(e_1) = \int_I e^{i2\pi\omega} d(\mathfrak{S}_\omega H) \leftrightarrow e_1\psi \leftrightarrow \mathbf{U}H$$

and therefore

$$\mathbf{U}H \simeq \int_I e^{i2\pi\omega} d(\mathfrak{S}_\omega H); \qquad (6.50)$$

this representation can be considered as a weak form of (6.47).

In similar fashion, if we take $H = F \leftrightarrow \psi$ and $g = e_k$, then

$$J_H(e_k) = \int_I e^{i2\pi k\omega} d(\mathfrak{S}_\omega H) \leftrightarrow e_k\psi \leftrightarrow \mathbf{U}^k H$$

and

$$\mathbf{U}^k H \simeq \int_I e^{i2\pi k\omega} d(\mathfrak{S}_\omega H). \qquad (6.51)$$

Taking $H = F$ we get (6.44) with $M_f(\omega) = \mathfrak{S}_\omega F$, in view of the equality $\mathbf{U}^k F \simeq F_k$.

Since $F \leftrightarrow \mathbf{1}$ and using (6.48) we obtain

$$\left[\mathfrak{S}_{\omega_1} F, \mathfrak{S}_{\omega_2} F\right] = \int_I \mathbf{1}_{\omega_1} \mathbf{1}_{\omega_2} dm_f = \Phi_f(\min(\omega_1, \omega_2))$$

and the proof is complete. $\square$

Remark 6.4 Properties (6.44) and (6.45) are similar to the spectral representation of a random stationary process in terms of the Fourier transformation of the stochastic orthogonal measure corresponding to the spectral measure of the process. Indeed, (6.45) means 'orthogonality', and $M_f(\omega) = \mathfrak{S}_\omega f$ corresponds to the cumulative distribution function of the 'stochastic orthogonal measure'.

Remark 6.5 The result of Theorem 6.1 remains valid for stationary sequences of the form $F = (f_0, f_1, \ldots, f_n, \ldots)$ if we define the inner product of the series $G_1, G_2 \in \mathcal{L}_f$ as

$$\left[G_1, G_2\right] = \lim_{N \to \infty} \frac{1}{N} \sum_{n=0}^{N-1} g_n^{(1)} \overline{g_n^{(2)}}. \qquad (6.52)$$

In this case, the operator $\mathbf{U}$ is isometric but not unitary. Since we do not use the spectral representation (6.47) directly, the final result is still valid.

Proposition 6.10 *Under the conditions of Theorem 6.1, for any $H \in \mathcal{H}_f$ and $\omega_0 \in I$ we have*

$$\frac{1}{N}\sum_{n=0}^{N-1} e^{-i2\pi n\omega_0} \mathbf{U}^n H \xrightarrow[N\to\infty]{\mathcal{H}_f} (\mathfrak{S}_{\omega_0} - \mathfrak{S}_{\omega_0-0})H. \tag{6.53}$$

Proof.
Let $H \leftrightarrow \psi$. By (6.51),

$$\frac{1}{N}\sum_{n=0}^{N-1} e^{-i2\pi n\omega_0} \mathbf{U}^n H$$

$$\simeq \int_I \left(\frac{1}{N}\sum_{n=0}^{N-1} e^{i2\pi n(\omega-\omega_0)}\right) d(\mathfrak{S}_\omega H)$$

$$= \int_I \frac{1 - e^{i2\pi N(\omega-\omega_0)}}{N(1 - e^{i2\pi(\omega-\omega_0)})} d(\mathfrak{S}_\omega H) \tag{6.54}$$

where the integrand is assumed to be 1 if $\omega = \omega_0$.
Since

$$(\mathfrak{S}_{\omega_0} - \mathfrak{S}_{\omega_0-0})H \simeq \int_{\{\omega_0\}} d(\mathfrak{S}_\omega H),$$

we obtain

$$\left\|\left\|\frac{1}{N}\sum_{n=0}^{N-1} e^{-i2\pi n\omega_0} \mathbf{U}^n H - (\mathfrak{S}_{\omega_0} - \mathfrak{S}_{\omega_0-0})H\right\|\right\|^2$$

$$= \int_{I\setminus\{\omega_0\}} \left|\frac{1 - e^{i2\pi N(\omega-\omega_0)}}{N(1 - e^{i2\pi(\omega-\omega_0)})}\right|^2 |\psi|^2(\omega) m_f(d\omega). \tag{6.55}$$

It is easy to see that

$$\left|\frac{1 - e^{i2\pi N(\omega-\omega_0)}}{N(1 - e^{i2\pi(\omega-\omega_0)})}\right|^2 = \frac{\sin^2(\pi N(\omega-\omega_0))}{N^2 \sin^2(\pi(\omega-\omega_0))} \leq \frac{C}{\varepsilon^2 N^2}$$

for $|\omega - \omega_0| \geq \varepsilon$, and the same expression is $O(1)$ uniformly on N for $|\omega - \omega_0| < \varepsilon$. Since $\psi \in \mathbf{L}_f^2$, the right side integral in (6.55) tends to zero as $N \to \infty$. □

Remark 6.6
1. In view of the equality

$$|||(\mathfrak{S}_{\omega_0} - \mathfrak{S}_{\omega_0-0})H|||^2 = |\psi|^2(\omega_0) m_f(\{\omega_0\}),$$

the left side of (6.53) tends to the zero element of $\mathcal{H}_f$ for any H if and only if $m_f(\{\omega_0\}) = 0$.

2. If we take $H = F$, then (6.53) can be rewritten as

$$\frac{1}{N}\sum_{n=0}^{N-1} e^{-i2\pi n\omega_0} F_n \xrightarrow[N\to\infty]{\mathcal{H}_f} M_f(\omega_0) - M_f(\omega_0 - 0). \tag{6.56}$$

The right side is equal to zero if and only if $m_f(\{\omega_0\}) = 0$. The convergence (6.56) means, in particular, that

$$\left|\left|\left|\frac{1}{N}\sum_{n=0}^{N-1} e^{-i2\pi n\omega_0} F_n\right|\right|\right|^2$$

$$= \lim_{T\to\infty} \frac{1}{2T-1} \sum_{k=-T+1}^{T-1} \left|\frac{1}{N}\sum_{n=0}^{N-1} e^{-i2\pi n\omega_0} f_{n+k}\right|^2$$

$$\xrightarrow[N\to\infty]{} |||M_f(\omega_0) - M_f(\omega_0 - 0)|||^2 = m_f(\{\omega_0\}). \tag{6.57}$$

3. If we take $H = F$ and $\omega_0 = 0$, then we come to the 'Law of Large Numbers in $\mathcal{H}_f$' result for stationary sequences: *the convergence*

$$\frac{1}{N}\sum_{n=0}^{N-1} F_n \xrightarrow[N\to\infty]{\mathcal{H}_f} 0 \tag{6.58}$$

takes place if and only if $m_f(\{0\}) = 0$.

4. It follows from (6.53) that

$$\left[\frac{1}{N}\sum_{n=0}^{N-1} e^{-i2\pi n\omega_1} \mathbf{U}^n H, \frac{1}{N}\sum_{n=0}^{N-1} e^{-i2\pi n\omega_2} \mathbf{U}^n H\right] \xrightarrow[N\to\infty]{} 0$$

if $\omega_1 \neq \omega_2$.

6.4.2 Classification of stationary sequences

(a) Periodic series

If a time series F has an (integer) period T, then its terms f_n can be expressed as

$$f_n = \sum_{k=-[T/2]+1}^{[T/2]} c_k e^{i2\pi kn/T} \tag{6.59}$$

with $[x]$ standing for the integer part of x. Since F is a real-valued series, it follows that $c_{-k} = \widetilde{c}_k$.

Using the equality

$$\frac{1}{N}\sum_{j=0}^{N-1} f_{n+j}f_{m+j} = \sum_{k,l=-[T/2]+1}^{[T/2]} c_k \bar{c}_l e^{i2\pi kn/T} e^{-i2\pi lm/T} I_N(k,l)$$

with

$$I_N(k,l) = \frac{1}{N}\sum_{j=0}^{N-1} e^{i2\pi kj/T} e^{-i2\pi lj/T} \xrightarrow[N\to\infty]{} \begin{cases} 1 & \text{for } l = k, \\ 0 & \text{for } l \neq k, \end{cases}$$

we can see that the periodic series F is stationary with covariance function

$$R_f(n) = \sum_{k=-[T/2]+1}^{[T/2]} |c_k|^2 e^{i2\pi kn/T}.$$

This function also has period T. The spectral measure is supported at the points $\omega_k = k/T$ with weights $|c_k|^2$. In terms of the spectral function,

$$\Phi_f(\omega) = \sum_{k/T \leq \omega} |c_k|^2.$$

The space $\mathcal{H}_f$ is finite-dimensional, and it is isometric to the complex vector space $\mathbf{L}_f^2 = \{\psi\} = \{(\alpha_{-[T/2]+1}, \ldots, \alpha_{[T/2]})\}$ equipped with the inner product

$$(\psi_1, \psi_2)_f = \sum_{k=-[T/2]}^{[T/2]} |c_k|^2 \alpha_k^{(1)} \overline{\alpha_k^{(2)}}.$$

On the other hand, if we denote

$$M_f^{(n)}(\omega) = \sum_{k/T \leq \omega} c_k e^{i2\pi kn/T},$$

then

$$f_{m+n} = \sum_k c_k e^{i2\pi mk/T} e^{i2\pi nk/T} = \int_I e^{i2\pi \omega m} dM_f^{(n)}(\omega).$$

If we set

$$M_f(\omega) = (\ldots, M_f^{(-1)}(\omega), M_f^{(0)}(\omega), M_f^{(1)}(\omega), \ldots),$$

then we come to the equality (6.44), which is now the pointwise equality.

(b) Almost periodic sequences

If the spectral measure m_f of a stationary sequence F is discrete, then this sequence will be called a *generalized almost periodic (GAP)* sequence. A generalized almost periodic sequence F is *generalized quasi-periodic* if the support Ω_f of the measure m_f is not a subset of any grid $\{k/T\}$, where T is an integer.

Let the measure m_f be supported on the set $\Omega_f = \{\omega_1, \omega_2, \ldots\} \subset (-1/2, 1/2]$, and suppose that $m_f(\{\omega_k\}) = d_k^2 > 0$. Since the measure m_f is finite,

$$m_f((-1/2, 1/2]) = \sum_k d_k^2 < \infty.$$

Due to the assumption that F is real-valued, the set $\Omega_f \cap (-1/2, 1/2)$ is symmetric around zero.

A simple example of a generalized quasi-periodic series with a spectral measure m_f supported on Ω_f can be described as follows. Denote by c_k a complex square root of d_k^2 and assume that $c_k = \overline{c_l}$ if $\omega_k = -\omega_l$. For $\omega_k = 1/2$ we take c_k as a real square root of d_k^2. If

$$|c_1| + \ldots + |c_k| + \ldots < \infty, \tag{6.60}$$

then we define

$$f_n = \sum_{\omega_k \in \Omega_f} c_k e^{i 2\pi \omega_k n}. \tag{6.61}$$

We can now check in the manner of the previous paragraph that this series is stationary with covariance function

$$R_f(n) = \sum_{\omega_k \in \Omega_f} |c_k|^2 e^{i 2\pi \omega_k n}. \tag{6.62}$$

If the set Ω_f is not a subset of the grid $\{\pm k/T\}$ with an integer T, then the series (6.61) is called *quasi-periodic*. (See Berge, Pomeau and Vidal, 1986, Chapter III.3.2, for examples and discussion.)

In addition to the series of the form (6.61) satisfying (6.60), there is another class of the stationary sequences with covariance function (6.62). By definition (see, for instance, Corduneanu, 1968, Chapter 1, or Osipov, 1988), a series F is called a *Bohr almost periodic* (briefly, *Bohr*) series if for any $\varepsilon > 0$ there exists a trigonometric polynomial series

$$T_\varepsilon = (\ldots, t_{-1}^{(\varepsilon)}, t_0^{(\varepsilon)}, t_1^{(\varepsilon)}, \ldots)$$

with

$$t_n^{(\varepsilon)} = \sum_{m=1}^{M} a_m e^{i 2\pi \omega_m n}, \quad \omega_m \in I,$$

such that $\sup_{n \in \mathcal{Z}} |f_n - t_n^{(\varepsilon)}| < \varepsilon$.

The class of Bohr series is closed under addition and term-by-term multiplication and includes the periodic series. Moreover, if F is a Bohr series, then for any $k \in \mathcal{Z}$, the shifted series F_k is also a Bohr series.

It can be proved (see Corduneanu, 1968, or Osipov, 1988) that Bohr sequences have the following properties:

1. For any $m \in \mathcal{Z}$ there exists a limit

$$\lim_{N \to \infty} \frac{1}{N} \sum_{k=0}^{N-1} f_{k+m} = \mathfrak{M}_f. \tag{6.63}$$

The number $\mathfrak{M}_f$ does not depend on m and the convergence in (6.63) is uniform in m;

2. It follows from (6.63) that the sequence

$$R_f^{(N)}(m,j) = \frac{1}{N} \sum_{k=0}^{N-1} f_{k+m} f_{k+j}$$

converges uniformly in $m - j$ to a number $R_f(m - j)$. Therefore, any Bohr sequence is a stationary sequence;

3. If F_1, F_2 are Bohr sequences, then the limit

$$[F_1, F_2] = \lim_{N \to \infty} \frac{1}{N} \sum_{k=0}^{N-1} f_k^{(1)} \overline{f_k^{(2)}}$$

is finite and can be used as the definition of an inner product of F_1 and F_2. If we take the closure of the set of Bohr sequences in the norm generated by this inner product, then we come to the Hilbert space $\mathcal{H}$ with the inner product $[\cdot, \cdot]$ and the norm $|||\cdot|||$.

For any Bohr sequence F, the space $\mathcal{H}_f$ is a closed linear subspace of $\mathcal{H}$;

4. Define the sequence

$$e_\omega = (\ldots, e^{-i2\pi 2\omega}, e^{-i2\pi\omega}, 1, e^{i2\pi\omega}, e^{i2\pi 2\omega}, \ldots).$$

Then there exists a countable set $\Omega_f = \{\omega_1, \omega_2 \ldots\} \subset I$ such that

$$c_m \stackrel{\text{def}}{=} [F, e_{\omega_m}] = \lim_{N \to \infty} \frac{1}{N} \sum_{k=0}^{N-1} f_k e^{-2\pi \omega_m k} \neq 0. \qquad (6.64)$$

For $\omega \notin \Omega_f$ the limit (6.64) exists and is equal to zero. The numbers c_m satisfy the inequality $|c_1|^2 + |c_2|^2 + \ldots < \infty$. The set Ω_f is called the *spectrum of the Bohr series* F.

5. For any $l \in \mathcal{Z}$

$$[F_l, e_\omega] = \begin{cases} 0 & \text{for } \omega \notin \Omega_f, \\ c_m e^{i2\pi \omega_m l} & \text{for } \omega = \omega_m. \end{cases} \qquad (6.65)$$

6. If F and G are Bohr series, then

$$[F, G] = \sum_{\omega \in \Omega_f \cap \Omega_g} [F, e_\omega] \overline{[G, e_\omega]}. \qquad (6.66)$$

7. The decomposition

$$f_n \sim \sum_k c_k e^{i2\pi \omega_k n} \qquad (6.67)$$

is valid in the sense that

$$\left|\left|\left| F - \sum_{k=1}^m c_k e_{\omega_k} \right|\right|\right| \xrightarrow[m \to \infty]{} 0.$$

It follows from (6.65) and (6.66) that

$$R_f(n) = \left[F_{k+n}, F_n\right] = \sum_m |c_m|^2 e^{i2\pi(k+n)\omega_m} e^{-i2\pi n\omega_m}$$

and therefore the spectral measure of any Bohr series F is a discrete measure.

(c) Aperiodic time series

A stationary series is called *aperiodic* (or, in another terminology, *chaotic*; see Schuster, 1995) if its spectral function Φ_f is continuous. Of course, the existence of the spectral density p_f is only a sufficient condition (and the most common example) of this situation. If the spectral density exists, the series is called *regularly aperiodic* or *regularly chaotic*. Since Φ_f is continuous, (6.58) is valid in this case.

Aperiodic series provide models for many chaotic dynamic processes. The chaotic behaviour of these processes can be checked by the convergence of the covariance function $R_f(n)$ to zero as $n \to \infty$.

If p_f = const, then the stationary series is called *white noise*. An example of a white noise series is the sequence $f_n = \{n\nu\} - 1/2$ with an irrational positive ν, where $\{a\}$ denotes the fractional part of a. (The proof can be easily accomplished with the help of Weyl's criterion of the uniform distribution of sequences, see for example Kuipers and Niederreiter, 1974, Chapter 1, §6.)

Other examples of regularly aperiodic time series can be constructed via a white noise series in the usual manner of the theory of random stationary processes. For example, if $F = (\ldots, f_{-1}, f_0, f_1, \ldots)$ is a white noise with $p_f \equiv 1$, then

$$g_n = \sum_{k=1}^{M} a_k f_{n-k}$$

will be the term of a regularly chaotic series G with the spectral density

$$p_g(\omega) = \left|\sum_{k=1}^{M} a_k e^{-i2\pi k\omega}\right|^2.$$

6.4.3 SVD for stationary series

Let $F = (f_0, \ldots, f_n, \ldots)$ be a stationary series and consider the collection of singular value decompositions generated by F. More precisely, for any $N > 1$ we fix a window length $L = L(N) < N$ and consider the trajectory matrix $\mathbf{X} = \mathbf{X}(N)$ of the series $F_N = (f_0, \ldots, f_{N-1})$ with its SVD

$$\mathbf{X} = \sum_j \sqrt{\lambda_j} U_j V_j^\mathrm{T}. \tag{6.68}$$

SSA FOR STATIONARY SERIES

Note that λ_j and U_j are the eigenvalues and eigenvectors of the matrix $\mathbf{S} = \mathbf{X}\mathbf{X}^\mathrm{T}$ with the elements

$$s_{km} = s_{km}(N) = \sum_{j=0}^{K-1} f_{k-1+j} f_{m-1+j},$$

where $K = N - L + 1$ and $k, m = 1, \ldots, L$. Therefore, $s_{km}/K \to R_f(k - m)$ and for fixed L and large K both eigenvalues and eigenvectors of the covariance matrix $\mathbf{C} = \mathbf{X}\mathbf{X}^\mathrm{T}/K$ are close to those of the Toeplitz matrix $\mathbf{R}_f = \mathbf{R}_f^{(L)}$ with the elements $r_{k,m} = R_f(k - m)$, where $k, m = 1, \ldots, L$.

Thus, if we consider the eigenvectors and eigenvalues of the matrix $\mathbf{R}_f$, then we also obtain information about the SVD (6.68) for large K.

(a) SVD for almost periodic sequences

Let us start with the case of GAP sequences, where

$$R_f(n) = \sum_{\omega_k \in \Omega_f} |c_k|^2 e^{i 2\pi \omega_k n}, \quad \Omega \subset (-1/2, 1/2].$$

Since $R_f(0) = \sum\limits_{\omega_k \in \Omega_f} |c_k|^2$ and

$$\frac{1}{L} \sum_{j=0}^{L-1} R_f(m - j) e^{i 2\pi \omega_k j}$$

$$= \sum_{\omega_l \in \Omega} |c_l|^2 e^{i 2\pi \omega_l m} \frac{1}{L} \sum_{j=0}^{L-1} e^{-i 2\pi \omega_l j} e^{i 2\pi \omega_k j}$$

$$\xrightarrow[L \to \infty]{} |c_k|^2 e^{i 2\pi \omega_k m},$$

we can claim that asymptotically (as $L \to \infty$) the eigenvalues ν_j of the matrix $\mathbf{R}_f^{(L)}/L$ become close to $|c_j|^2$ and the corresponding complex eigenvectors approximately have the form

$$E_{\omega_j}^{(L)} \stackrel{\text{def}}{=} \left(1, e^{i 2\pi \omega_j}, \ldots, e^{i 2\pi \omega_j (L-1)}\right)^\mathrm{T} / \sqrt{L}.$$

Note that $(E_{\omega_j}^{(L)}, E_{\omega_k}^{(L)}) \to 0$ for $\omega_j \neq \omega_k$ as $L \to \infty$.

In the case $\omega_j \neq 1/2$ the pair $(\omega_j, -\omega_j)$ generates two asymptotically orthogonal real harmonic eigenvectors with the frequency $|\omega_j|$.

Let us now fix the window length L large enough to achieve a good approximation of the eigenvalues ν_j by $|c_j|$ and the corresponding eigenvectors by $E_{\omega_j}^{(L)}$. Then for large $K = K(L)$ we obtain that the leading singular values of the SVD (6.68) are proportional to the largest absolute values of the amplitudes of the harmonic components of the GAP series, written in the form (6.67). Moreover, for large L and K the corresponding eigenvectors are close to the harmonic series with the frequencies associated with these amplitudes.

(b) SVD for aperiodic sequences

The situation is different if the series F is regularly chaotic. Assume that the spectral measure m_f of a stationary series F has a bounded spectral density p_f, that is

$$R_f(n) = \int_{-1/2}^{1/2} e^{i2\pi\omega n} p_f(\omega) d\omega.$$

Referring to Grenander and Szegö (1984, Chapters 5.2, 7.4, 7.7), we quote two important results. The first one is devoted to the asymptotic (as $L \to \infty$) distribution of the eigenvalues of the matrix $\mathbf{R}_f$.

Let $\mu_1, \ldots, \mu_L$ be the eigenvalues of the $L \times L$ Toeplitz matrix $\mathbf{R}_f$ with the elements $r_{mk} = R_f(m - k)$. Denote by P_L the discrete uniform distribution on the set $\{\mu_1, \ldots, \mu_L\}$.

Theorem 6.2 *As $L \to \infty$, the distribution P_L weakly converges to the distribution $\mathcal{D}_f$ of the random variable $p_f(\alpha)$, where α is a random variable uniformly distributed on $[-1/2, 1/2]$.*

The second theorem considers the asymptotic distribution of the eigenvectors of the matrix $\mathbf{R}_f$. For $m \geq 1$ we set

$$p_f^{(m)}(\omega) = \frac{1}{m} \int_{-1/2}^{1/2} \left(\frac{\sin \pi m(\omega - \omega')}{\sin \pi(\omega - \omega')} \right)^2 p_f(\omega') d\omega'.$$

The sequence of continuous functions $p_f^{(m)}$ approximates p_f as $m \to \infty$ in $\mathbf{L}^q((-1/2, 1/2))$ for $q \geq 1$. The approximation is uniform if p_f is continuous in $[-1/2, 1/2]$ (see, for example, Edwards, 1979, Chapter 6.1).

For $\omega_k = k/L$, $k = 0, 1, \ldots, [L/2]$, let us define the vectors $C_k, S_k \in \mathbf{R}^L$ by their components

$$c_L^{(j)} = \cos(2\pi\omega_k j), \quad s_L^{(j)} = \sin(2\pi\omega_k j), \quad -[L/2] < j \leq [L/2].$$

Finally, for any $0 \leq a < b$ we denote by $\Pi_L(a, b)$ and $\Pi_L^{(m)}(a, b)$ the orthogonal projection operators onto the linear spaces

$$\mathcal{L}_{a,b} = \operatorname{span}(U_k, \text{ such that } a \leq \mu_k \leq b)$$

and

$$\mathcal{L}_{a,b}^{(m)} = \operatorname{span}(C_k, S_k, \text{ such that } a \leq p_f^{(m)}(\omega_k) \leq b).$$

Theorem 6.3 *Let $L \to \infty$. Then there exists a sequence $m = m(L) \to \infty$ such that*

$$\left\| \Pi_L(a, b) - \Pi_L^{(m)}(a, b) \right\| \xrightarrow[L \to \infty]{} 0$$

for all $a < b$, the points of continuity of the distribution $\mathcal{D}_f$.

Let us discuss the results of the theorems starting with Theorem 6.3. Assume for simplicity that the spectral measure p_f is a continuous function, and for any c the number $n(c)$ of $\omega \in I$ such that $p_f(\omega) = c$ is finite. Then the the statement of Theorem 6.3 can be interpreted as follows:

1. In the limit the set $\{p_f(\omega), \omega \in I\}$ coincides with the set of eigenvalues of the matrix $\mathbf{R}_f$;

2. Every $\omega \neq 0, 1/2$ with $p_f(\omega) = c \neq 0$ asymptotically produces two real orthogonal harmonic eigenvectors with frequency ω; these eigenvectors correspond to the eigenvalue c.

The points $\omega = 0$ or $\omega = 1/2$ with $p_f(\omega) = c \neq 0$ produce the unique harmonic eigenvector of frequency ω corresponding to the eigenvalue c.

It follows from this description that for a monotone continuous spectral density, all the eigenvectors are asymptotically the harmonic series, and every frequency $\omega \in (0, 1/2)$ corresponds to a pair of equal eigenvalues. (This implies, in particular, that the phenomenon of frequency mixing in SSA is resolved in the limit.) Moreover, for a decreasing spectral density the leading eigenvalues correspond to low frequencies.

By contrast, if p_f is not a monotone function, then certain harmonic eigenvectors with different frequencies (asymptotically) correspond to the same eigenvalues.

Let us pass to Theorem 6.2. By definition, the weak convergence of P_L to the distribution $\mathcal{D}_f$ of the random variable $p_f(\alpha)$ means that for any bounded continuous function g we have

$$\frac{g(\mu_1) + \ldots + g(\mu_L)}{L} \xrightarrow[L \to \infty]{} \int_{-1/2}^{1/2} g(p_f(\omega))d\omega = 2 \int_0^{1/2} g(p_f(\omega))d\omega \qquad (6.69)$$

(here we have used the equality $p_f(-\omega) = p_f(\omega)$).

If $p_f \equiv a = \mathrm{const}$, then the stationary series is white noise. In this case, asymptotically in L, all the eigenvalues of the matrix $\mathbf{R}_f$ are equal to a. In the same manner, if p_f attains two values a_1 and a_2 on the sets A_1 and A_2 with $A_1 \cup A_2 = I$ and $\mathrm{meas}(A_1) = 1 - \mathrm{meas}(A_2) = p$, then asymptotically $100p\,\%$ of the eigenvalues are equal to a_1 and the rest of them are equal to a_2. The general case of piecewise constant p_f is similar.

For any $a < b$ let us denote by $\mathcal{N}_L(a, b)$ the number of eigenvalues μ_j such that $a \leq \mu_j \leq b$. Assume that p_f is a smooth function. Then the number $n(c)$ of roots of the equation $p_f(\omega) = c$ is finite for any c and the distribution $\mathcal{D}_f$ is continuous. Therefore, (6.69) is valid for any bounded Rieman integrable function and we can consider the local (asymptotic) density for the number of eigenvalues of the matrix $\mathbf{R}_f$.

Indeed, it follows from Theorem 6.2 that for any $a < b$

$$\frac{\mathcal{N}_L(a,b)}{L} \xrightarrow[L \to \infty]{} \int_I \mathbf{1}_{[a,b]}(p_f(\omega))d\omega, \qquad (6.70)$$

where $\mathbf{1}_A$ is the indicator of the set A. (The equality (6.70) corresponds to the choice $g = \mathbf{1}_{[a,b]}$ in (6.69).)

Let us denote by $\omega_1, \ldots, \omega_{\mathfrak{n}(c)}$ the roots of the equation $p_f(\omega) = c$. If we take $b = a + \varepsilon$, where ε is small positive, and assume that $\mathfrak{n}(a) = 1$ and $p'_f(\omega_1) \neq 0$, then (6.70) becomes

$$\frac{\mathcal{N}_L(a, a+\varepsilon)}{L} \sim \frac{\varepsilon}{|p'_f(\omega_1)|} + o(\varepsilon) \qquad (6.71)$$

as $L \to \infty$. In the same manner, if $p'_f(\omega_1) = 0$ and $p''_f(\omega_1) \neq 0$, then

$$\frac{\mathcal{N}_L(a, a+\varepsilon)}{L} \sim \sqrt{\frac{\varepsilon}{2|p''_f(\omega_1)|}} + o(\varepsilon^{1/2}), \qquad (6.72)$$

and so on. If $\mathfrak{n}(a) = 2$ and both $p'_f(\omega_1)$ and $p'_f(\omega_2)$ are nonzero, then

$$\frac{\mathcal{N}_L(a, a+\varepsilon)}{L} \sim \varepsilon \left(\frac{1}{|p'_f(\omega_1)|} + \frac{1}{|p'_f(\omega_2)|} \right) + o(\varepsilon). \qquad (6.73)$$

Other situations can be investigated in the same manner.

This means that for smooth p_f the eigenvalues are mainly concentrated around the numbers $p_f(\omega)$ such that $p'_f(\omega) = 0$. If $p'_f(\omega) \neq 0$ for all ω, then the function $1/|p'_f(\omega)|$ has the meaning of global asymptotic density of the number of eigenvalues.

Analogously, if we denote by $\Lambda_L(a,b)$ the sum of the eigenvalues μ_j such that $a \leq \mu_j \leq b$ and take

$$g(x) = g_{a,b}(x) = x\mathbf{1}_{[a,b]}(x), \quad m_1 \leq a < b \leq m_2$$

with $m_1 = \min p_f$, $m_2 = \max p_f$, then we obtain from (6.69) that

$$\frac{\Lambda_L(a,b)}{\mu_1 + \ldots + \mu_L} \xrightarrow[L \to \infty]{} \frac{1}{R_f(0)} \int_I p_f(\omega) \mathbf{1}_{[a,b]}(p_f(\omega))d\omega. \qquad (6.74)$$

For $a = m_1$ we obtain from (6.74) the function of the asymptotic eigenvalue share, standard in statistics. Taking $b = a + \varepsilon$ with small ε we can obtain the expressions for the local eigenvalue densities analogous to (6.71)-(6.73).

(c) Summary

Let us summarize the material of this subsection. For both almost periodic and regularly chaotic series, the asymptotic $(L, K \to \infty)$ singular vectors of the trajectory matrices have the form of harmonic series with frequencies belonging to the support of the spectral measure. In the almost periodic case this support is countable, while for the chaotic situation it is uncountable.

The corresponding eigenvalues are proportional to the weights of the discrete spectral measure for almost periodic sequences and to the values of the spectral density in the regularly aperiodic case.

Note that since we are considering the eigenvalues of the matrices $\mathbf{S} = \mathbf{X}\mathbf{X}^T$ and not that of the covariance matrices $\mathbf{C} = \mathbf{S}/K$, the asymptotic eigenvalues in the case of GAP sequences are proportional to LK. For regularly aperiodic sequences the corresponding coefficient is K. The difference can be used for checking whether or not long stationary-like sequences contain periodicities.

6.4.4 Separability of stationary series

Below we assume that the index set for the stationary sequences under consideration is $\mathcal{N} = \{0, 1, 2, \ldots\}$. According to Remark 6.5, the entire spectral theory is valid if we use the inner product in $\mathcal{H}_f$ determined by (6.52) instead of (6.41).

According to the definition of Section 6.1.2 (see (6.13)), two infinite time series $F^{(1)}, F^{(2)}$ are weakly pointwise regularly asymptotically separable if both window length L and $K = N - L + 1$ tend to infinity and

$$\frac{\sum_{k=0}^{N-1} f_{n+k}^{(1)} f_{m+k}^{(2)}}{\sqrt{\sum_{k=0}^{N-1} \left(f_{n+k}^{(1)}\right)^2} \sqrt{\sum_{k=0}^{N-1} \left(f_{m+k}^{(2)}\right)^2}} \xrightarrow[N \to \infty]{} 0$$

for any $m, n \geq 0$.

Let us consider the separability conditions for stationary series. If $F^{(1)}$ and $F^{(2)}$ are stationary sequences, then

$$\frac{1}{N} \sum_{k=0}^{N-1} \left(f_{n+k}^{(i)}\right)^2 \xrightarrow[N \to \infty]{} R_{f_i}(0) > 0, \quad i = 1, 2$$

for any n, and the asymptotic separability conditions are reduced to the requirement of the convergence

$$\frac{1}{N} \sum_{k=0}^{N-1} f_{n+k}^{(1)} f_{m+k}^{(2)} \xrightarrow[N \to \infty]{} 0.$$

Let us assume that $F \stackrel{\text{def}}{=} F^{(1)} + F^{(2)}$ is a stationary series with spectral measure m_f and $F^{(1)}, F^{(2)} \in \mathcal{H}_f$. In view of the results of Section 6.4.1, the Hilbert space $\mathcal{H}_f$ is isometric to $\mathbf{L}_f^2 \stackrel{\text{def}}{=} \mathbf{L}^2(dm_f)$, and the series F corresponds to the constant function $\mathbf{1} \in \mathbf{L}_f^2$, that is $F \leftrightarrow \mathbf{1}$.

Since $F^{(1)}, F^{(2)} \in \mathcal{H}_f$, there exist $\psi_1, \psi_2 \in \mathbf{L}_f^2$ such that $f_i \leftrightarrow \psi_i$, $i = 1, 2$, and $\psi_1 + \psi_2 = \mathbf{1}$ in $\mathbf{L}_f^2$. Therefore, the spectral measures of $F^{(1)}, F^{(2)}$ have the form $dm_{f_1} = |\psi_1|^2 dm_f$ and $dm_{f_2} = |\psi_2|^2 dm_f$, respectively. Moreover,

$$\frac{1}{N}\sum_{k=0}^{N-1} f^{(1)}_{n+k} f^{(2)}_{m+k}$$
$$\xrightarrow[N\to\infty]{} \left[F_n^{(1)}, F_m^{(2)}\right] = \int_I e^{i(n-m)\omega} \psi_1(\omega)\overline{\psi_2(\omega)} m_f(d\omega),$$

and the separability condition takes the form $\psi_1\overline{\psi_2} = 0$ m_f-almost everywhere. In other words, separation holds when the supports of the spectral measures m_{f_1} and m_{f_2} are disjoint.

For periodic series this result gives the opportunity to separate its various 'elementary' components: since m_f is concentrated on a grid $\{\pm k/T\}$ with integer k and T, each term $c_k e^{i2\pi nk/T} + c_{-k} e^{-i2\pi nk/T}$ on the right side of (6.59) can be asymptotically separated from the sum of the others.

For (generalized) quasi-periodic series everything is similar. Any chaotic series F can be (weakly) divided by SSA into two series belonging to $\mathcal{H}_f$ if their spectral densities have disjoint supports.

If $F^{(1)}$ is a GAP series ('signal') and $F^{(2)}$ is a chaotic series ('noise'), then their spectral measures always have disjoint supports. Thus, under the assumption that $F_N^{(1)}, F_N^{(1)} \in \mathcal{H}_f$, any generalized almost periodical signal is weakly pointwise regularly asymptotically separated from a stationary chaotic noise.

6.4.5 Periodograms

Let $F = (f_0, f_1, \ldots)$ be a stationary series. Then the series

$$\Pi^{(N)}_{k,f}(\omega) \stackrel{\text{def}}{=} \frac{1}{N}\left|\sum_{n=0}^{N-1} e^{-i2\pi\omega n} f_{n+k}\right|^2, \quad \omega \in I,$$

is called the *N-periodogram series* of the series F.

Periodogram series can be used as an approximation for the spectral measure of a stationary sequence. More precisely, we have the following result.

Theorem 6.4
1. *For any $\omega \in I$,*

$$\lim_{T\to\infty} \frac{1}{T} \sum_{k=0}^{T-1} \frac{\Pi^{(N)}_{k,f}(\omega)}{N} \xrightarrow[N\to\infty]{} m_f(\{\omega\}). \tag{6.75}$$

2. *For any bounded continuous $\Psi : I \mapsto \mathbf{R}$,*

$$\lim_{T\to\infty} \frac{1}{T} \sum_{k=0}^{T-1} \int_I \Psi(\omega) \Pi^{(N)}_{k,f}(\omega) d\omega \xrightarrow[N\to\infty]{} \int_I \Psi(\omega) m_f(d\omega). \tag{6.76}$$

3. *Let us assume that there exists a spectral density p_f that is continuous at the point ω. Then*

$$\lim_{T\to\infty} \frac{1}{T} \sum_{k=0}^{T-1} \Pi_{k,f}^{(N)}(\omega) \xrightarrow[N\to\infty]{} p_f(\omega). \qquad (6.77)$$

Proof.
1. According to Remark 6.5, the convergence (6.75) is exactly the convergence (6.57) of Remark 6.6.
2. Analogous to (6.54) and the first equality of (6.57), we have that the left side of (6.76) is equal to

$$\int_I \Psi(\omega) \left\|\frac{1}{\sqrt{N}} \sum_{n=0}^{N-1} e^{-i2\pi\omega n} F_n \right\|^2 d\omega$$

$$= \int_I \left(\int_I \Psi(\omega) \frac{\sin^2(\pi N(\omega - \omega_1))}{N \sin^2(\pi(\omega - \omega_1))} d\omega \right) m_f(d\omega_1).$$

Since the function

$$h_{\omega_1}^{(N)}(\omega) \stackrel{\text{def}}{=} \frac{\sin^2(\pi N(\omega - \omega_1))}{N \sin^2(\pi(\omega - \omega_1))}$$

is a density on I for any $\omega_1 \in I$ and, as $N \to \infty$, the associated distribution tends to the Dirac distribution concentrated at the point ω_1, we obtain

$$\int_I \Psi(\omega) h_{\omega_1}^{(N)}(\omega) \, d\omega \xrightarrow[N\to\infty]{} \Psi(\omega_1)$$

(see, for example, Edwards, 1979, Chapter 6.1.1). Therefore, the second assertion of the theorem is proved.
3. Similar to the proof of the previous assertion, we get

$$\lim_{T\to\infty} \frac{1}{T} \sum_{k=0}^{T-1} \Pi_{k,f}^{(N)}(\omega) = \int_I h_\omega^{(N)}(\omega_1) p_f(\omega_1) \, d\omega_1.$$

This completes the proof. □

Remark 6.7 If we consider the Bohr sequence F then, in view of (6.65),

$$\frac{\Pi_{k,f}^{(N)}(\omega)}{N} \xrightarrow[N\to\infty]{} m_f(\{\omega\})$$

for any k.

List of data sets and their sources

- Data 'Births', 'Coal sales', 'Demands', 'Drunkenness', 'Eggs', 'Fortified wine', 'Gold price', 'Hotels', 'Investment', 'Petroleum sales', 'Precipitation', 'Rosé wine', 'Sunspots', 'Tree rings', 'Unemployment', 'Wages': Time Series Data Library maintained by Rob Hyndman
  http://www-personal.buseco.monash.edu.au/~hyndman/TSDL
- Data 'England temperatures': The Meteorological Office (U.K.)
  http://www.meto.govt.uk/sec5/CR_data/Monthly/HadCET_act.txt
- Data 'Production': Economic Time Series Page
  http://www.economagic.com/em-cgi/data.exe/doeme/pnprbus
- Data 'White dwarf': The Santa Fe Time Series Competition Data
  http://www.stern.nyu.edu/~aweigend/Time-Series/SantaFe.html
- EEG Data: Dr. Dmitry Belov, Institute of Physiology, St. Petersburg University
- Data 'War': Table 10 in Janowitz and Schweizer (1989).

References

H.D.I. Abarbanel. *Analysis of Observed Chaotic Data*. Springer, New York, 1996.

M.R. Allen and L.A. Smith. Monte Carlo SSA: Detecting irregular oscillations in the presence of coloured noise. *Journal of Climate*, 9:3373–3404, 1996.

T.W. Anderson. *Statistical Analysis of Time Series*. Wiley, New York, 1994.

P. Berge, Y. Pomeau, and Ch. Vidal. *Order within Chaos: Towards a Deterministic Approach to Turbulence*. Wiley, New York, 1986.

M. Bouvet and H. Clergeot. Eigen and singular value decomposition technique for the solution of harmonic retrieval problems. In E.F. Deprettere (Ed.), *SVD and Signal Processing: Algorithms, Applications and Architechtures*, pp. 93–114. North-Holland, Amsterdam, 1988.

D. Brillinger. *Time Series. Data Analysis and Theory*. Holt, Rinehart and Winston, Inc., New York, 1975.

D.S. Broomhead and G.P. King. Extracting qualitative dynamics from experimental data. *Physica D*, 20:217–236, 1986a.

D.S. Broomhead and G.P. King. On the qualitative analysis of experimental dynamical systems. In S. Sarkar (Ed.), *Nonlinear Phenomena and Chaos*, pp. 113–144. Adam Hilger, Bristol, 1986b.

D.S. Broomhead, R. Jones, G.P. King, and E.R. Pike. Singular system analysis with application to dynamical systems. In E.R. Pike and L.A. Lugaito (Eds.), *Chaos, Noise and Fractals*, pp. 15–27. IOP Publishing, Bristol, 1987.

V.M. Buchstaber. Time series analysis and grassmannians. In S. Gindikin (Ed.), *Applied Problems of Radon Transform*, volume 162 of *AMS Transactions — Series 2*, pp. 1–17. AMS, Providence, 1994.

J.C. Clemens. Whole earth telescope observation of the white dwarf star PG1159-035. In A.S. Weigend and N.A. Gershenfeld (Eds.), *Time Series Prediction: Forecasting the Future and Understanding the Past*. Addison-Wesley, Reading, 1994.

C. Corduneanu. *Almost Periodic Functions*. Interscience Publishers, New York, 1968.

C.D. Cutler and D.T. Kaplan (Eds.), *Nonlinear Dynamics and Time Series: Building a Bridge between the Natural and Statistical Sciences*. American Mathematical Society, Providence, Rhode Island, 1997.

D. Danilov. Principal components in time series forecast. *Journal of Computational and Graphical Statistics*, 6:112–121, 1997a.

D. Danilov. The 'Caterpillar' method for time series forecasting. In D. Danilov and A. Zhigljavsky (Eds.), *Principal Components of Time Series: the 'Caterpillar' Method*, pp. 73–104. University of St.Petersburg, St.Petersburg, 1997b. (In Russian).

D. Danilov and A. Zhigljavsky (Eds.), *Principal Components of Time Series: the 'Caterpillar' Method*. University of St.Petersburg, St.Petersburg, 1997. (In Russian).

P. Diggle. *Time Series. A Biostatistical Introduction*. Clanderon Press, Oxford, 1998.

N.R. Draper and H. Smith. *Applied Regression Analysis*. Wiley, New York, third edition, 1998.

R.E. Edwards. *Fourier Series. A Modern Introduction*, volume 1. Springer-Verlag, New York, second edition, 1979.

B. Efron and R. Tibshirani. Bootstrap methods for standard errors, confidence intervals and other measures of statistical accuracy. *Statistical Science*, 1(1):54–75, 1986.

J.B. Elsner and A.A. Tsonis. *Singular Spectrum Analysis. A New Tool in Time Series Analysis*. Plenum Press, New York, 1996.

A.C. Fowler and G. Kember. Singular systems analysis as a moving-window spectral method. *European Journal of Applied Mathematics*, 9:55–79, 1998.

F.R. Gantmacher. *The Theory of Matrices*. Chelsea, Providence, RI, 1998.

A.O. Gelfond. *Finite Difference Calculus*. Nauka, Moscow, 1967. (In Russian).

M. Ghil and C. Taricco. Advanced spectral analysis methods. In G.C. Castagnoli and A. Provenzale (Eds.), *Past and Present Variability of the Solar-Terrestrial System: Measurement, Data Analysis and Theoretical Models*, pp. 137–159. IOS Press, Amsterdam, 1997.

M. Ghil and R. Vautard. Interdecadal oscillations and the warming trend in global temperature time series. *Nature*, 350:324–327, 1991.

U. Grenander and G. Szegö. *Toeplitz Forms and their Applications*. Chelsea, New York, 1984.

K.W. Hipel and A.I. McLeod. *Time Series Modelling of Water Resources and Environmental Systems*. Elsevier Science, Amsterdam, 1994.

M.F. Janowitz and B. Schweizer. Ordinal and percentile clustering. *Mathematical Social Sciences*, 18:135–186, 1989.

I.T. Jolliffe. *Principal Component Analysis*. Springer Series in Statistics. Springer-Verlag, New York, 1986.

H. Kantz and T. Schreiber. *Nonlinear Time Series Analysis*. Cambridge University Press, Cambridge, 1997.

M.G. Kendall and A. Stuart. *Design and Analysis, and Time Series*, volume 3 of *The Advanced Theory of Statistics*. Charles Griffin, London, third edition, 1976.

L. Kuipers and H. Niederreiter. *Uniform Distribution of Sequences*. Wiley, New York, 1974.

W. Ledemann and E. Lloyd (Eds.), *Statistics, Part B*, volume VI of *Handbook of Applied Mathematics*. Wiley, Chichester, 1984.

M. Loève. *Probability Theory*. D. Van Nostrand Company inc., Princeton, New Jersey, third edition, 1963.

V.K. Madisetti and D.B. Williams (Eds.), *The Digital Signal Processing Handbook*. CRC Press, Boca Raton, 1998.

S. Makridakis, S. Wheelwright, and R.J. Hyndman. *Forecasting: Methods and Applications*. Wiley, New York, third edition, 1998.

S.L. Marple-Jr. *Digital Spectral Analysis*. Prentice Hall, New Jersey, 1987.

R. McCleary and R.A. Hay. *Applied Time Series Analysis for the Social Sciences*. Sage Publication, London, 1980.

D.C. Montgomery and L.A. Johnson. *Forecasting and Time Series Analysis*. McGraw-Hill, New York, 1976.

REFERENCES

V.G. Moskvina and A.A. Zhigljavsky. Application of the singular-spectrum analysis for change-point detection in time series. *Journal of Time Series Analysis*, 2000. (submitted).

V.V. Nekrutkin. Decomposition of time series. In D. Danilov and A. Zhigljavsky (Eds.), *Principal Components of Time Series: the 'Caterpillar' Method*, pp. 194–227. University of St.Petersburg, St.Petersburg, 1997. (In Russian).

V.V. Nekrutkin. Approximation spaces and continuation of time series. In S.M. Ermakov and Yu.N. Kashtanov (Eds.), *Statistical Models with Applications in Econometrics and Neibouring Fields*, pp. 3–32. University of St.Petersburg, St.Petersburg, 1999. (In Russian).

G. Nicolis and I. Prigogine. *Exploring Complexity. An Introduction.* Freeman, New York, 1989.

T.M. O'Donovan. *Short Term Forecasting: An Introduction to the Box-Jenkins Approach.* Wiley, New York, 1983.

V.F. Osipov. *Borh-Frenel Analysis on Local-Compact Comutative Groups.* Leningrad State University, Leningrad, 1988. (In Russian).

R.K. Otnes and L. Enochson. *Applied Time Series Analysis*, volume 1 of *Basic Techniques*. Wiley, New York, 1978.

M. Paluš and D. Novotna. Detecting modes with nontrivial dynamics embedded in colored noise: enhanced Monte Carlo SSA and the case of climate oscillations. *Physics Letters A*, 248:191–202, 1998.

G. Plaut and R. Vautard. Spells of low-frequency oscillations and weather regimes in the northern hemisphere. *Journal of the Atmospheric Sciences*, 51:210–236, 1994.

M.B. Priestley. *Spectral Analysis and Time Series.* Academic Press, London, 1991.

D.L. Prothero and K.F. Wallis. Modelling macroeconomic time series (with discussion). *Journal of the Royal Statistical Society, A*, 139(Part 4):468–500, 1976.

T. Subba Rao. Canonical factor analysis and stationary time series models. *Sankhya: The Indian Journal of Statistics*, 38B:256–271, 1976.

T. Subba Rao and M.M. Gabr. *An Introduction to Bispectral Analysis and Bilinear Time Series Models.* Springer-Verlag, 1984.

R. Riesz and B. Sz.-Nagy. *Functional Analysis.* Dover Publications, New York, 1990.

Yu.A. Rozanov. *Stationary Random Processes.* Holden-Day, San-Francisco, 1967.

Y. Sauer, J.A. Yorke, and M. Casdagli. Embedology. *Journal of Statistical Physics*, 65:579–616, 1991.

H.G. Schuster. *Deterministic Chaos. An Introduction.* Physik-Verlag, Weinheim, third edition, 1995.

F. Takens. Detecting strange attractors in turbulence. In D.A.Rand and L.-S.Young (Eds.), *Dynamical Systems and Turbulence*, volume 898 of *Lecture Notes in Mathematics*, pp. 366–381. Springer-Verlag, Berlin, 1981.

H. Tong. *Non-linear Time Series: A Dynamical System Approach.* Oxford University Press, Oxford, 1993.

J.W. Tukey. *Exploratory Data Analysis.* Addison-Wesley, Reading, 1977.

F. Varadi, J.M. Pap, R.K. Ulrich, L. Bertello, and C.J. Henney. Searching for signal in noise by random-lag singular spectrum analysis. *The Astrophysical Journal*, 526:1052–1061, 1999.

R. Vautard and M. Ghil. Singular-spectrum analysis in nonlinear dynamics, with applications to paleoclimatic time series. *Physica D*, 35:395–424, 1989.

R. Vautard, P. Yiou, and M. Ghil. Singular-spectrum analysis: A toolkit for short, noisy chaotic signals. *Physica D*, 58:95–126, 1992.

W.W.S. Wei. *Time Series Analysis: Univariate and Multivariate Methods*. Addison-Wesley, New York, 1990.

A.S. Weigend and N.A. Gershenfeld (Eds.), *Time Series Prediction: Forecasting the Future and Understanding the Past*. Addison-Wesley, Reading, 1993.

P. Yiou, E. Baert, and M.F. Loutre. Spectral analysis of climate data. *Surveys in Geophysics*, 17:619–663, 1996.

P. Yiou, D. Sornette, and M. Ghil. Data-adaptive wavelets and multi-scale singular-spectrum analysis. *Physica D*, 142:254–290, 2000.

Index

base space, 151
base subseries (interval), 151
   starting, 154, 193
Basic SSA stages
   decomposition, 16
   reconstruction, 17
Basic SSA steps
   diagonal averaging, 17
   embedding, 16
   grouping, 17
   SVD, 17

change-point
   detection, 149
   moment, 157
confidence interval
   bootstrap, 118, 143
   empirical, 117
   Monte Carlo, 117

data
   'Births', 28, 43, 66, 70, 76, 91
   'Coal sales', 204
   'Demands', 211
   'Drunkenness', 29, 43
   'EEG', 92
   'Eggs', 27, 60, 61, 74, 135
   'Fortified wine', 138
   'Gold price', 142
   'Hotels', 63, 82
   'Investment', 82
   'Petroleum sales', 206
   'Precipitation', 136
   'Production', 25, 38, 42, 56, 71
   'Rosé wine', 39, 61, 88, 138
   'Sunspots', 209
   'Tree rings', 25, 57, 72, 86
   'Unemployment', 29, 42, 43, 65, 71, 89

'Wages', 131
'War', 31, 36, 38, 42, 43, 76
'White dwarf', 26, 34, 68, 72
detection
   change, 149
   change-point, 149
   function, 154
      column, 155
      diagonal, 155
      row, 155
      symmetric, 155

function
   detection, 154
   frequency-power (f-power), 203
   heterogeneity, 151
   root
      frequency, 200
      modulus, 200
   verticality, 146

Hankelization, 24
heterogeneity
   function, 151
      column, 152
      diagonal, 152
      row, 151
      symmetric, 152
   index, 150
      renormalized, 197
   matrix (H-matrix), 151
   permanent, 157
   temporary, 157

lagged (L-lagged) vectors, 16
linear recurrent formula (LRF), 97, 243
   characteristic polynomial of, 98, 250

extraneous root of, 100
main root of, 100
heterogeneous, 112
minimal, 97, 113, 243
linear space
  base, 151
  eigenspace, 23
  Schubert basis of, 246
  trajectory (L-trajectory), 21

matrix
  biorthogonal, 227
  diagonal averaging of, 17
  elementary (unit-rank), 17, 224
  Frobenius norm of, 20, 223
  Hankel, 16, 266
  Hankelization of, 24, 266
  heterogeneity, 151
  inner product of, 20, 223
  lag-covariance, 84
    Toeplitz, 85
  minimal decomposition of, 227
  orthogonal, 223
  periodogram, 202
  positive semidefinite, 219
  quasi-diagonal representation of, 223
  resultant, 17
  trajectory (L-trajectory), 16
  unitary (rotation), 222

periodogram, 34, 294
  matrix, 202
  power of, 35
  support of, 35
polynomial root
  modulus-frequency representation of, 200

reconstruction
  bootstrap average, 143
  by eigentriples, 24

singular value decomposition (SVD)
  double centring, 235
    second average triple, 236
    share of the 1st average triple, 236
    share of the 2nd average triple, 236
  eigenspace, 23, 219
  eigentriples, 17, 221
    base set of, 151
    rearrangement of, 176
    share of, 230
  eigenvalue, 17, 219
  eigenvector, 17, 219
  factor vector, 22
  principal component, 231
  principal direction (vector), 22, 231
  single centring, 233, 268
    first average triple, 233
    share of the 1st average triple, 234
  singular value, 20, 221
  singular vector, 20, 221
  vector of principal components, 22, 231
SSA
  Basic, 15
  Double centring, 80, 272
  Sequential, 89
  Single centring, 79, 268
  Toeplitz, 85
SSA forecasting
  bootstrap average, 118, 131
  centring recurrent, 112
    Basic, 113
    Toeplitz, 113
  centring vector, 112
    Basic, 113
    Toeplitz, 113
  Monte Carlo average, 119
  recurrent, 96
    Basic, 96
    Toeplitz, 111
  vector, 108
    Basic, 109
stationary time series, 33, 277
  almost periodic (Bohr), 34, 286
  aperiodic (chaotic), 34, 288
  covariance function of, 33, 277
  generalized almost periodic, 285
  generalized quasi-periodic, 285
  harmonic, 34
  periodic, 34, 284
  periodogram of, 294
  quasi-periodic, 34, 286

Index   305

   regularly aperiodic (chaotic), 288
   spectral density of, 34, 278
   spectral function of, 278
   spectral measure of, 33, 278
   white noise, 34, 288

test subseries (interval), 151
time series
   L-continuable, 252
   L-continuation of, 101, 252
   L-rank of, 98, 237
   w-orthogonal, 46, 268
   amplitude-modulated, 36
   continuation of
      multistart recurrent, 116
      recurrent, 93, 101
      recurrent approximate, 104
   correlation coefficient between, 261
      maximum, 47, 261
      spectral, 49
      weighted (w-correlation), 47
   finite-difference dimension of, 97, 243
   Fourier expansion of, 34
   frequency range of, 35
   governed by LRF, 243
   harmonic, 28
   homogeneous, 149
   inner product of, 267
   noise, 34
   rank of, 98, 237
   satisfying LRF, 243
   separable (L-separable), 22
      approximately, 22, 47, 261
      asymptotically, 47, 261
      regularly asymptotically, 261
      stably, 170
      stochastically, 264
      strongly, 51, 258
      under single centring, 269
      weakly, 45, 257
      weakly ε-separable, 261
      weakly pointwise regular, 263
   signal, 39
   stationary, 33, 277
   trend of, 38
Toeplitz
   lag-covariance matrix, 85

   singular values, 85
   SSA, 85
trajectory (L-trajectory) space, 21
transition interval, 157

verticality
   coefficient, 95, 247
   function, 146

window length, 16